POLYMER YEARBOOK

First Edition

POLYMER YEARBOOK

An annual desk reference featuring pertinent data, mini-reviews and generally useful information for polymer scientists.

Edited by Hans-Georg Elias and Richard A. Pethrick

ISSN 0738-1743

POLYMER YEARBOOK
First Edition

Edited by
HANS-GEORG ELIAS
Michigan Molecular Institute
and
RICHARD A. PETHRICK
University of Strathclyde

harwood academic publishers
chur . london . paris . new york

Harwood Academic Publishers

Poststrasse 22
7000 Chur
Switzerland

P.O. Box 197
London WC2 4DL
England

58, rue Lhomond
75005 Paris
France

P.O. Box 786
Cooper Station
New York, New York 10276
United States of America

First Published December 1983
Second Printing August 1984

Contents

Contributors

A. F. M. Barton	Murdoch University, Western Australia
L. G. Donaruma	Polytechnic Institute of New York, U.S.A.
W. Funke	University of Stuttgart, F.R.G.
N. Ise	Kyoto University, Japan
H. H. G. Jellinek	Clarkson College of Technology, Potsdam, U.S.A.
J. P. Kennedy	University of Akron, U.S.A.
H. R. Kricheldorf	University of Hamburg, Hamburg, F.R.G.
F. H. Otey	U.S. Department of Agriculture, Peoria, U.S.A.
R. M. Ottenbrite	Virginia Commonwealth University, Richmond, U.S.A.
S. Penczek	Polish Academy of Sciences, Lødz, Poland
E. H. Pryde	U.S. Department of Agriculture, Peoria, U.S.A.
H. K. Reimschuessel	Allied Corporation, Morristown, U.S.A.
A. Rudin	University of Waterloo, Canada
L. H. Tung	The Dow Chemical Company, Midland, U.S.A.
O. Vogl	Polytechnic Institute of New York, U.S.A.
B. A. Wolf	Johannes Gutenberg University, Mainz, F.R.G.

Preface

The editors wish to express their thanks to the contributors for their help and support in the production of this volume. It is hoped that future editions will contain more extensive tabulations of data and will develop in style and format so as to best serve the needs of the readers. Constructive comments and suggestions with regard to material for inclusion in future issues will be welcomed by the editors.

R. A. Pethrick
University of Strathclyde
Glasgow, Scotland

Introduction

Polymer science is one of the truly interdisciplinary scientific subjects spanning chemistry, biology, physics, mathematics, chemical and mechanical engineering and various elements of materials science. It is therefore not surprising to find that the practicing polymer scientist has to have an appreciation of a wide range of topics whilst still being a master of his own specialisation. A book which acts as a memory jogger is therefore a useful addition to any library collection. If, however, this text contains elements of current awareness of topics of interest then it becomes a must for the office or desk collection. The Polymer Yearbook we hope will fulfil those roles and become a must for every polymer scientist.

The first volume of Polymer Yearbook is an embryo of the future publication. It contains a collection of useful data, a number of reviews and a variety of current awareness features. The balance between these elements will vary over the next few years to provide a concise source and reference book for polymer scientists. It is not intended that the Yearbook should duplicate in any real way either the Polymer Handbook[1] or the Handbook of Chemistry and Physics.[2] Both of these texts provide invaluable information either in depth or breadth and it would be inappropriate to attempt to duplicate this material in the Yearbook. The data presented have been selected as being of everyday use and will change over the years to provide up-to-date information of immediate relevance.

The Polymer Yearbook contains four types of information:

● Factual information of general interest. In this section the reader will find a summary of the rules of nomenclature, lists of accepted notations, quantities and formulae in physics and mathematics as well as a short resume of some of the principles of polymer science. In the first edition two topics have been selected for special discussion: gel permeation chromatography and rubber elasticity.

● Reviews. Each year a number of short reviews will be published highlighting specific aspects of polymer science. No attempt will be made to produce comprehensive coverage of the subject area and the reader is referred to the Royal Society of Chemistry Specialist Periodic Reports on Macromolecular Chemistry[3] for an in-depth review of polymer chemistry.

● Current awareness. A special feature of the Yearbook will be a survey of recent publications in polymer science and also a summary of major review articles.

● Calendar of events. This section will attempt to summarise the subject area and location of forthcoming meetings and courses in polymer science.

It is hoped that the Polymer Yearbook will develop into a concise and widely used desk reference which will help to stimulate and develop polymer science in the future.

References

1. J. Brandorup and E. H. Immergut, Eds., *Polymer Handbook*, 2nd Ed. Wiley-Interscience, New York (1975).
2. R. C. Weast, Ed., *57th Handbook of Chemistry and Physics*, Chemical Rubber Publishing Company (1977).
3. A. Jenkins and J. Kennedy, *Specialist Periodic Reports Macromolecular Chemistry*, The Chemical Society, London (1980).

Nomenclature in Inorganic, Organic and Polymer Chemistry

In 1957 the International Union of Pure and Applied Chemists (IUPAC) Commission on Nomenclature presented a document entitled the 'Definitive Rules for the Nomenclature of Organic Chemistry'. This document was published by Butterworths Scientific Publications. Subsequently a number of additions have appeared addressing the problems of nomenclature in particular fields. A report entitled 'Nomenclature Dealing with Steric Regularity in High Polymers' appeared in 1962 and was produced by a sub-committee of the Commission on Macromolecules of the IUPAC. With the advent of high resolution nmr spectroscopy there has also arisen a need for a revision of the rules of nomenclature. Subsequent documents of the Commission and related publications proposing abbreviations and symbols to be used for the description of the conformation of polypeptide chains, have appeared as definitive documents. In 1979 IUPAC presented a document entitled 'Stereochemical Definitions and Notations relating to Polymers', which has subsequently been published in *Pure and Applied Chemistry*. This latter document attempts to produce a definitive set of definitions and notations for the description of the conformations of polymer molecules.

This section is planned to provide the reader with a brief summary of the rules of nomenclature as applied to inorganic, organic and polymer chemistry. References collected at the end of this section provide a more detailed description of the rules and definitions and the reader is referred to these for detail on a particular point. It is hoped that the abbreviated discussion of the topic presented in the following sections will be adequate for general reference purposes.

The last part of this section provides a summary of certain conventions used in the presentation of information and includes a section on the Wiswesser Chemical Line–Notation and a condensate of the rules governing the Chemical Abstracts Service (CAS) Registry Numbers.

NOMENCLATURE IN INORGANIC CHEMISTRY

The elements should have names and symbols which differ as little as possible among the different common languages used in the scientific literature. The following list represents the nomenclature in current common usage in English and French literature.

Name	Symbol	Atomic Number	Atomic Weight
Actinium	Ac	89	—
Aluminum	Al	13	26.98154
Americium	Am	95	—
Antimony	Sb	51	121.75
Argon	Ar	18	39.948
Arsenic	As	33	74.9216
Astatine	At	85	—
Barium	Ba	56	137.34
Berkelium	Bk	97	—
Beryllium	Be	4	9.01218
Bismuth	Bi	83	208.9804
Boron	B	5	10.81
Bromine	Br	35	79.904
Cadmium	Cd	48	112.40
Calcium	Ca	20	40.08
Californium	Cf	98	—
Carbon	C	6	12.011
Cerium	Ce	58	140.12
Cesium	Cs	55	132.9054
Chlorine	Cl	17	35.453
Chromium	Cr	24	51.996
Cobalt	Co	27	59.9332
Copper	Cu	29	63.546
Curium	Cm	96	—
Dysprosium	Dy	66	162.50
Einsteinium	Es	99	—
Erbium	Er	68	167.26
Europium	Eu	63	151.96
Fermium	Fm	100	—
Fluorine	F	9	18.99840
Francium	Fr	87	—
Gadolinium	Gd	64	157.25
Gallium	Ga	31	69.72
Germanium	Ge	32	72.59
Gold	Au	79	196.9665
Hafnium	Hf	72	178.49
Helium	He	2	4.00260
Holmium	Ho	67	164.9304
Hydrogen	H	1	1.0079
Indium	In	49	114.82
Iodine	I	53	126.9045
Iridium	Ir	77	192.22
Iron	Fe	26	55.847
Krypton	Kr	36	83.80
Lanthanum	La	57	138.9055
Lawrencium	Lr	103	—
Lead	Pb	82	207.2
Lithium	Li	3	6.941
Lutetium	Lu	71	174.97
Magnesium	Mg	12	24.305

Name	Symbol	Atomic Number	Atomic Weight
Manganese	Mn	25	54.9380
Mendelevium	Md	101	—
Mercury	Hg	80	200.59
Molybdenum	Mo	42	95.94
Neodymium	Nd	60	144.24
Neon	Ne	10	20.179
Neptunium	Np	93	—
Nickel	Ni	28	58.71
Niobium	Nb	41	92.9064
Nitrogen	N	7	14.0067
Nobelium	No	102	—
Osmium	Os	76	190.2
Oxygen	O	8	15.9994
Palladium	Pd	46	106.4
Phosphorus	P	15	30.97376
Platinum	Pt	78	195.09
Plutonium	Pu	94	—
Polonium	Po	84	—
Potassium	K	19	39.098
Praseodymium	Pr	59	140.9077
Promethium	Pm	61	—
Protactinium	Pa	91	—
Radium	Ra	88	—
Radon	Rn	86	—
Rhenium	Re	75	186.2
Rhodium	Rh	45	102.9055
Rubidium	Rb	37	85.4678
Ruthenium	Ru	44	101.07
Samarium	Sm	62	150.4
Scandium	Sc	21	44.9559
Selenium	Se	34	78.96
Silicon	Si	14	28.086
Silver	Ag	47	107.868
Sodium	Na	11	22.98977
Strontium	Sr	38	87.62
Sulfur	S	16	32.06
Tantalum	Ta	73	180.9479
Technetium	Tc	43	—
Tellurium	Te	52	127.60
Terbium	Tb	65	158.9254
Thallium	Tl	81	204.37
Thorium	Th	90	232.0381
Thulium	Tm	69	168.9342
Tin	Sn	50	118.69
Titanium	Ti	22	47.90
Tungsten	W	74	183.85
Unnilquadium	Un	104	—
Uranium	U	92	238.029
Vanadium	V	23	50.9414
Xenon	Xe	54	131.30
Ytterbium	Yb	70	173.04
Yttrium	Y	39	88.9059
Zinc	Zn	30	65.38
Zirconium	Zr	40	91.22

Names of groups of elements

The following collective names are approved:

Group name	Elements
Halogens	F, Cl, Br, I and At
Chalcogens	O, S, Se, Te and Po
Alkalie metals	Li, Na, K, Rb, Cs, Fr
Alkaline earth metals	Ca, Sr, Ba, Ra
Inert gas	He, Ne, Ar, Kr, Xe, Rn
Rare earth metals	Sc, Y and La to Lu
Lanthanum series	La (51) to Lu (71)
Lanthanides	Ce (58) to Lu (71)
Actinium series	Ac (89) to Lr (103)

Designation of mass, charge etc. on atomic symbols

The mass number, atomic number, number of atoms and ionic charge of an element may be indicated by means of four indices placed around the symbol. The positions occupied are to be thus:

left upper index   — mass number
right lower index — number of atoms
left lower index   — atomic number
right upper index — ionic charge

Example of use of these rules—$^{32}_{16}S_2^{2+}$ represents a doubly ionized molecule containing two atoms of sulphur, each of which has the atomic number 16 and mass number 32.

It has been the practice of American chemists and physicists to put the mass number at the upper right of the symbol, however the nomenclature committee recognizes the advantage of putting the symbol in the upper left position so that the upper right is available for the ionic charge.

Isotopically labelled compounds may be described by adding to the name of the compound, the symbol for the isotrope in parentheses. Example:

$H^{36}Cl$ Hydrogen chloride^{36}(Cl) or hydrogen chloride-36

Allotropes

A number of trivial names are to be found in common usage indicating an element occurring in a particular modification. Using systematic nomenclature each modification should be based on the number of atoms involved.

Atom	Trivial name	Systematic name
H	atomic hydrogen	monohydrogen
O_2	oxygen gas	dioxygen
O_3	ozone	trioxygen
P_4	white phosphorus	tetra phosphorus
S_8	λ-sulphur	cyclooctasulphur or octasulphur

Minerals occurring in nature with similar compositions, but having different crystal structures, often are given different names; zinc blende, wurtzite, quartz, tridymite and cristobalite. These polymorphic modifications are systematically designated by a Greek letter or with a Roman numeral—α-iron, ice-I etc.

Formulae and names of compounds

Formulas

The empirical formula is formed by juxtaposition of the atomic symbols to give the simplest possible formula expressing the stoichiometric composition of the compound in question. The formula should be consistent with the correct molecular weight—e.g. N_2H_4 and not NH_2. When the molecular weight varies with temperature, e.g., NO_2 in equilibrium with N_2O_4, the simplest possible formula should be chosen.

Systematic names

Compounds should be named according to the following rules:
- the name of the electropositive constituent will not be modified.
- the electronegative constituent if monatomic or if a binary compound of a nonmetal is formed then the name should terminate in the designate -ide.
- if the electronegative constituent is polyatomic then the name should be terminated in -ate. There are certain exceptions which may be terminated either in the phrase -ide or -ate.

Polyatomic anions

OH^-	hydroxide ion	peroxide ion	O_2^{2-}
O_2^-	hyperoxide ion	ozonide ion	O_3^-
S_2^{2-}	disulphide	triiodide	I_3^-
HF_2^-	hydrogen difluoride ion	azide ion	N_3^-
NH^{2-}	imide ion	amide ion	NH_2^-
$NHOH^-$	hydroxylamide ion	hydrazide ion	$N_2H_3^-$
CN^-	cyanide ion	acetylide	C_2^{2-}

- in inorganic polyatomic groupings it is usually possible to identify a characteristic atom and the group is designated by the phrase -ate, -phosphate; $(PO_4)^{3-}$—tetraoxophosphate.

- the stoichiometric proportions may be denoted by means of Greek numerical prefixes (mono, di, tri, tetra, penta, hexa, hepta, octa, ennea, deca, hendeca and dodeca) preceding without hyphen the names of the elements to which they refer, e.g. Fe_3O_4 triiron tetraoxide. An alternative approach which is sometimes used is the Stock notation. It is usually applied to cations and anions but should not be applied to nonmetals, e.g.:

$$FeCl_3 \text{—iron(III) chloride.}$$

- in certain instances it is convenient to use a trivial name rather than the systematic one since this can also convey information on the stoichiometric nature of the compound. For example: sodium tetraoxosulphate will normally be called sodium sulphate.

Trivial names

A number of atomic groupings are known by their trivial names:

water	H_2O	ammonia	NH_3	hydrazine	N_2H_4
diboran	B_2H_6	silane	SiH_4	phosphine	PH_3
arsine	AsH_3	stilbine	SbH_3	disilane	Si_2H_6
diphosphine	P_2H_4	diarsine	As_2H_4		

Names for ions and radicals

Cations

Monoatomic cations should be named in the same manner as the corresponding element, without change or suffix, e.g. Cu^+ the copper(I) ion.

A similar principle should also apply to the naming of monoatomic cations e.g., NO^+ the nitosyl cation, NO_2^+ the nitryl cation.

Polyatomic cations formed from the addition of an ion to another atom or cation will be regarded as complex and named as follows:

$$Al(H_2O)_6^{3+} \text{—hexaaquoaluminum ion.}$$

The addition of a proton to monatomic anions lead to the formation of -onium cationic species. Examples of these would be: iodonium, oxonium, sulphonium, selenium, tellurium, arsonium, phosphonium, stilbonium ions.

The main group of exceptions to the above rules are the *nitrogen bases*. The name ammonium for the NH_4^+ ion does not conform to the above rules but is accepted by common usage. Substituted ammonium ions derived from nitrogen bases with names ending in -amine will receive names formed by changing -amine to -ammonium. E.g., $HONH_3^+$, the hydroxylammonium ion.

Anions

The names for monoatomic anions shall consist of the name of the element and end in the termination -ide. Thus

H^-	hydride ion	D^-	deuteride ion
F^-	fluoride ion	Br^-	bromide ion
Cl^-	chloride ion	I^-	iodide ion
O^{2-}	oxide ion	S^{2-}	sulphide ion
Se^{2-}	selenide ion	Te^{2-}	telluride ion
N^{2-}	nitride ion	P^{3-}	phosphide ion
As^{3-}	arsenide ion	Sb^{3-}	antimonide ion
C^{4-}	carbide ion	Si^{4-}	silicide ion
B^{3-}	boride ion		

In the case of polyatomic the termination -ite is used to denote a low oxidation state anion. A number of trivial names are in common use.

NO_2^-	nitrite	ClO^-	hypochlorite
NOO_2^-	peroxonitrite	$N_2O_2^{2-}$	hyponitrite
$P_2H_2O_5^{2-}$	diphosphite (pyrophosphite)	PHO_3^{2-}	phosphite
$PH_2O_2^-$	hypophosphite	AsO_3^{3-}	arsenite
SO_3^{2-}	sulphite	$S_2O_5^{2-}$	disulphite (pyrosulphite)
$S_2O_4^{2-}$	dithionite	$S_2O_2^{2-}$	thiosulphite
SeO_3^{2-}	solenite	ClO_2^-	chlorite

NB: note that European literature uses sulphate for sulfate—American usage.

Radicals

A radical is a group of atoms which can be identified as a discrete entity and often occur as such in a number of compounds. The following table contains a selection of the more common radicals. The radical name for a species containing oxygen or other chalcogenide will end in -yl.

HO	hydroxyl	CO	carbonyl
NO	nitrosyl	SO	sulphinyl
SO_2	suphonyl	S_2O_5	pyrosulphuryl
SeO	seleninyl	SeO_2	selenonyl
CrO_2	chromyl	UO_2	uranyl
NpO_2	neptunyl	NO_2	nitryl
PO	phosphoryl	VO	vanadyl
PuO_2	plutonyl	ClO	chlorosyl
ClO_2	chloryl	ClO_2	perchloryl

In the case where radicals have different valences, the oxidation number of the characteristic element should be indicated by means of the Stock notation. For example, the uranyl group UO_2 may refer either to the ion UO_2^{2+} or to the ion UO_2^+; these can be distinguished as uranyl (VI) and uranyl (V) respectively. These polyatomic radicals are always considered as forming the positive part of the compound.

$COCl_2$	carbonyl chloride	NOS	nitrosyl sulphide
PON	phosphoryl nitride	$PSCl_2$	thiophosphoryl chloride
POCl	phosphoryl(III) chloride	IO_2F	iodyl fluoride
SO_2NH	sulphonyl imide	$SO_2(N_3)_2$	sulphonyl azide
$NO_2HS_2O_7$	nitryl hydrogen disulphide	S_2O_5ClF	pyrosulphurylchloride-fluoride

Acids and salts

Many compounds which now according to convention are called acids do not conform to rules classically defining an acid. In this area the common usage consists of a mixture of trivial and systematic names. In order to clarify the situation we will summarize the rules for various groups of acids and salts.

Binary and pseudobinary acids

Acids are normally designated by the termination -ide e.g. hydrogen chloride, hydrogen cyanide.

Polyatomic anions and associated acids

Acids giving rise to anions bearing names ending in -ate and -ite may also be designated using the termination -ic and -ous for the corresponding acid related to the anion -ate and -ite, respectively. For example chloric acid will give rise to chlorate, sulphuric acid to sulphate and phosphorous acid to phosphite. The same nomenclature can be used to describe the less common acids of hexacyanoferrate salts—hexacyanoferric acid.

Oxo acids

Many common acids contain oxygen and it is a long established custom not to indicate these oxygen atoms. Certain of these acids can be distinguished by their oxidation state and the characteristic termination -ic. The -ous names are reserved for acids corresponding to the -ite anions listed previously.

To distinguish between different acids prefixes are used. Hypo- denotes a lower oxidation state.

$H_4B_2O_4$	hypoboric acid	$H_2N_2O_2$	hyponitrous acid
$H_4P_2O_6$	hypophosphoric acid	HPH_2O_2	hypophosphorous acid
$HOCl$	hypochlorous acid similarly for other halogens		

Similarly the prefix -ic is used to designate a higher oxidation state. $HBrO_4$ perbromic acid. The prefix per- used to designate a higher oxidation state should not be confused with the prefix peroxo- which indicates the presence of the —O—O— grouping.

HNO_4	peroxonitric acid	H_2SO_5	peroxosulphuric acid
H_3PO_5	peroxophosphoric acid	$H_4P_2O_8$	peroxodiphosphoric acid
$H_2S_2O_8$	peroxosulphuric acid		

The prefixes ortho- and meta- have been designated for acids containing different 'water contents'. For acids derived from removing water from the corresponding ortho acid the systematic name is tetraoxo acid.

H_3BO_3	orthoboric acid	H_4SiO_4	orthosilicic acid
H_3PO_4	orthophosphoric acid	H_5IO_6	orthoperiodic acid
H_6TeO_6	orthotelluric acid	$(HBO_2)_n$	metaboric acid
$(H_2SiO_3)_n$	metasilicic acid	(HPO_3)	metaphosphoric acid

The prefix pyro- has been used to designate the acid formed from the elimination of water from two molecules of ortho acid.

The prefix thio- is used to indicate that oxygen has been replaced by sulphur in the corresponding oxo-acid.

$H_2S_2O_2$	thiosulphurous acid	$H_2S_2O_3$	thiosulphuric acid
$HSCN$	thiocyanic acid	H_3PO_3S	monothiophosporic acid
H_3AsS_3	trithioarsenious acid	$H_3PO_2S_2$	dithiophosphoric acid
H_3AsS_4	tetraarsenic acid	H_2CS_3	trithiocarbonic acid

The prefixes seleno- and telluro- may be used in a similar manner.

Names of oxo-acids

H_3BO_3	orthoboric acid	$(HBO_2)_n$	metaboric acid
$(HBO_2)_3$	trimetaboric acid	$H_4B_2O_4$	hypoboric acid
$HBrO_3$	bromic acid	$HBrO_2$	bromous acid
$HBrO$	hypobromous acid	H_2CO_3	carbonic acid
$HClO_4$	perchloric acid	$HClO_3$	chloric acid
$HClO_2$	chlorous acid	$HClO$	hypochlorous acid
H_2CrO_4	chromic acid	$H_2Cr_2O_7$	dichromic acid
H_5IO_4	(ortho)periodic acid	HIO_3	iodic acid
HIO	hyperiodic acid	$HMnO_4$	permanganic acid
H_2MnO_4	manganic acid	HNO_3	nitric acid
HNO_4	peroxonitric acid	HNO_2	nitrous acid
$HOONO$	peroxonitrous acid	$H_2N_2O_2$	hyponitrous acid

oxo-acids (continued)

HNCO	isocyanic acid	HOCN	cyanic acid
HONC	fulminic acid	H_3PO_4	(ortho)phosphoric acid
$H_4P_2O_7$	diphosphoric or pyrophosphoric acid	$H_{n+2}P_nO_{3n+1}$	polyphosphoric acid
		$(HPO_3)_3$	trimetaphosphoric acid
$H_5P_3O_{10}$	triphosphoric acid	$H_4P_2O_8$	peroxodiphosphoric acid
$(HPO_3)_n$	metaphosphoric acid	$H_2PH_2O_2$	phosphorous acid
$(HPO_3)_4$	tetrametaphosphoric acid	$(HO)_2P-$	
H_3PO_5	peroxo(mono)phosphoric acid	$O-PO(OH)_2$	diphosphoric(III,V) acid
$(HO)_2OP-$		H_3AsO_4	arsenic acid
$PO(OH)_2$	hypophosphoric acid	$HReO_4$	perrhenic acid
$H_4P_2O_5$	diphosphorous or pyrophosphorous acid	$HSb(OH)_6$	hexahydroxoantimonic acid
		H_4SiO_4	orthosilicic acid
HPH_2O_2	hypophosphorous acid	$H_2S_2O_7$	disulphuric or pyrosulphuric acid
H_2AsO_3	arsenious acid		
H_2ReO_4	rhenic acid	$H_2S_2O_8$	peroxodisulphuric acid
H_4SiO_4	orthosilicic acid	$H_2S_2O_6$	disulphurous or pyrosulphurous acid
H_2SO_4	sulphuric acid		
H_2SO_5	peroxo(mono)sulphuric acid	$H_2S_2O_4$	dithionous acid
$H_2S_2O_3$	thiosulphuric acid	$H_2S_xO_6$	
$H_2S_2O_2$	thiosulphurous acid	$(x = 3, 4, ...)$	polythionic acid
H_2SO_2	sulphoxylic acid	H_2SeO_3	selenious acid
H_2SeO_4	selenic acid	$HTcO_4$	pertechnetic acid
$HTeO_6$	(ortho)telluric acid		
H_2TcO_4	technic acid.		

Salts containing acid hydrogens

The addition of the word hydrogen in the name of a salt denotes the presence of replaceable hydrogen. For example:

$NaHCO_3$ sodium hydrogen carbonate
NaH_2PO_4 sodium dihydrogen phosphate
$NaH(PHO_3)$ sodium hydrogen phosphite

Double salts

When hydrogen is present it should be considered as a cation and would follow the nomenclature indicated above.

Salts can also be obtained containing more than one cation and they are named as follows:

$KMgF_3$ potassium magnesium fluoride
$KNaCO_3$ potassium sodium carbonate
$NaNH_4HPO_4 . 4H_2O$ sodium ammonium hydrogen phosphate tetrahydrate

Basic salts

The designation oxy- or hydroxy- indicates that the salt is basic and are usually named as follows:

Mg(OH)Cl	magnesiumhydroxidechloride
$CuCl_2 3Cu(OH)_2$	dicoppertrihydroxidechloride

The subject of the nomenclature for coordination complexes and inorganic ligands will not be considered here. There are a number of detailed discussions of this topic and the reader is referred to the original literature.

ORGANIC CHEMICAL NOMENCLATURE

Hydrocarbons

Acyclic hydrocarbons (linear)

The nomenclature used for the acyclic hydrocarbons is the backbone upon which the designation of all other saturated unbranched acyclic molecules are based. In this area a number of trivial names are retained, however a more systematic nomenclature exists for the higher members of this homologous series. The generic name of this group of molecules is 'alkane' and their common termination is '-ane'.

Code for alkanes

The value of *n* in the following table indicates the length of the hydrocarbon chain and the molecules conform to the general formula

$$C_nH_{2n+2}$$

n	Name	n	Name	n	Name	n	Name
1	Methane	11	Undecane	21	Heneiocosane	31	Hentriacontane
2	Ethane	12	Dodecane	22	Docosane	32	Dotriacontane
3	Propane	13	Tridecane	23	Tricosane	33	Tritiacontane
4	Butane	14	Tetradecane	24	Tetracosane	40	Tetracontane
5	Pentane	15	Pentadecane	25	Pentacosane	50	Pentacontane
6	Hexane	16	Hexaecane	26	Hexacosane	60	Hexacontane
7	Heptane	17	Heptadecane	27	Heptacosane	70	Heptacontane
8	Octane	18	Octadecane	28	Octacosane	80	Octacontane
9	Nonane	19	Nonadecane	29	Nonacosane	90	Nonacontane
10	Decane	20	Eisosane	30	Triacontane	100	Hectane
112	Dodecconta-hectane	122	Dodoconta-hectane	191	Unnonaconta-hectane		

The corresponding radical generated by the loss of a hydrogen atom and having the formula:

$$C_nH_{2n+1}^{.}$$

will have the same generic designation as the hydrocarbon from which it is derived but will be terminated in '-yl'.

Examples $CH_3^{\cdot 3}$ methyl C_6H_{13} hexyl

Acyclic hydrocarbons (branched)

The nomenclature for branched chain hydrocarbons is based on the name of the equivalent hydrocarbon with the longest chain. The branch chain structure is designated as numerated prefix. For example:

2-methylpentane

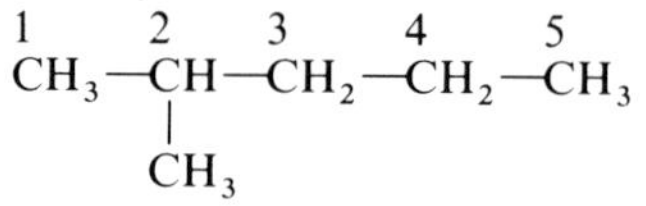

$$\overset{1}{CH_3}-\overset{2}{CH}-\overset{3}{CH_2}-\overset{4}{CH_2}-\overset{5}{CH_3}$$
$$\underset{CH_3}{|}$$

2,4-dimethylpentane

$$\overset{1}{CH_3}-\overset{2}{CH}-\overset{3}{CH_2}-\overset{4}{CH}-\overset{5}{CH_3}$$
$$\underset{CH_3}{|}\qquad\underset{CH_3}{|}$$

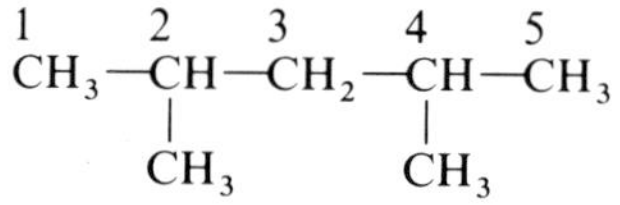

The numbering convention for branched chains requires that the sum of the numerical values should be a minimum value. For example:

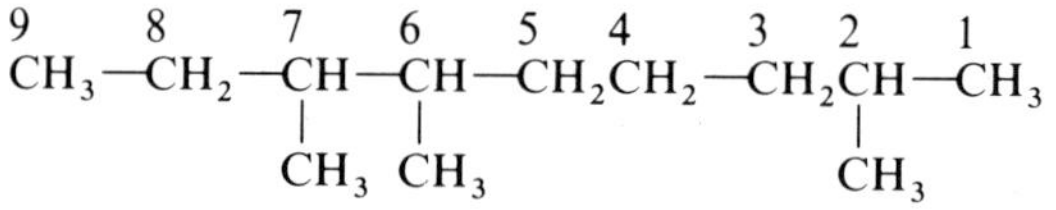

$$\overset{9}{CH_3}-\overset{8}{CH_2}-\overset{7}{CH}-\overset{6}{CH}-\overset{5}{CH_2}\overset{4}{CH_2}-\overset{3}{CH_2}\overset{2}{CH}-\overset{1}{CH_3}$$
$$\underset{CH_3}{|}\ \underset{CH_3}{|}\qquad\qquad\underset{CH_3}{|}$$

This should be designated 2,6,7-trimethylnonane rather than 3,4,8-trimethylnonane.

Multibranched chains

If the molecule contains two or more side chains the accepted designation will be the one which is either least complex or lowest alphabetical order, the lowest numerical rule being also operative.

Example:

Order based on complexity

5-methyl-6-methyldecane

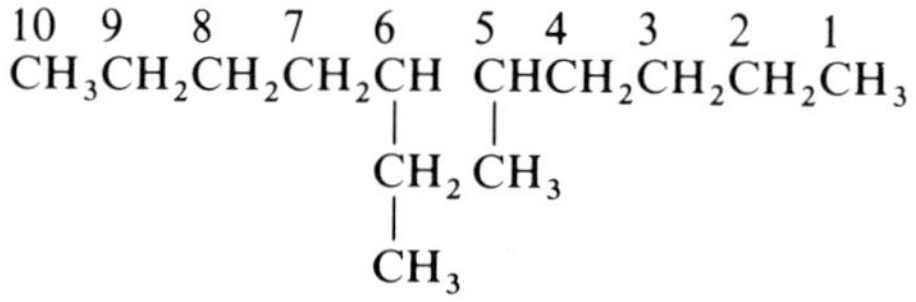

$$\overset{10}{CH_3}\overset{9}{CH_2}\overset{8}{CH_2}\overset{7}{CH_2}\overset{6}{CH}\ \overset{5}{CH}\overset{4}{CH_2}\overset{3}{CH_2}\overset{2}{CH_2}\overset{1}{CH_3}$$
$$\qquad\qquad\underset{CH_2}{|}\ \underset{CH_3}{|}$$
$$\qquad\qquad\underset{CH_3}{|}$$

5-propyl-6-isopropyldecane

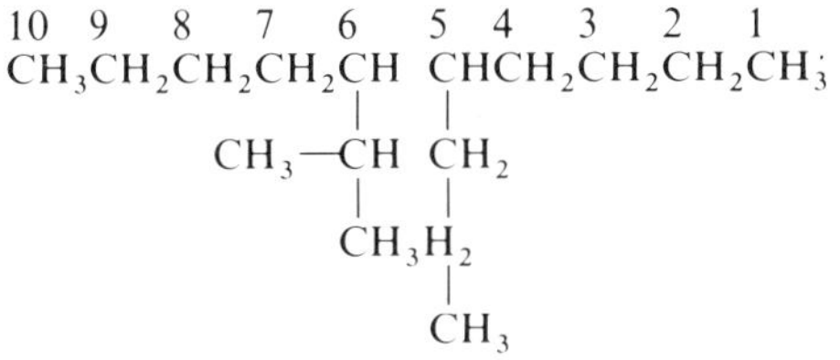

```
10  9    8    7    6    5   4    3    2    1
CH₃CH₂CH₂CH₂CH  CHCH₂CH₂CH₂CH₃
              |     |
        CH₃—CH  CH₂
              |     |
            CH₃H₂
                   |
                  CH₃
```

Alphabetical order

5-ethyl-6-methyldecane

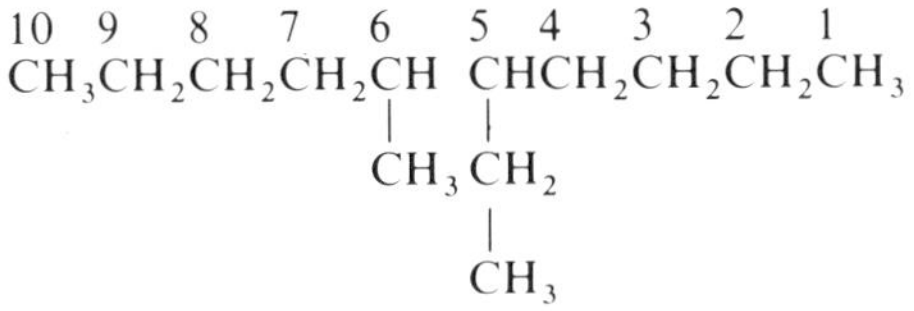

```
10  9    8    7    6    5   4    3    2    1
CH₃CH₂CH₂CH₂CH  CHCH₂CH₂CH₂CH₃
              |     |
           CH₃CH₂
                   |
                  CH₃
```

5-isopropyl-6-propyldecane

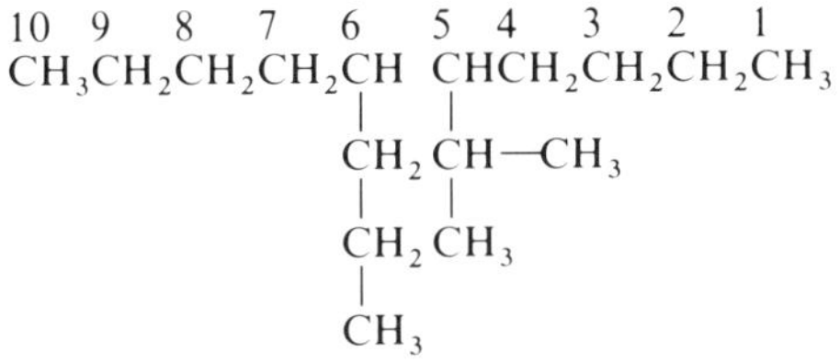

```
10  9    8    7    6    5   4    3    2    1
CH₃CH₂CH₂CH₂CH  CHCH₂CH₂CH₂CH₃
              |     |
           CH₂ CH—CH₃
              |     |
           CH₂ CH₃
              |
             CH₃
```

Radicals   Univalent branched radicals are designated by prefixing the side chains to the name of the unbranched alkyl radical possessing the longest possible chain starting from the carbon atom with the free valence, the starting atom being numbered as 1. Examples:

```
                 7    6    5    4    3    2    1
1-methylheptyl   CH₃CH₂CH₂CH₂CH₂CH₂CH(CH₃)—
```

```
                 7    6    5    4    3    2         1
2-methylhelptyl  CH₃CH₂CH₂CH₂CH₂CH(CH₃)CH₂—
```

```
                 7    6    5         4    3    2    1
5-methylheptyl   CH₃CH₂CH(CH₃)CH₂CH₂CH₂—
```

If there are two possible names which can be used the accepted one will be the least complex of the alternatives.

Acyclic hydrocarbons—double or triple bonds

The presence of a single double bond is designated by changing the ending to '-ene' of the equivalent hydrocarbon molecule. If there are two or more double bonds, the ending will be '-adiene', 'atriene' etc. The presence of a triple bond is designated by the ending '-yne'. If a molecule contains both double and triple bonds then the numbering designates the order of the nomenclature. This may in certain cases lead to a molecule ending in '-yne' rather than 'ene', when the alternative designation would lead to a higher number for the double bond. Examples:

$$\underset{\text{3-hexene}}{} \quad \overset{6}{C}H_3\overset{5}{C}H_2\overset{4}{C}H=\overset{3}{C}H\overset{2}{C}H_2\overset{1}{C}H_3$$

$$\underset{\text{1,4-butadiene}}{} \quad \overset{4}{C}H_2=\overset{3}{C}H-\overset{2}{C}H=\overset{1}{C}H_2$$

$$\underset{\text{1,3-hexadiene-5-yne}}{} \quad \overset{6}{C}H\equiv\overset{5}{C}H-\overset{4}{C}H=\overset{3}{C}H-\overset{2}{C}H=\overset{1}{C}H_2$$

$$\underset{\text{1-Hexen-4-yne}}{} \quad \overset{6}{C}H_3\overset{5}{C}\equiv\overset{4}{C}-\overset{3}{C}H_2-\overset{2}{C}H=\overset{1}{C}H_2$$

Monocyclic hydrocarbons

The generic name is derived from the equivalent linear hydrocarbon and prefixed with 'cyclo'. If the ring is saturated then the nomenclature will follow the alkane series and hence will be called 'cycloalkane'. If the molecule contains double bonded units then the generic name is obtained by changing the ending from '-ane' to 'yl', the position of the double bond being associated with carbon atom 1. This series will therefore be called 'cycloalkyl'. The introduction of more than one double bond leads to the designates 'adiene', 'atriene' etc. As before the correct nomenclature corresponds to the one with the lowest possible numerical value for the designated double and or triple bonds. The latter will be designated by changing the ending to '-yne' for the corresponding cycloalkane. Examples:

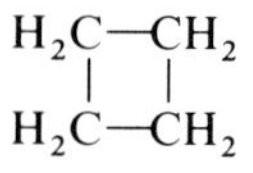

cyclobutane

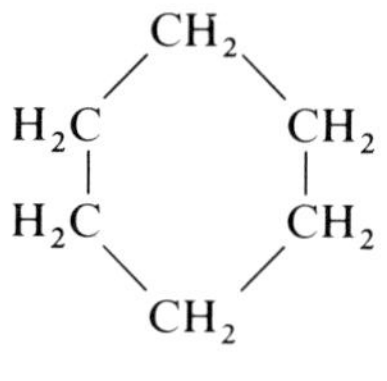

cyclohexane

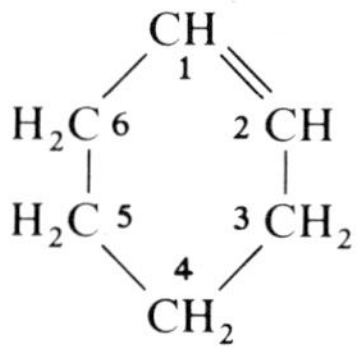

cyclohexene

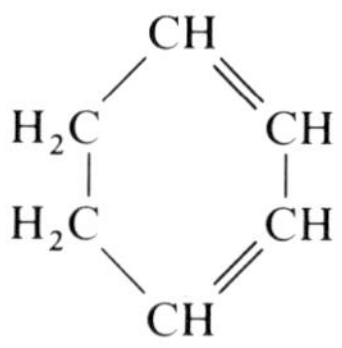

1,3 cyclohexadiene

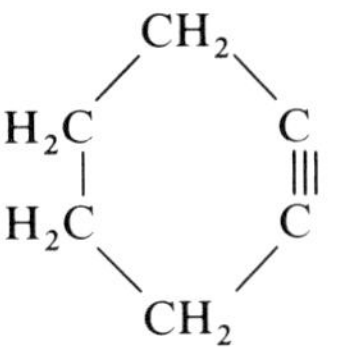

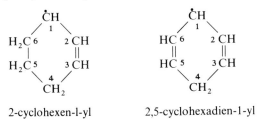

$$CH_2$$
$$H_2C \qquad C$$
$$H_2C \qquad C$$
$$CH_2$$

cyclohexyne

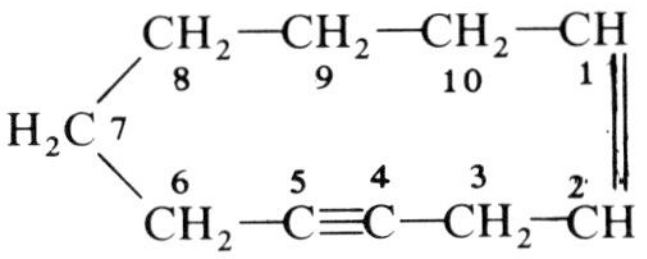

$$CH_2{-}CH_2{-}CH_2{-}CH$$
$$\quad 8 \qquad 9 \qquad 10 \qquad 1$$
$$H_2C\ 7$$
$$\quad 6 \qquad 5 \quad 4 \quad 3 \quad 2$$
$$CH_2{-}C{\equiv}C{-}CH_2{-}CH$$

1-cyclohexene-4-yne

Certain trivial names are accepted e.g. fulvene for methylene-cyclopentadiene and benzene for 1,3,5 cyclohexatrine.

Radicals The name of the radical is generated from the equivalent designation of its saturated homologue the termination being changed to 'yl', the carbon atom with the free valence being numbered as 1. in the case of saturated ring systems. In the case of rings which contain unsaturated linkages double or triple bonds, the point of attachment of the radical is designated as 1. With these latter systems the double and triple bonds are numbered to give the lowest numerical value. The radical of benzene retains its trivial name 'phenyl'. Examples:

$$\overset{\bullet}{C}H$$
$$\quad 1$$
$$H_2C\ 6 \qquad 2\ CH$$
$$H_2C\ 5 \qquad 3\ CH$$
$$\quad 4$$
$$CH_2$$

2-cyclohexen-l-yl

$$\overset{\bullet}{C}H$$
$$\quad 1$$
$$HC\ 6 \qquad 2\ CH$$
$$HC\ 5 \qquad 3\ CH$$
$$\quad 4$$
$$CH_2$$

2,5-cyclohexadien-1-yl

Aromatic molecules

The nomenclature follows the rules already indicated except that a number of trivial names are allowed. The benzene ring may be numbered so as to give the lowest possible numerical value. For disubstituted benzenes the o-(orth), m-(meta) and p-(para) convention for the 1,2-, 1,3- and 1,4-substituted ring systems is acceptable. When equivalent designates exist alphabetical order should take precedence.

Trivial names

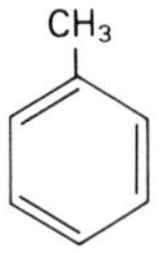

$$CH_3$$

Toluene

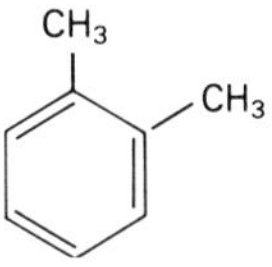

$$CH_3$$
$$CH_3$$

σ Xylene

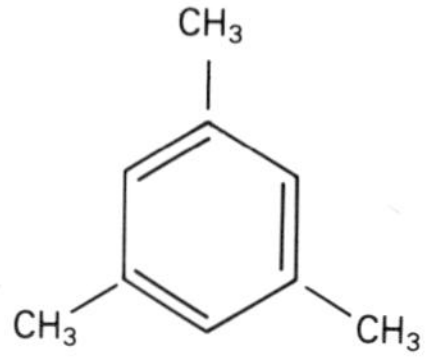

Mesitylene

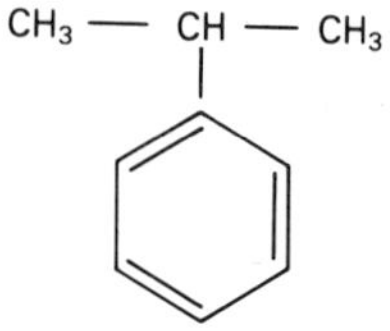

Cumene

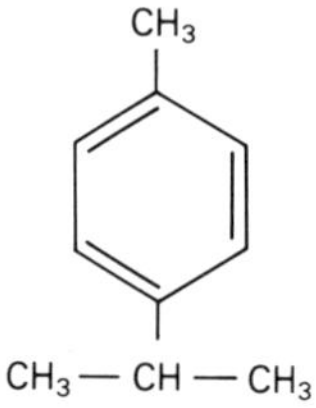

p-Cymene

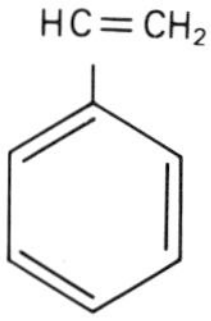

Styrene

Examples

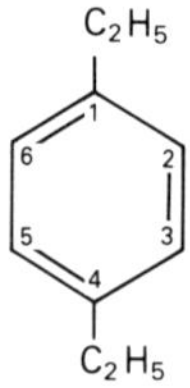

1,4 Diethyl benzene

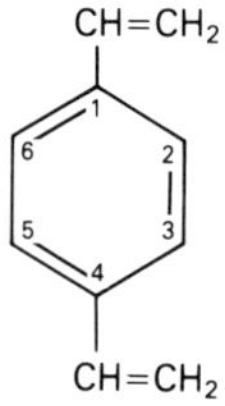

1,4 Divinyl benzene

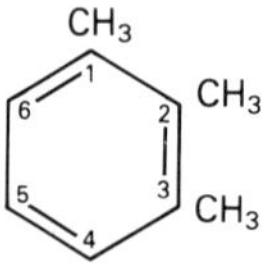

1,2,3 Trimethyl benzene

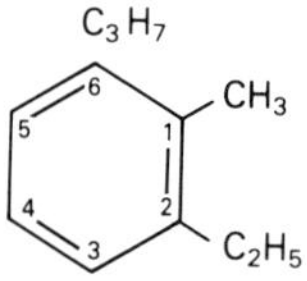

1 Methyl-2-ethyl-6-propyl
benzene

The generic name of monocyclic and polycyclic aromatic hydrocarbons is 'arene'.

Radicals Univalent radicals derived from monocyclic aromatic hydro-carbons and having the free valence at a ring atom are designated according to the appropriate ring system. In each case the position of the radical is indicated by the number 1. If the radical is resident on the side chain then the designate is the same as that for the linear chains except that certain trivial names are retained. The generic names of univalent and bivalent aromatic hydrocarbon radicals are 'aryl' and 'arylene' respectively.

Trivial names (site of valence electron)

Ring systems

Phenyl

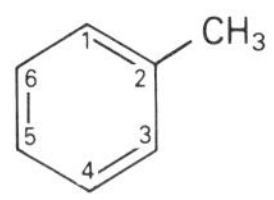

σ Tolyl

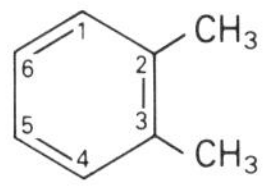

2,3-Xylyl

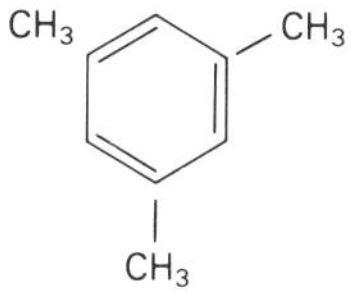

Mesityl

Side chain

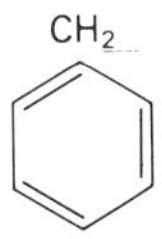

Benzyl

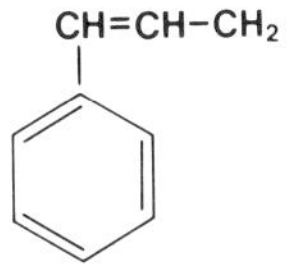

Cinnamyl

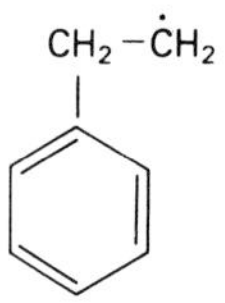

$$CH_2-\overset{\cdot}{C}H_2$$

Phenethyl

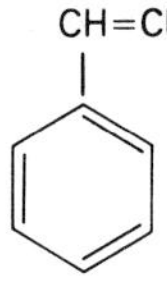

$$CH=CH$$

Styryl

NOMENCLATURE AND NOTATIONS IN POLYMER SCIENCE

Recent advances in spectroscopic techniques have provided the polymer scientist with the means of identifying and determining the stereochemistry of macromolecules in considerable detail. As a result the rules to be used for the description of a polymer structure have themselves required revision. A IUPAC produced report entitled 'Nomenclature Dealing with Steric Regularity in High Polymers' in 1962 and this has been revised several times. The most recent version, produced in 1979 has been used as the basis for the present document. In this section, stereochemical formula for polymer chains are shown as Fischer projections rotated through $90°$ or as hypothetical extended zigzag chains. This latter representation gives a clear indication of the three dimensional arrangement of the chain. The use of the Fischer projections is best illustrated by an example:

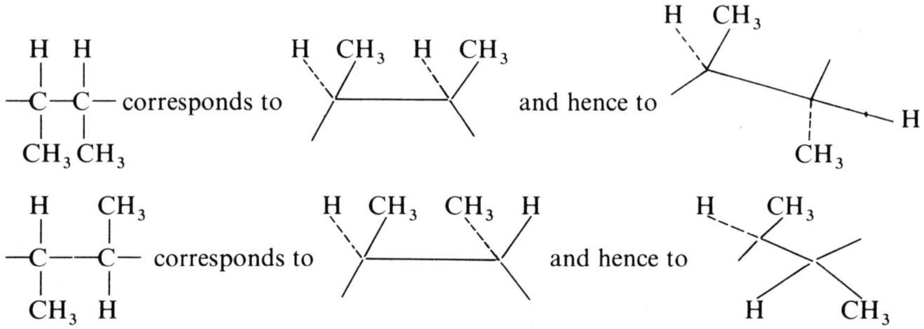

The above structures provide information on the spatial distribution of atoms and can be used to describe configurational placements, stereo-regularity etc. It is often more convenient to use the first representation of the chemical structure shown above to describe repeat units. For instance, the result of a vinyl polymerization can lead to a structure with different repeat unit sequences or alternatively two bracketted constitutional units may be regarded as different even though the repetition of either one of them would give the same regular polymer.

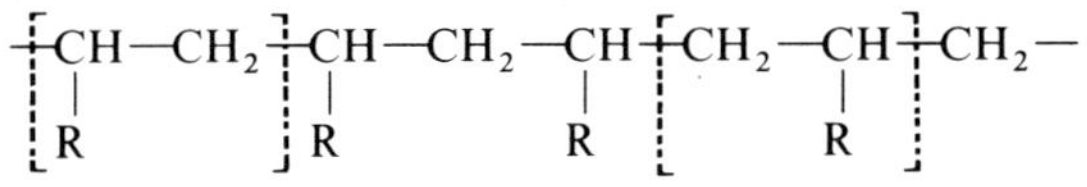

$$-CH-CH_2-CH-CH_2-CH-CH_2-CH-CH_2-$$
$$\quad\ |\qquad\quad\ |\qquad\quad\ |\qquad\quad\ |$$
$$\quad\ R\qquad\quad R\qquad\quad R\qquad\quad R$$

In order that definitions should be unambiguous it is necessary that the structures to which they refer are idealised. However in reality one has to face the fact that a polymer may not conform totally to one particular stereochemical structure. The presence of other structural forms although present in small concentrations can be best indicated by the phrases 'almost completely isotactic' or 'highly syndiotactic'. Although such expressions are imprecise and hence subject to criticism by purists, experienced polymer scientists will appreciate the value of using such terms.

Basic definitions

Constitutional repeating unit (CRU)   This is the designate repeating unit which is essentially independent of the method of preparation and is characteristic of the particular polymer species. The unit will usually be bivalent, higher valencies are possible but rarely encountered in practice, except in ladder polymers. The naming within the CRU depends on the seniority of the constituent groupings. The descending order of seniority among the types of bivalent groups is

—heterocyclic rings
—chains containing heteroatoms
—carbocyclic rings
—chains containing only carbon atoms.

This order is unaffected by the presence of rings, atoms or groups that are not part of the main chain. Example: $-(O-CH_2CH_2)_n-$ poly(oxyethylene) not poly(ethyleneoxy).

Configurational base unit   The elements of the polymer back bone can be sub-divided into one or more explicit stereoisomeric units. A constitutional repeating unit may, dependant upon its structure, be able to exist in one or more stereoisomeric forms.

Stereorepeating unit   A configurational repeating unit having defined configuration at all sites of stereoisomerism in the main chain of a polymer molecule.

Enantiomeric and distereoisomeric units   Two configurational units when joined can form a structure which can exist in two optically active forms;

enantometric units. These units will be non-superposable mirror images. Two non-superposable configurational units that correspond to the same constitutional unit are considered to be diastereoisomeric if they are not mirror images.

In a polymer such as polyvinyl chloride $-(CHCl-CH_2)_n-$, the constitutional repeating unit is $-CHCl-CH_2-$ and the corresponding configurational base units are

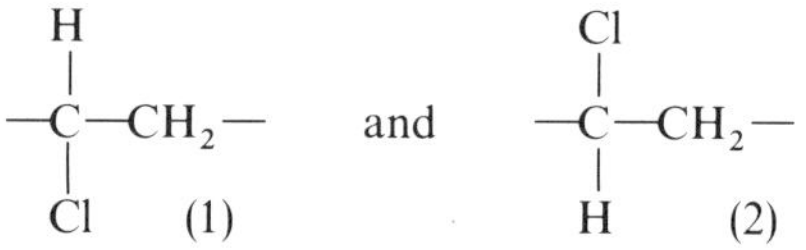

$$
\begin{array}{cccc}
& H & & Cl \\
& | & & | \\
-C&-CH_2- & \text{and} & -C-CH_2- \\
& | & & | \\
& Cl \quad (1) & & H \quad (2)
\end{array}
$$

The configurational base units (1) and (2) are enantiomeric, while the configurational units (1) and (3) cannot be enantiomeric because the constitutional units are different species, according to this nomenclature.

$$
\begin{array}{cccc}
& H & & H \\
& | & & | \\
-C&-CH_2- & & -CH_2-C- \\
& | & & | \\
& Cl \quad (1) & & Cl \quad (3)
\end{array}
$$

It does not matter whether (1) or (2) is taken as the configurational repeating unit and stererepeating unit of polyvinylchloride; since the infinite chains obtained from each are *not* enantiomeric and differ in the chain orientation. Within each pair of units, such as:

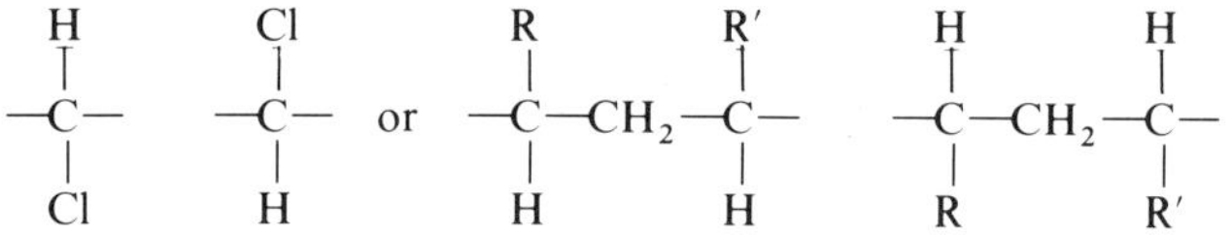

$$
\begin{array}{cccccc}
H & Cl & & R & R' & & H & H \\
| & | & & | & | & & | & | \\
-C- & -C- & \text{or} & -C-CH_2-C- & & & -C-CH_2-C- \\
| & | & & | & | & & | & | \\
Cl & H & & H & H & & R & R'
\end{array}
$$

the components are enantiometric since they are non-superposable mirror images, as defined above. However, with the constitutional unit $-CHR-CHR'-CH_2-CH_2-$ the two corresponding configurational units

$$
\begin{array}{c}
H \quad H \\
| \quad | \\
-C-C-CH_2-CH_2- \\
| \quad | \\
R \quad R'
\end{array}
$$

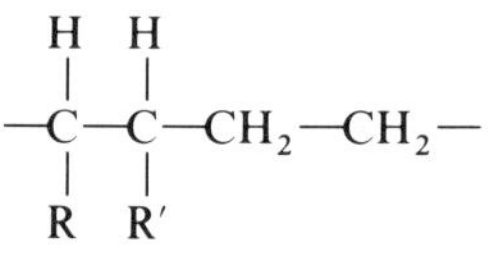

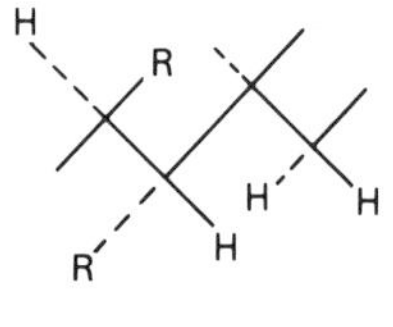

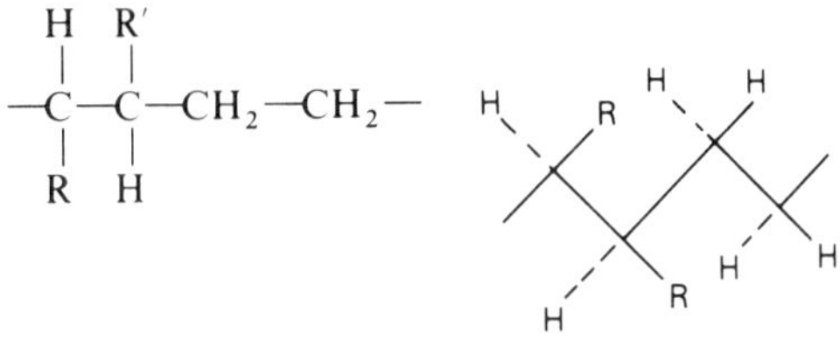

$$-\overset{\underset{\displaystyle R}{|}}{\underset{}{C}}-\overset{\underset{\displaystyle H}{|}}{\overset{\displaystyle R'}{C}}-CH_2-CH_2-$$

are diastereoisomeric. The units

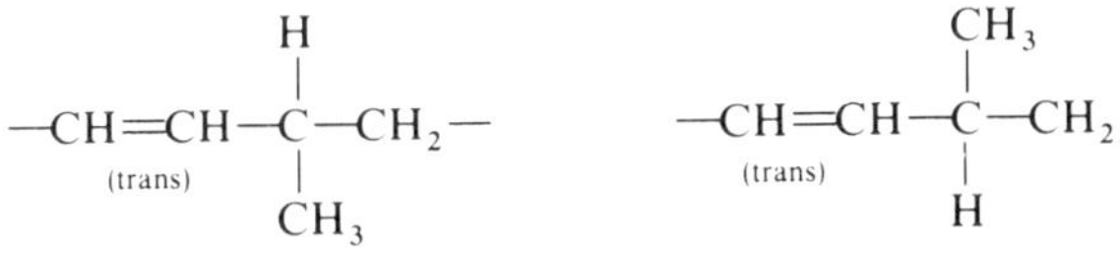

$$-CH=CH-\overset{\underset{\displaystyle CH_3}{|}}{\overset{\displaystyle H}{C}}-CH_2-$$ (trans)

$$-CH=CH-\overset{\underset{\displaystyle H}{|}}{\overset{\displaystyle CH_3}{C}}-CH_2$$ (trans)

are enantiomeric, while the units

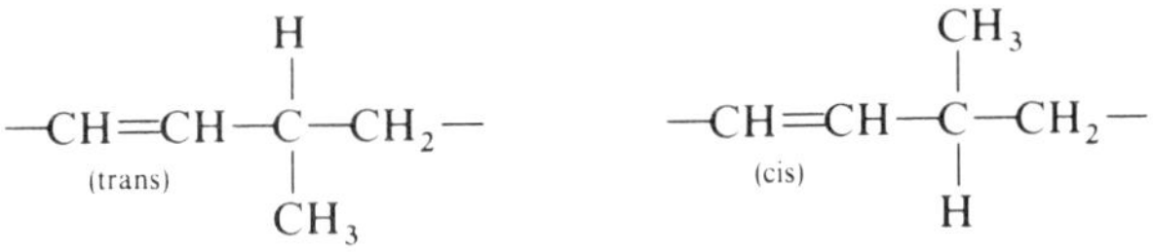

$$-CH=CH-\overset{\underset{\displaystyle CH_3}{|}}{\overset{\displaystyle H}{C}}-CH_2-$$ (trans)

$$-CH=CH-\overset{\underset{\displaystyle H}{|}}{\overset{\displaystyle CH_3}{C}}-CH_2-$$ (cis)

are not enantiometric, but diastereoisomeric. The simplest possible stereorepeating units in a stereoregular polyvinylchloride are

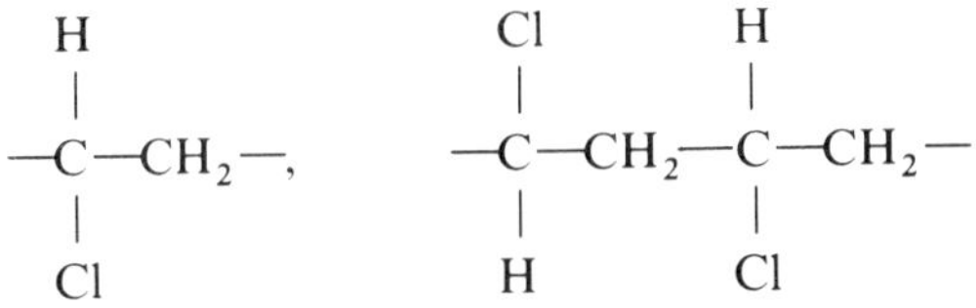

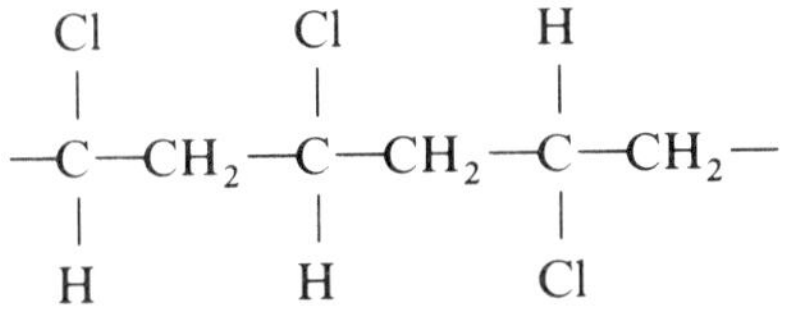

$$-\overset{\underset{\displaystyle Cl}{|}}{\overset{\displaystyle H}{C}}-CH_2-,$$

$$-\overset{\underset{\displaystyle H}{|}}{\overset{\displaystyle Cl}{C}}-CH_2-\overset{\underset{\displaystyle Cl}{|}}{\overset{\displaystyle H}{C}}-CH_2-$$

$$-\overset{\underset{\displaystyle H}{|}}{\overset{\displaystyle Cl}{C}}-CH_2-\overset{\underset{\displaystyle H}{|}}{\overset{\displaystyle Cl}{C}}-CH_2-\overset{\underset{\displaystyle Cl}{|}}{\overset{\displaystyle H}{C}}-CH_2-$$

and the corresponding stereoregular polymers are

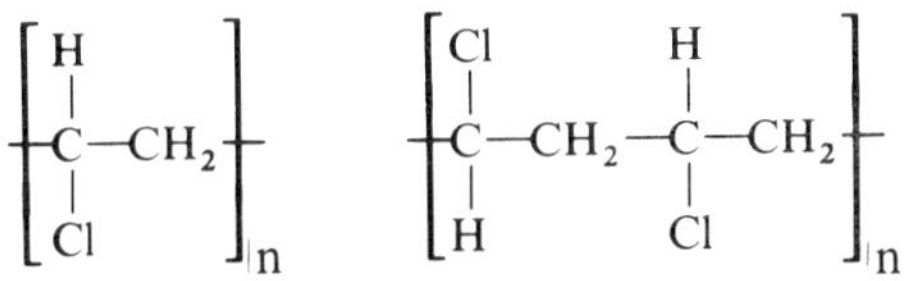

$$\left[\overset{\underset{\displaystyle Cl}{|}}{\overset{\displaystyle H}{C}}-CH_2\right]_n$$

Isotactic

$$\left[\overset{\underset{\displaystyle H}{|}}{\overset{\displaystyle Cl}{C}}-CH_2-\overset{\underset{\displaystyle Cl}{|}}{\overset{\displaystyle H}{C}}-CH_2\right]_n$$

Syndiotactic

$$\left[\begin{array}{c} \underset{|}{\overset{|}{\text{Cl}}} \\ \text{C} \\ | \\ \text{H} \end{array}\!\!=\!\!\text{CH}_2\!\!-\!\!\underset{|}{\overset{|}{\underset{\text{H}}{\overset{\text{Cl}}{\text{C}}}}}\!\!-\!\!\text{CH}_2\!\!-\!\!\underset{|}{\overset{|}{\underset{\text{Cl}}{\overset{\text{H}}{\text{C}}}}}\!\!-\!\!\text{CH}_2\!\!\right]_n$$

Heterotactic

Tacticity and tactic polymer   A stereoregular polymer is designated as being tactic, its stereochemical structure is described by the tacticity. Three tactic forms are defined on the basis of the arrangement in space of consecutive monomer units.

a) *Isotactic* (same side). A regular polymer in which the chiral or prochiral atoms in the main chain are identical in their arrangement.

b) *Syndiotactic* (alternating sides). A regular polymer the monomer units of which alternate in terms of their conformational base, they are enantiomeric. In a syndiotactic polymer, the configurational repeating unit consists of two configurational base units that are enantiometric.

c) *Atactic*. A regular polymer, the monomer units of which are randomly distributed so as not to form a particular configurational sequence. The polymer should ideally contain equal numbers of the possible configurational base units.

Isotactic and syndiotactic polymers are designated as being stereoregular polymers. The above definitions do not indicate the range of order. Spectroscopic techniques are often sensitive to interactions ranging over diad, triads and higher-mers. Stereoregular polymers may therefore be designated in .terms of the concentration of particular stereochemical sequences.

Degrees of triad iso-, syndio- and hetero-tacticity   The fractions of the triads in a regular vinyl polymer that are of the mm, rr and mr=rm types can be assigned to their stereochemical designates. In cases where triad analysis is not attainable, the diad isotactic and syndiotactic analysis will give the concentrations of m and r types respectively.

Nomenclature for chain backbones

The nomenclature for a polymer is simply to add the prefix poly- to the appropriate constitutional repeating unit–monomer equivalent. Example:

$$CH_2\!=\!CH_2 \qquad \text{―}(CH_2)_n\text{―} \qquad \text{poly(methylene)}$$

the value of 'n' should in theory designate the degree of polymerization, the number of monomers units in the polymer. For a regular single stranded polymer constructed from bivalent units the name conforms to the structure poly(bivalent constitutional repeating unit). In those cases in which a choice is possible between a bivalent and a higher valent CRU, the bivalent is always preferred. The principle of minimizing the number of free valences in the CRU supersedes all orders of seniority. The application of this principle is best illustrated by the following examples:

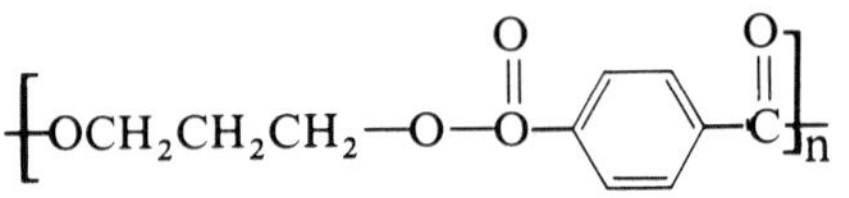

$$\left[OCH_2CH_2CH_2-O-\overset{\overset{\displaystyle O}{\|}}{O}-\left\langle \bigcirc \right\rangle-\overset{\overset{\displaystyle O}{\|}}{C} \right]_n$$

poly(oxypropyleneoxyterephthaloyl)

The CRU does not imply the method of preparation of the polymer, the name is simply that of the largest identifiable repeating unit.

Copolymer systems

Random In this type of polymer the sequence of the two monomer units forming the polymer varies randomly down the chain. Example a copolymer of A and B can be designated as:

A—B—B—A—B—A—A—A—B—B—A—B   (Random)

Alternating Certain polymerizations lead to an alternation of the sequence structure and the term refers to the structure:

—A—B—A—B—A—B—A—B—A—

Block copolymers A sequence structure in which the constituent units are separated into defined blocks is referred to as a diblock copolymer and has structure:

A—A—A—A—A—B—B—B—B—B

This is a diblock copolymer of A and B. Another possible sequence is

A—A—A—A—A—B—B—B—B—B—B—B—A—A—A—A—A

This is an ABA triblock copolymer.

Graft copolymers If the side chain of a polymer is different from that of the back bone it can be designated as a graft copolymer. Hence if the back bone is B and the side chain A, then we have an A graft copolymer of B.

Identification of the simplest constitutional repeating unit

The preferred nomenclature should be chosen on the basis of the lowest numbering of a particular substituent. Hence for deuterated polymer we might have:

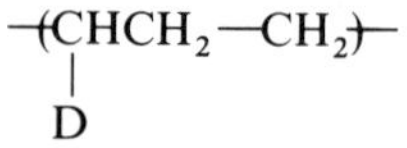

$$-\!\!\!\!-(\text{CHCH}_2\!\!-\!\!\text{CH}_2)\!\!-\!\!\!\!-$$

poly(trimethylene-d_1)

with D substituent

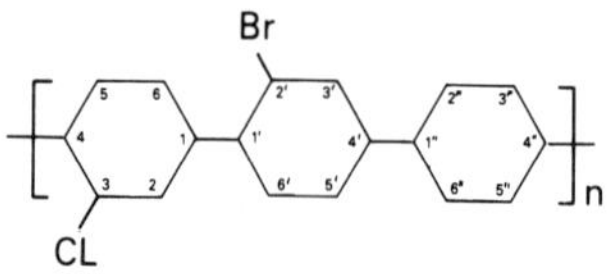

poly(3-chloro-2'-bromo-p-terphenyl-4,4''-ylene)

If a bivalent group has a numbering system explicitly determined by the monomer and formed backbone structure then the polymer will follow that system. Examples of the application of this rule to ring structures are shown below.

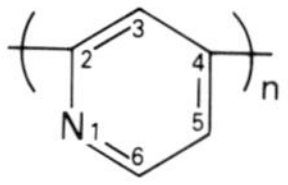

Poly(2,4-pyridinediyl)

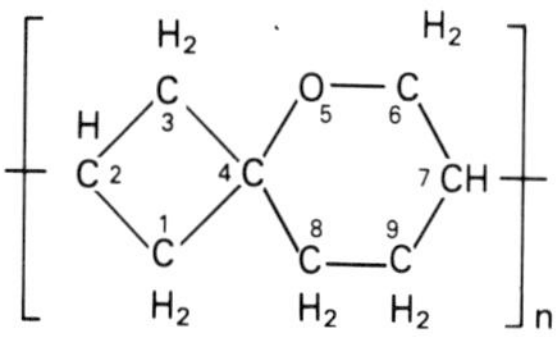

Poly(5-oxaspiro[3,5]non-2,7-ylene)

Bivalent constitutional repeating unit having two or more subunits

These polymer are regular alternating copolymers, however, they are often a consequence of a particular chemistry and as such are not designated as copolymers. The name accepted is that of the largest possible sub unit and the elements are named left to right in the order in which they appear in the polymer. If there is more than one

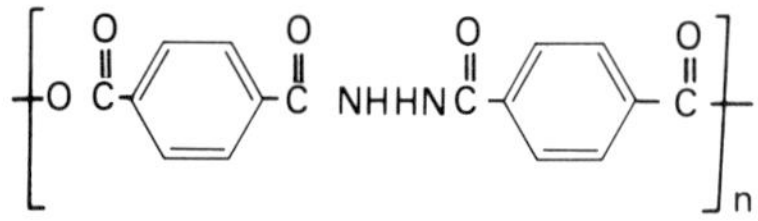

Poly(oxyterephthaloylhydrazoterephthaloyl)

possible way of naming the sub-unit then the following rules should be applied. The order of seniority among the types of bivalent groups is (1) heterocyclic rings, followed by (2) chains containing hetero atoms, (3) carbocyclic rings, and finally chains containing only carbon. When two rings contain the same number of heteroatoms then the one with the least number of hydrogen atoms should take precedence. The order of seniority for heteroatoms is O, S, Se, Te, N, As, Sb, Bi, Si, Ge, Sn, Pb, and Hg, other atoms may be placed in the order in which they appear in the periodic table.

Substituents

In many polymers the substituent can be included in the name of the CRU without requiring application of any particular order. If this is not the case then substituents can be indicated according to the position which they occupy either on the side chain or within the CRU. The name is included in that of the sub-unit by means of prefixes appended to the name of the sub-unit to which they are bound. Example:

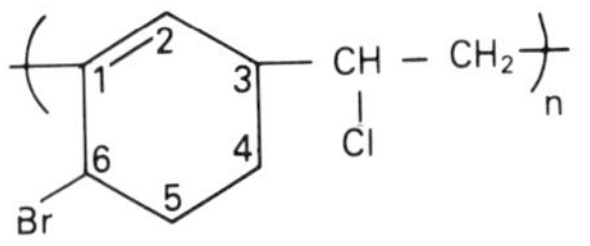

Poly[(6-bromo-1-cyclohexen-1,3-ylene)(1-chloro ethylene)]

End groups

The end group may be designated by placing a prefix to the name of the polymer. The left hand head group is designated α and the right hand is designated ω. In practice this is rarely used except in the definition of oligomeric species where the nature of the end group may have significant effects on the properties of the polymer. Example:

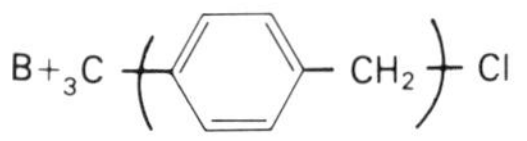

α-(tribromomethyl)-ω-chloro-poly(1,4-phenylene methylene)

Trivial names

Although the above rules allow an unambiguous definition of a polymer in terms of its IUPAC nomenclature, in practice trivial names are widely used and often indicate the source from which the polymeric material is generated.

Name	Structure	Trivial name
poly(methylene)	$-(CH_2CH_2)_n-$	polyethylene
poly(propylene)	$-(CH-CH_2)-$ $\quad\vert$ $\quad CH_3$	polypropylene
poly(1,1-dimethylethylene)	CH_3 $\quad\vert$ $-(C-CH_2)_n-$ $\quad\vert$ $\quad CH_3$	polyisobutylene
poly(1-methyl-1-butenylene)	$-(C=CHCH_2CH_2)_n-$ $\quad\vert$ $\quad CH_3$	polyisoprene
poly(1-butenylene)	$-(CH=CHCH_2CH_2)_n-$ `	polybutadiene
poly(1-phenylethylene)	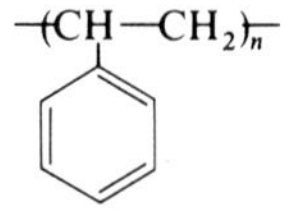 $-(CH-CH_2)_n-$ (phenyl)	polystyrene
poly(1-cyanoethylene)	$-(CHCH_2)_n-$ $\quad\vert$ $\quad CN$	polyacrylonitrile
poly(1-hydroxyethylene)	$-(CH-CH_2)_n-$ $\quad\vert$ $\quad OH$	poly(vinylalcohol)
poly(1-chloroethylene)	$-(CH-CH_2)_n-$ $\quad\vert$ $\quad Cl$	poly(vinylchloride)
poly(1-acetoxyethylene)	$-(CH-CH_2)_n-$ $\quad\vert$ $\quad OOCCH_3$	poly(vinylacetate)
poly(1,1-difluoroethylene)	F $\quad\vert$ $-(C-CH_2)_n-$ $\quad\vert$ $\quad F$	poly(vinylidenefluoride)
poly(difluoromethylene)	$F\quad F$ $\;\vert\quad\vert$ $-(C-C)_n-$ $\;\vert\quad\vert$ $F\quad F$	poly(tetrafluoroethylene)
poly((2-propyl-1,3-dioxane-4,6-diyl)-methylene)	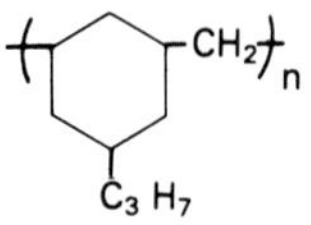 (ring structure with $-CH_2-$) $_n$ C_3H_7	poly(vinylbutryl)

]

Name	Structure	Trivial name
poly(1-(methyoxycarbonyl) ethylene)	$-(CHCH_2)_n-$ $\|$ $COOCH_3$	poly(methylacrylate)
poly(1-(methoxycarbonyl)-1-methyl-ethylene)	CH_3 $\|$ $-(C-CH_2)_n-$ $\|$ $COOCH_3$	poly(methyl methacrylate)
poly(oxymethylene)	$-(OCH_2)_n-$	poly(formaldehyde)
poly(oxyethylene)	$-(OCH_2CH_2)_n-$	poly(ethylene oxide) (NB: sometimes called polyethylene glycol)
poly(phenyleneoxide)	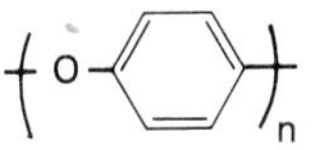	poly(phenyleneoxide)
poly(oxyethylẹneoxytereph-thaloyl)	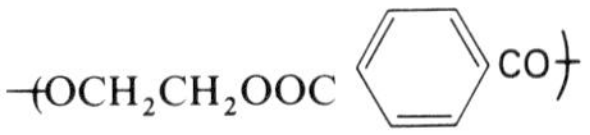 $-(OCH_2CH_2OOC$ $CO)-$	poly(ethylene terephthalate)
poly(iminohexamethyleneimi-noadipoyl)	$-(NH(CH_2)_6NHCO(CH_2)_4CO)_n-$	poly(hexamethylene adipamide)

Wiswesser chemical line notation

This notation has become important with recent attempts at simplifying nomenclature to assist with computer retrieval of information. The system establishes both the topology and make up of the molecule by means of a unique and unambiguous code. The symbols used are standard characters available on most data processing equipment; the 26 capital letters (A to Z), the 10 digits ($\emptyset$ to 9), the zero is written using the normal computer notation $\emptyset$ to distinguish it from O, four punctuation marks (-*/&) and the blank space.

The WLN retains unchanged most of the familiar atomic symbols, setting off two letters between hyphens—lithium is -LI-. There are a number of important exceptions; K, U, V, W and Y and certain of the halogen atoms. Adoption of this approach allows an economy of use and the symbols and their designates are presented in Table I. The carbon atom is only designated by a C when it is an unbranched carbon atom multiply bonded to a heteroatom or doubly bonded to two other carbons—allenes. Unsaturation is designated by the letter U—singly designating a double bond and UU designating a triple bond. Examples of the application of the above coding are illustrated below:

$$CH_3CH_2CH_2N=NCH_3, \qquad CH_2=C=CH_2, \qquad N\equiv CCH_2CH_2CH_3$$

$$3NUN1 \qquad\qquad 1UCU1 \qquad\qquad NC3$$

Encoding straight chain compounds

There are three general rules to be followed:

a) translate the two dimensional structural diagram into a 'graphic formula' using the symbols in Table I.

b) choose a path which leads from one end of the molecule to the other.

c) convert the 'graphic formula' into a linear notation.

The application of the above rules are best illustrated with an example.

TABLE I
WLN specific symbols

Atomic symbol	WLN symbol	Moiety	WLN symbol
Br	E	$\diagdown N^+ \diagup$ / $\diagup$ $\diagdown$	K
Cl	G	—NH—	M
K	-KA-	—OH	Q
U	-UR-	·=·	U
V	-VA-	·≡·	UU
W	-WO-	—NH₂	Z
Y	-YT-	$\diagdown$ C=O $\diagup$	V
		$\diagdown \diagup$ C $\diagup \diagdown$	X
		$\diagdown$ C= or $\diagdown \diagup$ CH $\diagup$	Y
		O O $\diagdown \diagtwoup$	W
		⬡	R

Note the · on the ·=· symbol may indicate any suitable atom which is designated and sandwich the symbol U. Hence —N=N— is designated as NUN.

Structural diagram	Graphic formula	Notation
Cl—CH₂CH=CHCNH₂ ‖ O	G2U1VZ	ZV1U2G

For a straight chain compound the proper path begins at the termination of the chain having the later alphameric position. The symbols are encoded according to their ASCII numerical values, with the exception of * which is quoted last:

$$\& - / \emptyset\ 1\ 2\ 3 \ldots 9\ A\ B\ C\ D\ E\ F \ldots X\ Y\ Z\ *$$

First position last cited.   Last position first cited

Encoding branched chain compounds

The presence of branching points in an atom chain complicates the selection of the preferred path through the symbols and then requires that the side chain be folded into the linear representation of the molecular structure. The main path will therefore be the longest continuous path in the translated structure which passes through the maximum number of side chain structures. The priority for numbering side chains is governed by the following rules:

- the side chain with the smallest number of branch symbols
- the side chain with the smallest number of total symbols
- the side chain with the latest alphanumeric notation.

The points at which the side chain are folded into the main structure are indicated by an ampersand ($\&$). Example

Structure

$$CH_3CH_2\diagdown$$
$$CH_2OH$$
$$N-CH_2CH_2CH-CH_2-C-CH_3$$
$$CH_3CH_2CH_2\diagup \qquad CF_3 \qquad CH_2CH_2OH$$

<table>
<tr><td>Graphic formula</td><td>Notation</td></tr>
</table>

```
            F   Q
  2      F  X  F  I
         N  2  Y  I  X  I
  3               2
                  Q
```

Notation: Q2X1&IQIYXFFF2N3&1

It will be noted that separate symbols for hydrogen are not normally required and hence if no indication to the contrary exists it is assumed that the free valences are occupied by hydrogen.

Encoding aromatic derivatives

The phenyl ring is designated by a single R. Topographically the phenyl ring can either be a terminus, link or branch point in a symbol chain. For a disubstituted aromatic the point of attachment is indicated by a letter which is preceded in the notation by a space. The encoding within the ring follows the usual rules of precedence. When one of the substituents is branched then the ampersand is used to show the constitution of the side chain. Example:

Structure

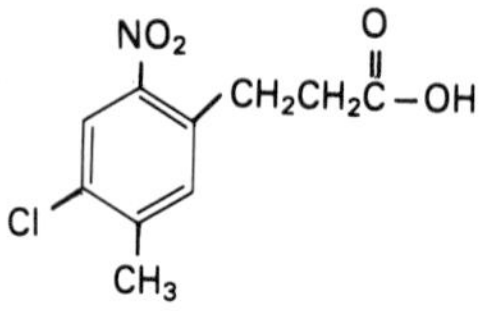

NO$_2$

CH$_2$CH$_2$C—OH (with O above C as double bond)

Cl

CH$_3$

Graphic formula

W
N
G R 2 V Q
I

Locant assignment

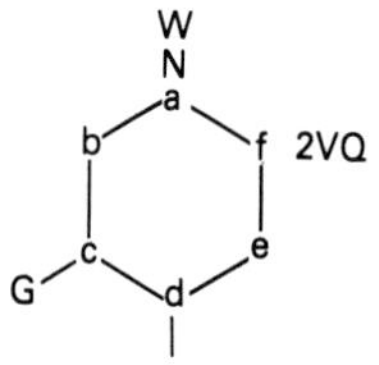

W
N
a
b   f   2VQ

c   e
G   d
I

Notation

WNR CQ D1 F2VQ

Encoding monocyclic compounds

The rules outlined above are insufficient to allow coding of a monocyclic system. The WLN notation indicates a carbocyclic by the format L...J with the requisite information contained between these symbols. Similarly for a chelate the symbols D...J and heterocyclic T...J are used. An initial L, D or T indicates the beginning of the ring description and the final J shows that the ring description is complete. Locants and symbols for the side chains follow the final J.

The topology and composition of a monocyclic system are described by citing after the initiating L, D or T symbol and before the final J the following:

a) a numeral corresponding to the number of atoms in the ring system. If the ring contains ten or more atoms then the numerals should be enclosed between hyphens.

b) ring segment symbols, these indicate groups directly attached to the ring for instance

the symbol Y for C= or V for C=O.

Examples

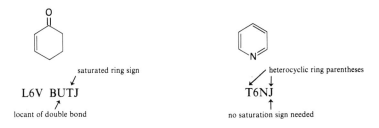

saturated ring sign
↙
L6V BUTJ
↗
locant of double bond

heterocyclic ring parentheses
↙ ↓
T6NJ
↑
no saturation sign needed

Encoding polycyclic ring systems

The WLN system is capable of describing complex ring systems, even as in the case of spiro compounds where the bridge is complex. The principle rule which underlines the naming is the choice of topological paths that form closed loops. The features are organized on the basis of the order of priority indicated above: ring segments, saturation signs and substituent information being indicated in the usual order.

Example

5-member ring
↙
T56 BNJ
↖
6-member ring

Bicyclic system

The above outline summarizes the main points of the WLN method. A number of particular points have not been discussed and the reader is referred to detailed discussions on this topic.

References

E. G. Smith and P. A. Baker, *The Wiswesser Line Formula Chemical Notation* (WLN) CIMI, P.O. Box 2740, Cherry Hill N.J., 08034 (1976).

A. Zamora and T. Ebe, Pathfinder II. A computer program that generates wiswesser line notations for complex polycylic structures, *J. Chem. Info. Comput Sci.*, **16**, 36–39 (1976).

T. Ebe and A. Zamora, Wiswesser line notation processing at chemical abstracts service, *J. Chem. Info. Comput. Sci.*, **16**, 33–35 (1976).

C. E. Granito and E. Garfield, Substructure search and correlation in the management of chemical information, *Naturwissenschaften*, **60**, 189–197 (1973).

D. A. Koniver, W. J. Wiswesser and E. S. Usdin, Wiswesser line notation: simplified techniques for converting chemical structures to WLN, *Science*, **176**, 1437–1439 (1972).

C. E. Granito, S. Roberts and G. W. Gibson, The conversion of WLN to ring codes: I. The conversion of ring systems, *J. Chem. Doc.*, **12**, 190–196 (1972).

Chemical abstracts registry handbook and registry number update

The REGISTRY HANDBOOK–Number Section provides a comprehensive method of relating Registry Numbers to the specific chemical substances they identify. This method of identifying a particular chemical by a coded number dates from 1965 and was constructed to provide an ambigious computer-language description of a compounds molecular structure including all stereochemical details. The way this system has been evolved leads to there being no chemical significance to its value; it being simply a machine checkable number assigned in sequential order to each substance as it enters the Registry System. This is in contrast to the Wiswesser Notation discussed in the previous section where the code is governed by the molecular structure. Chemical Abstracts Service (CAS) Registry Numbers appear in Chemical Abstracts issues and volume indexes as well as in the many CAS computer readable files. These registry numbers are also to be found in the National Library of Medicine's TOXLINE and CHEMLINE computer based information retrieval services. Thus despite the non systematic base we will consider the method used to identify a particular entry.

The Registry Handbook lists, in Registry Number (A) order, systematic Chemical Abstracts index names (B) and molecular formulas (C) for all substances recorded in the files of CAS, and was originally based on the files as they existed on December 31, 1971. Annual supplements to the RH–Number Section appear each year and identify all new additions for the preceding calendar year.

$$Ⓐ \qquad\qquad Ⓑ \qquad\qquad Ⓒ$$

$$93\text{–}89\text{–}0 \qquad \text{Benzoic acid, ethylester} \qquad C_9H_{10}O_2$$

Subsequently CA have produced a Registry Handbook–Registry Number Update which allows the reader to check whether the number in a particular list is the currently valid one to be used. As a consequence of errors having occurred and the assignment of numbers to materials identified by their formula names and trade identifications, multiple registration numbers exist for the same material. The Registry Update identifies these and provides the current values for particular materials.

The CAS Registry Number may be written in general form as $N_i \ldots N_4 N_3\text{–}N_2 N_1\text{–}R$ in which R represents the check digit and N_i through N_1 represents a fundamental sequential number. The check digit is derived from the following formula

$$\frac{iN_i + \cdots 4N_4 + 3N_3 + 2N_2 + 1N_1}{10} = Q + \frac{R}{10}$$

in which Q represents an integer which is discarded.

For example, 2-chloroethanol has the total CAS Registry Number 107–07–3, the validity of which is checked on input as follows:

$$\frac{(5 \times 1) + (4 \times 0) + (3 \times 7) + (2 \times 0) + (1 \times 7)}{10} = \frac{33}{10} = 3 + \frac{3}{10}$$

Hence, $Q = 3$ (disregarded) and R (the check digit) $= 3$.

For further details consult:

Chemical Abstracts Service
   Registry Handbook–Number Section (1965–71)
   Registry Handbook–Registry Number Update (1965–78)

Also the Cumulative lists issued annually. The above notes have been constructed from data provided in the above publications.

Acknowledgement

Extensive reference has been made to the following documents in the preparation of the section on nomenclature:

A. D. Jenkins, *Pure and Applied Chem.*, **51**, 1101–1121 (1979).
M. L. McGlashan, M. A. Paul and D. H. Wiffen, *Pure and Applied Chem.*, **51**, 1–41 (1979).

Mathematical Symbols, Units and Formulae

The ISO 31 International Standard Series, 'Part XI Mathematical signs and symbols for use in physical sciences and technology'; has recommended that the following convention with regards mathematical symbols should be generally adopted. Mathematical operators (for example d and Δ) and mathematical constants (for example e and π) should always be printed in Roman (upright) type. Letter symbols for numbers other than mathematical constants should be printed in italic type.

Definition	Symbol	Definition	Symbol
equal to	$=$	not equal to	$\neq$
identically equal to	$\equiv$	corresponds to	$\hat{=}$
approximately equal to	$\approx$	approaches	$\rightarrow$
asymptotically equal to	$\simeq$	proportional to	$\alpha \sim$
infinity	∞	less than	$<$
greater than	$>$	less than or equal to	$\leqslant$
greater than or equal to	$\geqslant$	much less than	$\ll$
much greater than	$\gg$		
plus	$+$	minus	$-$
multiplied by	$\times \ \ddagger$	a divided by b	$\dfrac{a}{b} \ \ a/b \ \ ab^{-1}$
magnitude of a (modulus)	$\lvert a \rvert$	a raised to the power n	a^n
square root of a	$a^{\frac{1}{2}} \ \sqrt{a}$	n'th root of a	$a^{1/n} \ \sqrt[n]{a}$
mean value of a	$\langle a \rangle \ \bar{a}$	natural logarith of a	$\ln a \ \log_e a$
decadic logarith of a	$\lg a \ \log_{10} a$	binary logarith of a	$\operatorname{lb} a \ \log_2 a$
exponential of a	$\exp a \ \mathrm{e}^a$		

Mathematical constants

$$\pi = 3.141\ 592\ 653\ 589\ 793\ 238\ 462\ 643\ 383\ 279$$
$$\mathrm{e} = 2.718\ 281\ 828\ 459\ 045\ 235\ 360\ 287\ 713\ 526$$
Euler's constant $\quad \gamma = 0.577\ 215\ 664\ 901\ 532\ 860\ 61$
Golden Ratio $\quad \phi = 1.618\ 033\ 988\ 749\ 848\ 204\ 586$

MATHEMATICAL RELATIONSHIPS

Many of the mathematical problems encountered in polymer science involve the use of fundamental formulae which are only available in elementary texts. The following table presents a list of essential formulae.

Algebraic

Quadratic equation:
$ax^2 + bx + c = 0$ will have roots α_1 and α_2, where

$$\alpha_1 = \frac{-b + \sqrt{b^2 - 4ac}}{2a}$$

$$\alpha_2 = \frac{+b - \sqrt{b^2 - 4ac}}{2a}$$

Roots real if $b^2 > 4ac$, coincident if $b^2 = 4ac$, imaginary if $b^2 < 4ac$ and rational if $b^2 - 4ac$ is a complete square.

Arithmetical progression:

$$a, a + d, a + 2d, \ldots$$

The nth term is $a + (n - 1)d$, and the sum to n terms

$$= \frac{n}{2} 2a + (n - 1)d = \frac{n}{2}(a + 1),$$

where 1 is the last term.

Geometrical progression:

$$a, ar, ar^2, ar^3, \ldots$$

The nth term is ar^{n-1}, and the sum to n terms is

$$= \frac{a(1 - r^n)}{1 - r}.$$

If the modulus of r is less than unity, then the sum to infinity

$$= \frac{a}{1 - r}.$$

Natural numbers:

$$1, 2, 3, 4, \ldots, n$$

The sum S_1, sum of squares S_2 and sum of cubes S_3 are given by

$$S_1 = \frac{n}{2}(n + 1), \quad S_2 = \frac{n(n + 1)(2n + 1)}{6} \quad S_3 = \frac{n(n + 1)^2}{2}$$

Binomial theorem:

$$(a + x)^n = a^n + na^{n-1}x + \frac{n(n - 1)}{2!}a^{n-2}x^2 + \cdots$$

for all rational values of n.

Trigonometrical functions

Limits of function:

$$-1 \leqslant \sin x \leqslant 1 \qquad\qquad -1 \leqslant \cos x \leqslant 1$$

$$-\infty < \tan x < +\infty \qquad\qquad -\infty < \cot x < +\infty$$

Equalities:

$$\sin(-x) = -\sin x \qquad\qquad \cos(-x) = \cos x$$

$$\tan(-x) = -\tan x \qquad\qquad \cot(-x) = -\cot x$$

Relations between trigonometric functions:

$$\sin^2 x + \cos^2 x = 1 \qquad\qquad \tan x = \sin x / \cos x$$

$$\cot x = \cos x / \sin x \qquad\qquad \cosec x = 1/\sin x$$

$$1 + \tan^2 x = 1/\cos^2 x \qquad\qquad 1 + \cot^2 x = 1/\sin^2 x$$

Addition, multiple and half angle formulae:

$$\sin(x + y) = \sin x \cos y + \cos x \sin y$$

$$\sin(x - y) = \sin x \cos y - \cos x \sin y$$

$$\cos(x + y) = \cos x \cos y - \sin x \sin y$$

$$\cos(x - y) = \cos x \cos y + \sin x \sin y$$

$$\tan(x + y) = \frac{\tan x + \tan y}{1 - \tan x \tan y}$$

$$\tan(x - y) = \frac{\tan x - \tan y}{1 + \tan x \tan y}$$

$$\sin x + \sin y = 2 \sin \frac{x + y}{2} \cos \frac{x - y}{2}$$

$$\sin x - \sin y = 2 \cos \frac{x + y}{2} \sin \frac{x - y}{2}$$

$$\cos x + \cos y = 2 \cos \frac{x + y}{2} \cos \frac{x - y}{2}$$

$$\cos x - \cos y = 2 \sin \frac{x + y}{2} \sin \frac{x - y}{2}$$

$$\sin 2x = 2 \sin x \cos x = \frac{2 \tan x}{1 + \tan^2 x}$$

$$\cos 2x = \cos^2 x - \sin^2 x = 2\cos^2 x - 1 = 1 - 2\sin^2 x$$

$$\cos^2 x = \tfrac{1}{2}(1 + \cos 2x) \qquad\qquad \sin^2 x = \tfrac{1}{2}(1 - \cos 2x)$$

$$\sin 3x = 3\sin x - 4\sin^3 x \qquad\qquad \cos 3x = 4\cos^3 x - 3\cos x$$

$$\tan 3x = \frac{3\tan x - \tan^3 x}{1 - 3\tan x}$$

$$2\sin x \cos y = \sin(x + y) + \sin(x - y)$$

$$2\cos x \sin y = \sin(x + y) - \sin(x - y)$$

$$2\cos x \cos y = \cos(x + y) + \cos(x - y)$$

$$2\sin x \sin y = \cos(x - y) - \cos(x + y)$$

Hyperbolic functions:

$$\sinh x = \tfrac{1}{2}(e^x - e^{-x}) \qquad \cosh x = \tfrac{1}{2}(e^x + e^{-x})$$

$$\tanh x = \frac{e^x - e^{-x}}{e^x + e^{-x}} = \sinh x / \cosh x$$

$$\coth x = \frac{e^x + e^{-x}}{e^x - e^{-x}} = \frac{1}{\tanh x}$$

$$\operatorname{sech} x = 1/\cosh x \qquad\qquad \operatorname{cosech} x = 1/\sinh x$$

$$\cosh^2 x - \sinh^2 x = 1 \qquad\qquad \cosh x + \sinh x = e^x$$

$$\cosh x - \sinh x = e^{-x} \qquad\qquad \text{where } i = \sqrt{-1}$$

$$\sin ix = i\sinh x \qquad\qquad \cos ix = i\cosh x$$

$$\tan ix = i\tanh x \qquad\qquad \cot ix = \approx i\coth x$$

Inverse trigonometric functions:

$$\arcsin x = \sin^{-1} x \qquad\qquad \arccos x = \cos^{-1} x$$

$$\arctan x = \tan^{-1} x \qquad\qquad \operatorname{arccot} x = \cot^{-1} x$$

Relationships in coordinate geometry

Equation of	Equation	Comment
straight line	$y = mx + c$	m is slope c is intercept
circle	$(x - x_1^2) + (y - y_1^2) = r^2$	centre x_1, y_1 radius $x_1^2 + y_1^2 = r^2$
parabola	$y^2 = 4ax$	a is a constant
ellipse	$\dfrac{x^2}{a^2} + \dfrac{y^2}{b^2} = 1$	where $b^2 = a^2(1 - e)$ and e is the eccentricity (e < 1) and 2a and 2b are respectively the major and minor axes.
hyperbola	$\dfrac{x^2}{a^2} + \dfrac{y^2}{b^2} = 1$	where $b^2 = a^2(1 - e)$ and e is the eccentricity (e > 1) and 2a and 2b are respectively the major and minor axes.

Vector algebra

Consider a vector in three dimensional space and defined by three real numbers. A vector is designated by a bold letter $\mathbf{a} = (a_1, a_2, a_3)$.

Definitions:
a) equality: $(a_1, a_2, a_3) = (b_1, b_2, b_3)$ only if $a_1 = b_1$, $a_2 = b_2$, $a_3 = b_3$
b) sum: $(a_1, a_2, a_3) + (b_1, b_2, b_3) = (a_1 + b_1, a_2 + b_2, a_3 + b_3)$.

Vectors satisfy the following laws:
a) the commutative law: $\mathbf{a} + \mathbf{b} = \mathbf{b} + \mathbf{a}$
b) the associative law: $(\mathbf{a} + \mathbf{b}) + \mathbf{c} = \mathbf{a} + (\mathbf{b} + \mathbf{c})$
c) distributive laws: $k(\mathbf{a} + \mathbf{b}) = k\mathbf{a} + k\mathbf{b}$; $(k_1 + k_2)\mathbf{a} = k_1\mathbf{a} + k_2\mathbf{a}$ where k_1 and k_2 are real numbers.

Length of vector

$$\mathbf{a}(a_1, a_2, a_3) = \sqrt{(a_1^2 + a_2^2 + a_3^2)}$$

Principle of coordinate vectors:
Unit vectors $\mathbf{i}(1, 0, 0)$, $\mathbf{j}(0, 1, 0)$, $\mathbf{k}(0, 0, 1)$
Every vector can be expressed in the form

$$\mathbf{a} = a_1\mathbf{i} + a_2\mathbf{j} + a_3\mathbf{k}$$

Scalar product of vectors:

$$\mathbf{a} \cdot \mathbf{b} = \mathbf{b} \cdot \mathbf{a} \qquad \mathbf{c} \cdot (\mathbf{a} + \mathbf{b}) = \mathbf{a} \cdot \mathbf{c} + \mathbf{b} \cdot \mathbf{c} \qquad \mathbf{a} \cdot \mathbf{a} = \mathbf{a}^2$$

Two vectors are perpendicular is $\mathbf{a} \cdot \mathbf{b} = 0$.

Vector product satisfies the relation:

$$\mathbf{a} \times \mathbf{b} = -(\mathbf{b} \times \mathbf{a}) \qquad k\mathbf{a} \times \mathbf{b} = k(\mathbf{a} \times \mathbf{b})$$
$$\mathbf{a} \times (\mathbf{b} + \mathbf{c}) = \mathbf{a} \times \mathbf{b} + \mathbf{a} \times \mathbf{c}$$

Denoting the direction angles of a non-zero vector $a(a_1, a_2, a_3)$ as α, β, γ the following relations hold for the direction cosines of the vector a.

$$\cos \alpha = a_1/|\mathbf{a}|; \qquad \cos \beta = a_2/|\mathbf{a}|; \qquad \cos \gamma = a_3/|\mathbf{a}|$$
$$\cos^2 \alpha + \cos^2 \beta + \cos^2 \gamma = 1$$

Operations with unit vectors:

$$\mathbf{i} \times \mathbf{j} = \mathbf{k}, \qquad \mathbf{j} \times \mathbf{k} = \mathbf{i} \qquad \mathbf{k} \times \mathbf{i} = \mathbf{j}$$
$$\mathbf{i} \times \mathbf{i} = 1, \qquad \mathbf{j} \times \mathbf{j} = 1 \qquad \mathbf{k} \times \mathbf{k} = 1$$

Vector analysis:
Consider that a vector is a function of a scalar variable t.

$$\mathbf{a}(t) = a_1(t)\mathbf{i} + a_2(t)\mathbf{j} + a_3(t)\mathbf{k} = r$$

$$i \cdot j = j \cdot i = 0 \qquad j \cdot k = k \cdot j = 0$$

$$k \cdot i = i \cdot k = 0$$

$$r^2 = r \cdot r = \cos \alpha^2 + \cos \beta^2 + \cos \gamma = 1^2 + m^2 + n^2 = 1$$

Likewise if $\mathbf{r}_1$ and $\mathbf{r}_2$ are two vectors inclined at an angle θ then

$$\mathbf{r}_1 \cdot \mathbf{r}_2 = |\mathbf{r}_1||\mathbf{r}_2|\cos \theta$$

$$\text{since } r_1 = |\mathbf{r}_1|\hat{\mathbf{r}}_1 \qquad r_2 = |\mathbf{r}_2|\hat{\mathbf{r}}_2$$

$$\cos \theta = \hat{\mathbf{r}}_1 \cdot \hat{\mathbf{r}}_2 = \cos \alpha_1 \cos \alpha_2 + \cos \beta_1 \cos \beta_2 + \cos \gamma_1 \cos \gamma_2$$

$$= l_1 l_2 + m_1 m_2 + n_1 n_2$$

Hence $\mathbf{r}_1$ and $\mathbf{r}_2$ are perpendicular if

$$l_1 l_2 + m_1 m_2 + n_1 n_2 = 0$$

and if parallel

$$l_1 l_2 + m_1 m_2 + n_1 n_2 = \pm 1$$

Scalar and vector fields

We will associate with every point (x, y, z) of a region R a vector $a(x, y, z)$. Similarly if a scalar $\phi(x, y, z)$ is defined at every point (x, y, z) of R then a scalar field is said to exist in R. Suppose a scalar function $\phi(x, y, z)$ is differentiable in R. Then we define the gradient of $\phi(x, y, z)$ denoted by grad ϕ as

$$\text{grad } \phi = \mathbf{i}\frac{d\phi}{dx} + \mathbf{j}\frac{d\phi}{dy} + \mathbf{k}\frac{d\phi}{dz}$$

which is usually written as

$$\text{grad } \phi = \left(\mathbf{i}\frac{d}{dx} + \mathbf{j}\frac{d}{dy} + \mathbf{k}\frac{d}{dz}\right)\phi$$

the quantity in the bracket is a vector differential operator and is denoted by the symbol ∇ (del or nabla)

$$\nabla = \mathbf{i}\frac{d}{dx} + \mathbf{j}\frac{d}{dy} + \mathbf{k}\frac{d}{dz}$$

Consequently

$$\nabla \phi = \text{grad } \phi = \mathbf{i}\frac{\partial \phi}{\partial x} + \mathbf{j}\frac{\partial \phi}{\partial y} + \mathbf{k}\frac{\partial \phi}{\partial z}$$

Using the operator ∇ we may define the three dimensional Laplacian as the scalar product

$$\nabla^2\phi = \nabla\cdot\nabla\phi = \left(i\frac{d}{dx} + j\frac{d}{dy} + k\frac{d}{dz}\right)\cdot\left(i\frac{d\phi}{dx} + j\frac{d\phi}{dy} + k\frac{d\phi}{dz}\right)$$

$$= \frac{d^2\phi}{dx^2} + \frac{d^2\phi}{dy^2} + \frac{d^2\phi}{dz^2}$$

Furthermore the operation ∇ may be applied to any vector function $\mathbf{a}(x, y, z)$ to give the divergence of $\mathbf{a}(x, y, z)$—written as div $\mathbf{a}$. Clearly

$$\nabla\cdot\mathbf{a} = \text{div}\,\mathbf{a} = \left(i\frac{d}{dx} + j\frac{d}{dy} + k\frac{d}{dz}\right)\cdot(ia_x + ja_y + ka_z)$$

$$= \frac{da_x}{dx} + \frac{da_y}{dy} + \frac{da_z}{dz}$$

where a_x, a_y, a_z are general functions of x, y and z. Unlike grad ϕ which is a vector function, div $\mathbf{a}$ is a scalar function.

The quantity $\nabla^2\phi$ can be written symbolically as

$$\nabla^2\phi = \nabla\cdot\nabla\phi = \text{div grad }\phi.$$

Besides the operation ∇ as applied to a vector function $\mathbf{a}(x, y, z)$ we may also consider operation $\nabla \wedge$ on $\mathbf{a}(x, y, z)$. The vector product $\nabla \wedge \mathbf{a}$ defines the curl of $\mathbf{a}$—written as curl $\mathbf{a}$ and is such that

$$\nabla \wedge \mathbf{a} = \text{curl}\,\mathbf{a} = \left(i\frac{\partial}{\partial x} + j\frac{\partial}{\partial y} + k\frac{\partial}{\partial z}\right) \wedge (ia_x + ja_y + ka_z)$$

$$= \begin{vmatrix} i & j & k \\ \frac{\partial}{\partial x} & \frac{\partial}{\partial y} & \frac{\partial}{\partial z} \\ a_x & a_y & a_z \end{vmatrix}$$

$$= i\left(\frac{\partial a_z}{\partial y} - \frac{\partial a_y}{\partial z}\right) + j\left(\frac{\partial a_x}{\partial z} - \frac{\partial a_z}{\partial x}\right) + k\left(\frac{\partial a_y}{\partial x} - \frac{\partial a_x}{\partial y}\right)$$

Curl $\mathbf{a}$ is a vector function. From the definitions of div $\mathbf{a}$ and curl $\mathbf{a}$ we now see that if $\mathbf{a}$ and $\mathbf{b}$ are two vector functions then

$$\text{div}(\mathbf{a} + \mathbf{b}) = \text{div}\,\mathbf{a} + \text{div}\,\mathbf{b}$$

or

$$\nabla\cdot(\mathbf{a} + \mathbf{b}) = \nabla\cdot\mathbf{a} + \nabla\cdot\mathbf{b}$$

and

$$\text{curl}(\mathbf{a} + \mathbf{b}) = \text{curl } \mathbf{a} + \text{curl } \mathbf{b}$$

or

$$\nabla \wedge (\mathbf{a} + \mathbf{b}) = \nabla \wedge \mathbf{a} + \nabla \wedge \mathbf{b}$$

Consequently

$$\nabla \cdot (\nabla \wedge \mathbf{a}) = \text{div curl } \mathbf{a} = 0 \qquad \nabla \wedge \nabla \phi = \text{curl grad } \phi = 0.$$

Mathematical formulae used in polymer science

The following list is not intended to be comprehensive but rather to provide the reader with a selection of the more commonly used equations and mathematical relationships.

Integrals—elementary forms

$$\int a \, \partial x = ax$$

$$\int (u + v) \, \partial x = \int u \, \partial x + \int v \, \partial x$$

$$\int u \frac{\partial v}{\partial \chi} \partial x = uv - \int v \frac{\partial u}{\partial x} \partial x$$

$$\int e^{ax} \, \partial x = e^{ax}/a$$

$$\int \frac{\partial x}{x} = \log x$$

$$\int a \cdot f(x) \, \partial x = a \int f(x) \, \partial x$$

$$\int u \, \partial v = u \int \partial v - \int v \, \partial u = uv - \int v \, \partial u$$

$$\int x^n \, \partial x = \frac{x^{n+1}}{n+1}$$

$$\int \log x \, \partial x = x \log x - x$$

$$\int a^x \log a \, \partial x = a^x$$

Logarithmic forms:

$$\int (\log x)\, \partial x = x \log x - x$$

$$\int \frac{(\log x)^n}{x}\, \partial x = \frac{1}{n+1}(\log x)^{n+1}$$

$$\int x^n(\log ax)\, \partial x = \frac{x^{n+1}}{n+1}\log ax - \frac{x^{n+1}}{(n+1)^2}$$

$$\int \frac{\partial x}{x \log x} = \log(\log x)$$

Exponential forms:

$$\int e^x\, \partial x = e^x$$

$$\int \frac{\partial x}{1+e^x} = \log \frac{e^x}{1+e^x}$$

$$\int e^{-x}\, \partial x = -e^{-x}$$

$$\int \frac{e^{ax}\, \partial x}{x} = \log x + \frac{ax}{1!} + \frac{a^2 x^2}{2.2!} + \cdots$$

Hyperbolic forms:

$$\int (\sinh x)\, \partial x = \cosh x$$

$$\int (\tanh x)\, \partial x = \log \cosh x$$

$$\int (\operatorname{sech} x)\, \partial x = \tan^{-1}(\sinh x)$$

$$\int x(\sinh x)\, \partial x = x \cosh x - \sinh x$$

$$\int (\cosh x)\, \partial x = \sinh x$$

$$\int (\coth x)\, \partial x = \log \sinh x$$

$$\int \cosh x \, \partial x = \log \tanh\left(\frac{x}{2}\right)$$

$$\int x(\cosh x) \, \partial x = x \sinh x - \cosh x$$

Trigonometric functions:

$$\int (\sin ax) \, \partial x = \frac{1}{a} \cos ax$$

$$\int (\tan ax) \, \partial x = -\frac{1}{a} \log \cos ax$$

$$\int (\cos ax) \, \partial x = \frac{1}{a} \sin ax$$

$$\int (\cos ax) \, \partial x = \frac{1}{a} \log \sin ax$$

$$\int (\sin^n ax) \, \partial x = -\frac{\sin^{n-1} ax \cos ax}{na} + \frac{n-1}{n} \int (\sin^{n-2} ax) \, \partial x$$

$$\int (\cos^n ax) \, \partial x = \frac{1}{na} \cos^{n-1} ax \sin ax + \frac{n-1}{n} \int (\cos^{n-2} ax) \, \partial x$$

$$\int (\sin ax)(\cos ax) \, \partial x = \frac{1}{2a} \sin^2 ax$$

Definite integrals:

$$\int_0^\infty x^{n-1} e^{-x} \, \partial x = \int_0^1 \left(\log \frac{1}{x}\right)^{n-1} \partial x = \frac{1}{n} \prod_{m=1}^\infty \frac{\left(1 + \frac{1}{m}\right)^n}{1 + \frac{n}{m}}$$

$$= \Gamma(n), n \neq 0, -1, -2, \ldots$$

(Gamma Function)

$$\int_0^1 x^{m-1}(1-x)^{n-1} \, \partial x = \int_0^\infty \frac{x^{m-1}}{(1+x)^{m+n}} \, \partial x$$

$$= \frac{\Gamma(m)\Gamma(n)}{\Gamma(m+n)} = B(m,n)$$

(Beta Function) where m and n are positive real numbers.

$$\int_0^\infty \cos(x^2)\,\partial x = \int_0^\infty \sin(x^2)\,\partial x = \frac{1}{2}\sqrt{\frac{\pi}{2}}$$

$$\int_0^\infty \frac{\sin x}{\sqrt{x}}\,\partial x = \int_0^\infty \frac{\cos x}{\sqrt{x}}\,\partial x = \sqrt{\frac{\pi}{2}}$$

$$\int_0^\infty \frac{\partial x}{x^m} = \frac{1}{m-1}\,[m > 1]$$

$$\int_0^\infty \frac{\partial x}{(1+x)\sqrt{x}} = \pi$$

$$\int_0^\infty \frac{\sin mx\,\partial x}{x} = \frac{\pi}{2}$$

if $m > 0$; 0, if $m = 0$; $-\dfrac{\pi}{2}$, if $m < 0$.

$$\int_0^\infty \frac{\cos x\,\partial x}{x} = \infty$$

$$\int_0^\infty \frac{\tan x\,\partial x}{x} = \frac{\pi}{2}$$

$$\int_0^\pi \sin ax \cdot \sin bx\,\partial x = \int_0^\pi \cos ax \cos bx\,\partial x = 0$$

$(a \neq b;\ a,\ b$ integers$)$

$$\int_0^\infty e^{-ax}\,\partial x = \frac{1}{a}(a > 0)$$

$$\int_0^\infty x\,e^{-x^2}\,\partial x = \tfrac{1}{2}$$

Statistics

Definitions:

Sample average

$$\bar{x} = \frac{1}{n}\sum_{i=1}^n x_i$$

where n is the number of points in the data set.

Sample variance

$$s_x^2 = \frac{1}{n-1} \sum_{i=1}^{n} (x_i - \bar{x})^2 = \frac{1}{n-1} \left[\sum_{i=1}^{n} x_i^2 - \frac{1}{n} \left(\sum_{i=1}^{n} x_i \right)^2 \right]$$

Sample covariance

$$s_{xy} = \frac{1}{n-1} \sum_{i=1}^{n} (x_i - \bar{x})(y_i - \bar{y}) = \frac{1}{n-1} \left[\sum_{i=1}^{n} x_i y_i - \frac{1}{n} \sum_{i=1}^{n} x_i \sum_{i=1}^{n} y_i \right]$$

Sample standard deviation

$$s_x = \sqrt{s_x^2}$$

Sample coefficient of variance

$$c_x = \frac{s_x}{\bar{x}}$$

Sample correlation coefficient

$$r_{xy} = \frac{s_{xy}}{s_x s_y}$$

Sample regression coefficient of y on x

$$b = \frac{s_{xy}}{s_x^2}$$

Principle of least squares

Assume that the variables x and y are related functions

$$y = f(x)$$

The equation of the experimental data is

$$y = f(x; b_1, b_2, \ldots, b_n)$$

The sum of squared differences s are given by

$$S = \sum_{i=1}^{n} [y_i - f(x_i, b_1^*, b_2^*, \ldots, b_p^*)]^2$$

the estimates b_j^* ($j = 1, 2, \ldots, p$) will be determined from the condition

$$S = \min$$

suppose that the x and y are related by the equation

$$y = a + bx$$

where a and b are unknown parameters

$$S = \sum_{i=1}^{n} (y_i - a* - b*x_i)^2 = \min$$

Therefore the solutions of the equation are

$$\frac{\partial S}{\partial a*} = 0 \qquad \frac{\partial S}{\partial b*} = 0$$

We have

$$-\frac{1}{2}\frac{\partial S}{\partial a*} = \sum_{i=1}^{n} (y_i - a* - b*x_i) = \sum_{i=1}^{n} y_i - na* - \left(\sum_{i=1}^{n} x_i\right) b*$$

$$-\frac{1}{2}\frac{\partial S}{\partial b*} = \sum_{i=1}^{n} (y_i - a* - b*x_i)x_i$$

$$= \sum_{i=1}^{n} x_i y_i - \left(\sum_{i=1}^{n} x_i\right) a* - \left(\sum_{i=1}^{n} x_i^2\right) b*$$

Thus

$$na* + \left(\sum_{i=1}^{n} x_i\right) b* = \sum_{i=1}^{n} y_i$$

$$\left(\sum_{i=1}^{n} x_i\right) a* + \left(\sum_{i=1}^{n} x_i^2\right) b* = \left(\sum_{i=1}^{n} x_i y_i\right)$$

Linear regression analysis of standard functions:
   a) $y = a$

Least squares condition:

$$\sum_{i=1}^{n} (y_i - a*)^2 = \min$$

Estimate of the parameter a:

$$a* = \bar{y}$$

Residual sum of squares and degrees of freedom:

$$S_0 = \sum_{i=1}^{n} y_i^2 - \frac{\left(\sum_{i=1}^{n} y_i\right)^2}{n} \qquad v = n - 1$$

Estimates of standard errors:

$$s = \sqrt{\frac{S_0}{n-1}} \qquad S_{a*} = \frac{S}{\sqrt{n}}$$

b) $y = bx$

Least squares condition

$$\sum_{i=1}^{n} (y_i - b^*x_i)^2 = \min$$

Normal equations:

$$\left(\sum_{i=1}^{n} x_i^2 \right) b^* = \sum_{i=1}^{n} x_i y_i$$

Estimate of the parameter b:

$$b^* = \frac{\sum_{i=1}^{n} x_i y_i}{\sum_{i=1}^{n} x_i^2}$$

Residual sum of squares and degrees of freedom:

$$S_0 = \sum_{i=1}^{n} y_i^2 - \frac{\left(\sum_{i=1}^{n} x_i y_i \right)^2}{\sum_{i=1}^{n} x_i^2} \qquad v = n - 1$$

Estimates of standard errors:

$$S = \sqrt{\frac{S_0}{n-1}} \qquad S_{b*} = \frac{S}{\sqrt{\left(\sum_{i=1}^{n} x_i^2 \right)}}$$

c) $y = a + bx$

Least squares condition:

$$\sum_{i=1}^{n} (y_i - a^* - b^*x_i)^2 = \min.$$

Normal equations:

$$na^* + \left(\sum_{i=1}^{n} x_i \right) b^* = \sum_{i=1}^{n} y_i$$

$$\left(\sum_{i=1}^{n} x_i \right) a^* + \left(\sum_{i=1}^{n} x_i^2 \right) b^* = \sum_{i=1}^{n} x_i y_i$$

Estimates of parameters a and b:

$$a^* = \bar{y} - b^* \bar{x} \qquad b^* = \frac{\displaystyle\sum_{i=1}^{n} x_i y_i - \frac{\left(\sum_{i=1}^{n} x_i \right)\left(\sum_{i=1}^{n} y_i \right)}{n}}{\displaystyle\sum_{i=1}^{n} x_i^2 - \frac{\left(\sum_{i=1}^{n} x_i \right)^2}{n}}$$

Residual sum of squares and degrees of freedom:

$$S_0 = \sum_{i=1}^{n} y_i^2 - \frac{\left(\sum_{i=1}^{n} y_i \right)^2}{n} - b^* \left[\sum_{i=1}^{n} x_i y_i - \frac{\left(\sum_{i=1}^{n} x_i \right)\left(\sum_{i=1}^{n} y_i \right)}{n} \right]$$

$$v = n - 2$$

Estimates of standard errors:

$$S = \sqrt{\frac{S_0}{n-2}} \qquad S_{a^*} = S \sqrt{\frac{\frac{1}{n} \sum_{i=1}^{n} x_i^2}{\sum_{i=1}^{n} x_i^2 - \frac{\left(\sum_{i=1}^{n} x_i \right)^2}{n}}}$$

$$S_{b^*} = \frac{S}{\sqrt{\left[\sum_{i=1}^{n} x_i^2 - \frac{\left(\sum_{i=1}^{n} x_i \right)^2}{n} \right]}}$$

d) $y = a + bx + cx^2$

Least squares condition:

$$\sum_{i=1}^{n} (y_i - a^* - b^* x_i - c^* x_i^2)^2 = \min$$

Normal equation

$$S_m = \sum_{i=1}^{n} x_i^m$$

$$na* + S_1 b* + S_2 c* = \sum_{i=1}^{n} y_i$$

$$S_1 a* + S_2 b* + S_3 c* = \sum_{i=1}^{n} x_i y_i$$

$$S_2 a* + S_3 b* + S_4 c* = \sum_{i=1}^{n} x_i^2 y_i$$

Estimates of the parameters a, b, c:

$$a* = \frac{(S_2 S_4 - S_3^2) \sum_{i=1}^{n} y_i + (S_2 S_3 - S_1 S_4) \sum_{i=1}^{n} x_i y_i + (S_1 S_3 - S_2^2) \sum_{i=1}^{n} x_i^2 y_i}{\Delta}$$

$$\cdot \; \Delta = n(S_2 S_4 - S_3^2) + S_1(S_2 S_3 - S_1 S_4) + S_2(S_1 S_3 - S_2^2)$$

$$b* = \frac{(S_2 S_3 - S_1 S_4) \sum_{i=1}^{n} y_i + (nS_4 - S_2^2) \sum_{i=1}^{n} x_i y_i + (S_1 S_2 - nS_3) \sum_{i=1}^{n} x_i^2 y_i}{\Delta}$$

$$c* = \frac{(S_1 S_3 - S_2^2) \sum_{i=1}^{n} y_i + (S_1 S_2 - nS_3) \sum_{i=1}^{n} x_i y_i + (nS_2 + S_1^2) \sum_{i=1}^{n} x_i^2 y_i}{\Delta}$$

Residual sum of squares and degrees of freedom:

$$S_0 = \sum_{i=1}^{n} y_i^2 - a* \sum_{i=1}^{n} y_i - b* \sum_{i=1}^{n} x_i y_i - c* \sum_{i=1}^{n} x_i^2 y_i \qquad v = n - 3$$

Estimates of standard errors:

$$S = \sqrt{\frac{S_0}{n-3}} \qquad S_{a*} = S \sqrt{\left(\frac{S_2 S_4 - S_3^2}{\Delta}\right)}$$

$$S_{b*} = S \sqrt{\left(\frac{nS_4 - S_2^2}{\Delta}\right)} \qquad S_{c*} = S \sqrt{\left(\frac{nS_2 - S_1^2}{\Delta}\right)}$$

Special functions

A number of special functions are used in the analysis and transformation of data. The most commonly used functions and certain of their operations are summarized below. The following remarks refer to an orthogonal system.

Fourier integral

An orthogonal system of trigonometric functions is complete in the range (π, $-\pi$). For the pointwise convergence of a Fourier series the following relationships will be obeyed. Let $f(x)$ be a periodic function with period 2 (i.e. $f(x + 2) = f(x)$ and let $f(x)$ and $f'(x)$ be piecewise continuous in the range (π, $-\pi$). Then

$$f(x) = \frac{a_0}{2} + \sum_{n=1}^{\infty} (a_n \cos nx + b_n \sin nx)$$

The coefficients are defined by:

$$a_n = \frac{1}{\pi} \int_{-\pi}^{\pi} f(x) \cos nx \, dx \qquad (n = 0, 1, 2, \ldots)$$

$$b_n = \frac{1}{\pi} \int_{-\pi}^{\pi} f(x) \sin nx \, dx \qquad (n = 0, 1, 2, \ldots)$$

the above is true at each point x, where $f(x)$ is continuous, while at each point of discontinuity:

$$\tfrac{1}{2}f(x + 0) + f(x - 0) = \frac{a_0}{2} + \sum_{n=1}^{\infty} (a_n \cos nx + b_n \sin nx)$$

The symbols $f(x + 0)$ and $f(x - 0)$ denote the limits from the right and left respectively of the function $f(x)$ at the point x.

Bessel functions

The bessel function of the first kind and order (index) n is defined by the equation

$$J_n(x) = \left(\frac{x}{2}\right)^n \sum_{k=1}^{\infty} \frac{(-1)^k}{k!\Gamma(n + k + 1)} \left(\frac{x}{2}\right)^{2k}$$

the series converges for any real n and all x.

If $k! = 1$ for $k = 0$. The symbol $\Gamma(u)$ denotes the gamma function defined by

$$\Gamma(x) = \int e^{-t} t^{x-1} \, dt \qquad \text{Gamma function}$$

For $x = 0, -1, -2, \ldots$ we define $1/(x) = 0$. For integral x we have

$$J_{-n}(x) = (-1)^n J_n(x)$$

For a negative integer n it is not necessary to use the above formula directly, but to put $n = -m$ (m a positive integer) and compute $J_m(\chi)$.

The function $y = J_n(x)$ satisfies the (Bessel) differential equation

$$x^2 y'' + xy' + (x^2 - n^2)y = 0$$

Bessel functions with integral orders are coefficients in the expansion of the so-called generating function, that is, for all x and all $t \neq 0$ we have

$$e^{(x/2)(t - 1/t)} = J_n(x)t^n = J_0(x) + J_1(x)t + J_2(x)t^2 + \cdots$$

$$+ J_{-1}(x)t^{-1} + J_{-2}(x)t^{-2} + \cdots$$

For $n = 0, 1, 2, \ldots$ we have

$$J_n(x) = \frac{1}{\pi} \int_0^\pi \cos(x \sin v - nv)\,dv$$

The functions of orders zero and one

$$J_0(x) = 1 - \frac{x^2}{2^2} + \frac{x^4}{(2.4)^2} - \frac{x^6}{(2.4.6)^2} + \cdots$$

$$J_1(x) = \frac{x}{2}\left(1 - \frac{x^2}{2.4} + \frac{x^4}{2.4^2.6} - \frac{x^6}{2.(4.6)^2.8} + \cdots\right)$$

Spherical Bessel functions

$$J_{\frac{1}{2}}(x) = \sqrt{\left(\frac{2}{\pi x}\right)} \sin x, \qquad J_{\frac{3}{2}}(x) = \sqrt{\left(\frac{2}{\pi x}\right)}\left(\frac{\sin x}{x} - \cos x\right),$$

$$J_{-\frac{1}{2}}(x) = \sqrt{\left(\frac{2}{\pi x}\right)} \cos x, \qquad J_{-\frac{3}{2}}(x) = -\sqrt{\left(\frac{2}{\pi x}\right)}\left(\frac{\cos x}{x} + \sin x\right).$$

Fourier–Bessel expansion

Let $J_n(\lambda_1, x)$, $J_n(\lambda_2, x)$, $J_n(\lambda_3, x)$, $\ldots$ have positive roots, $\lambda_1, \lambda_2, \lambda_3, \ldots$ and , satisfy the equation $J_n(\lambda c) = 0$ (n fixed, real, non negative; $c > 0$). The functions further constitute an orthogonal system in the interval $(0, c)$ with the weight functions x, $J_n(\lambda_1, x)$, $J_n(\lambda_2, x)$, hence

$$\int_0^c x J_n(\lambda_1, x)J_n(\lambda_k x)\,dx = 0 \quad \text{for} \quad i \neq k.$$

Let $f(x)$ and $f'(x)$ be piecewise continuous functions in the interval $(0, c)$.

Then

$$\sum_{k=1}^{\infty} a_k J_n(\lambda_k x) = \begin{array}{l} f(x) \quad \text{at each point continuity of} \\ \frac{1}{2}f(x+0) - f(x-0) \quad \text{at each point of} \\ \text{discontinuity of the function } f(x) \end{array}$$

$(0 < x < c)$ and $f(x+0)$, $f(x-0)$ are the limits from the right and left, respectively of the function $f(x)$ at the point x. The above series converges uniformly in every closed interval interior to an interval in which $f(x)$ is continuous. If the λ_k are the positive roots of the equation $J_n(\lambda, c) = 0$, we have

$$a_k = \frac{2}{c^2 J_{n+1}^2(\lambda_k c)} \int_0^c x J_n(\lambda_k x) f(x) \, \partial x \qquad (k = 1, 2, 3, \ldots)$$

if the λ_k's are the non-negative roots we have

$$a_k = \frac{2\lambda_k^2}{(\lambda_k^2 c + h^2 - n^2)J_n^2(\lambda_k c)} \int_0^c x J_n(\lambda_k x) f(x) \, \partial x \quad (k = 1, 2, 3, \ldots)$$

Bessel functions of the second kind are defined by the relationship

$$Y_n(x) = \frac{J_n(x)\cos n\pi - J_{-n}(x)}{\sin n\pi} \qquad (n \text{ non-integral})$$

For integral n, $Y_n(x)$ may be defined by the equation

$$Y_n(x) = \lim_{\substack{r-n \\ r \neq n}} Y_r(x)$$

where $Y_n(x)$ is approximated by functions $Y_r(x)$ with indices r close to the given integer n.

Legendre polynomial

The Legendre polynomial is defined by the equation

$$P_n(x) = \frac{1.3.5\ldots(2n-1)}{n!}$$

$$\times \left[x^n - \frac{n(n-1)}{2(2n-1)} x^{n-2} + \frac{n(n-1)(n-2)(n-3)}{2.4(2n-1)(2n-3)} x^{n-4} - \cdots \right]$$

where n indicates that it is of the nth degree.
    The function $y = P_n(x)$ satisfies the (Legendre) differential equation

$$(1 - x^2)y'' - 2xy' + n(n+1)y = 0$$

c

The polynomial $P_n(\cos \theta)$ satisfies the equation

$$\frac{1}{\sin \theta}\frac{\partial}{\partial \theta}\left(\sin \theta \frac{\partial y}{\partial \theta}\right) + n(n + 1)y = 0$$

which is obtained from the differential equation by the substitution $x = \cos \theta$.

The first five Legendre polynomials in the variables x and θ are:

$P_0(x) = 1$

$P_1(x) = x = \cos \theta$

$P_2(x) = \frac{3}{2}x^2 - \frac{1}{2} = \frac{1}{2}(3 \cos^2 \theta - 1) = \frac{1}{4}(3 \cos 2\theta + 1)$

$P_3(x) = \frac{5}{2}x^3 - \frac{3}{2}x = \frac{1}{2}(5 \cos^3 \theta - 3 \cos \theta) = \frac{1}{8}(5 \cos 3\theta + 3 \cos \theta)$

$P_4(x) = \frac{35}{8}x^4 - \frac{15}{4}x^2 + \frac{3}{8}$

$\quad = \frac{1}{8}(35 \cos^4 \theta - 30 \cos^2 \theta + 3) = \frac{1}{64}(35 \cos 4\theta + 20 \cos 2\theta + 9)$

Fundamental properties

$$P_n(-x) = (-1)P_n(x)$$
$$P_n(1) = 1$$

The generating function. For $|x| \leqslant 1, |t| < 1$ we have

$$\frac{1}{\sqrt{(1 - 2xt + t^2)}} = P_0(x) + P_1(x)t + P_2(x)t^2 + \cdots$$

The above uniquely determines the Legendre polynomials.

Bessel functions of the third kind—Hankel functions—are defined by the formulae

$$H_n^{(1)}(x) = J_n(x) + iY_n(x)$$
$$H_n^{(2)}(x) = J_n(x) - iY_n(x)$$

where i is the imaginary unit.

Error function

The error function erf x is defined as

$$\text{erf } x = \frac{2}{\sqrt{\pi}}\int_0^x e^{-u^2} du$$

which is the area under the curve e^{-u^2} from $u = 0$ to $u = x$; the limits erf(0) $= 0$ and erf(∞) $= 1$. Closely related to this function are the functions $C(x)$

and $S(x)$—the Fresnel integrals defined by

$$C(x) = \int_0^x \cos\frac{\pi u^2}{2}\, du$$

$$S(x) = \int_0^x \sin\frac{\pi u^2}{2}\, du$$

Consequently

$$C(x) - iS(x) = \int_0^x e^{-i(\pi/2)u^2}\, du$$

Now assuming that

$$\int_0^\infty e^{-au^2}\, du = \tfrac{1}{2}\sqrt{\frac{\pi}{a}}$$

holds when a is complex.

References

H. Bateman, *Partial Differential Equations of Mathematical Physics*, Cambridge University Press, London – New York (1959).

K. Courant and D. Hilbert, *Methods of Mathematical Physics*, Vol. 1, 2, New York, Interscience (1962).

P. M. Morse and H. Feshbach, *Methods of Mathematical Physics*, Parts I, II, New York–Toronto–London, McGraw-Hill (1963).

K. Rektorys, *Survey of Applicable Mathematics*, Iliffe Books Ltd, London (1969).

B. Spain, *Tensor Calculus*, Edinburgh, Oliver and Boyd (1953).

B. Spain, *Vector Analysis*, von Nostrand, London (1965).

Physical Quantities–International System of Units (SI)

Physical quantities—international system of units (SI)

Nearly all physical quantities may be described in terms of seven basic dimensional quantities. These are length (l), mass (m), time (t), temperature (T), electric current (I), light intensity (I_v) and the amount of substance (n). There is no generally accepted name for this latter unit, however the mole is becoming relatively widely accepted as the unit of substance in physical chemistry.

The International Union of Pure and Applied Chemistry (IUPAC) has recommended that the International System of Units, summarized in the document 'Le Systeme International d'Unites (SI),' published by the Bureau des Poids et Measures in a second edition in 1973.

SI definition of physical quantities

The SI base units are the metre, kilogram, second, kelvin, ampere, mole and candela.

Metre (m) is equal to 1,650,763.73 wavelengths in vacuum of the radiation corresponding to the transition between the levels $2p_{10}$ and $5d_5$ of the Krypton-86 atom.

Kilogram (kg) is the unit of mass and is defined by the international prototype of the kilogram.

Second (s) is the duration of 9,192,631,770 periods of the radiation corresponding to the transition between the two hyperfine levels of the ground state of the ceasium-133 atom.

Kelvin (K) is the thermodynamic unit of temperature and is defined as $1/273.16$ of the triple point of water.

Ampere (I) is the constant current which will produce a force equal to 2×10^{-7} newton metre between two straight parallel conductors of infinite length, of negligible cross section and placed 1 metre apart in vacuum.

Mole (mol) is the amount of substance which contains as many elementary entities as there are atoms in 0.012 kilogram of carbon-12. When the mole is used the elementary entities must be specified and may be atoms, molecules, ions, electrons, other particles or specified groups of such particles.

Candela (cd) is the luminous intensity, in the perpendicular direction of a surface of 1/600000 square metre of a black body at the temperature of freezing platinum under a pressure of 101,325 newtons per square metre.

Physical quantities, units and symbols

The value of a physical quantity is equal to the product of a numerical value and a unit:

physical quantity = numerical value × unit.

Neither any physical quantity, nor the symbol used to denote it, should imply a particular choice of unit. Operations on equations involving physical quantities, units and numerical values, should follow the ordinary rules of algebra. Headings in tables and labels on the axes of graphs should be dimensionless.

Printing of symbols for units

The symbol for a unit should be printed in Roman (upright) type, should remain unaltered in the plural and should not be followed by a full stop except when it occurs at the end of a sentence in text.
Example: 2 cm but not 2 cms.
A proper name should begin with a capital Roman (upright) letter.
Example: J for Joule or T for Tesla.

Presentation of symbols and units

Compound prefixes should always be avoided e.g. ns but not mμs for 10^{-9} s. A product of two units may be represented in any of the following forms:

N m   or   N.m   or   N × m

The representation Nm is not recommended. When division is required no more than one solidus (/) should be used unless parentheses are employed to avoid ambiguity. For example: $J K^{-1} mol^{-1}$ or $J/(K mol)$ but never J/K/mol.

The symbols for physical quantities are single letters of the Latin or Greek alphabet except for certain transport properties like the Reynolds 'number'. Symbols for physical quantities are printed in the italic type. Symbols for vector quantities should be printed in bold type.

The symbols for physical quantity may be modified by subscripts and superscripts. Such subscripts and superscripts should be printed in italic type if they are themselves symbols for physical quantities. All other subscripts and superscripts should be printed in Roman (upright) type.
Examples: Heat capacity at constant pressure: C_p
Specific heat capacity of substance B: C_B

The words 'specific' and 'molar' have restricted meanings. The term 'specific' being used for 'divided by mass' and 'molar' always means 'divided by amount of substance.' When the extensive physical quantity is represented by a capital letter, the corresponding specific quantity may be symbolized by the corresponding lower case letter. Molar quantities may be denoted by the superscripts ^{m} to the symbol for the extensive quantity.

Example: Volume $\quad V$
specific volume $\quad V/m = v$
Molar volume $\quad V/n = V^m$

Definition of derived SI quantities

Physical quantity	SI Unit		
	Name	Unit	Definition
Force	Newton	N	$m\,kg\,s^{-2}$
Pressure	Pascal	Pa	$m^{-1}kg\,s^{-2}(=N\,m^{-2})$
Stress	Pascal	Pa	$m^{-1}kg\,s^{-2}(=N\,m^{-2})$
Energy	Joule	J	$m^2kg\,s^{-2}(=N\,m)$
Power	Watt	W	$m^2kg\,s^{-3}(=J\,s^{-1})$
Luminous flux	lumen	lm	$cd\,sr*$
Illuminance	lux	lx	$cd\,sr\,m^{-2}$
Frequency	hertz	Hz	s^{-1}
Activity (of radio-active source)	becquerel	Bq	s^{-1}

	Unit		Symbol
Area	square metre		m^2
volume	cubic metre		m^3
density	kilogram per cubic metre		$kg\,m^{-3}$
velocity	metre per second		$m\,s^{-1}$
kinetic viscosity	square metre per	$m^2\,s^{-1}$	
diffusion coefficient	second		

Physical quantity	SIUnit		
	Name	Unit	Definition
Electric charge	Coulomb	C	$s\,A$
Electrical potential	Volt	V	$m^2kg\,s^{-3}A^{-1}(=J\,A^{-1}s^{-1})$
Electric resistance	Ohm	Ω	$m^2kg\,s^{-3}A^{-2}(=V\,A^{-1})$
Electric conductance	siemens	S	$m^{-2}kg^{-1}s^3A^2(=A\,V^{-1})$
Electric capacitance	farad	F	$m^{-2}kg^{-1}s^4A^2(=A\,V^{-1}s)$
Inductance	henry	H	$m^2kg\,s^{-2}A^{-2}(=V\,A^{-1}s)$
Magnetic flux	weber	Wb	$m^2kg\,s^{-2}A^{-1}(=V\,s)$
Magnetic flux density	tesla	T	$kg\,s^{-2}A^{-1}(=V\,s\,m^{-1})$
Absorbed dose (of radiation)	gray	Gy	$m^2s^{-2}(=J\,kg^{-1})$

	Unit	Symbol
Electric field strength	volt per metre	$V\,m^{-1}$
Magnetic field strength	ampere per metre	$A\,m^{-1}$
luminance	candela per square metre	$cd\,m^{-2}$
molar entropy	joule per kelvin mole	$J\,K^{-1}mol^{-1}$
concentration	mole per cubic metre	$mol\,m^{-3}$
heat capacity	joule per kelvin mole	$J\,K^{-1}mol^{-1}$
dynamic viscosity	newton second per square metre	$N\,s\,m^{-2}$

Accepted SI prefixes

SI units may be prefixed by decimal multiples and sub-multiples constructed using the following table.

Fraction	Prefix	Symbol	Multiple	Prefix	Symbol
10^{-1}	deci	d	10	deca	da
10^{-2}	centi	c	10^2	hecto	h
10^{-3}	milli	m	10^3	kilo	k
10^{-6}	micro		10^6	mega	M
10^{-9}	nano	n	10^9	giga	G
10^{-12}	pico	p	10^{12}	tera	T
10^{-15}	femto	f	10^{15}	peta	P
10^{-18}	atto	a	10^{18}	exa	E

Recommended format for numbers

Numbers should be printed in upright type. The decimal sign between digits in a number should be a comma (,) or a point (.). To facilitate the reading of long numbers the digits may be grouped in threes but no comma or point should ever be used except for the decimal sign.

Example: 1 234,567 891 or 1 234.567 891 but not 1,234.562,891.

When the decimal sign is placed before the first digit of a number a zero should always be placed before the decimal sign.

Example: $0.123\ 456 \times 10^4$ but not $.123\ 456 \times 10^4$.

Multiplication and division

The multiplication sign between numbers should be a cross ($\times$) or (but never when a point is used as the decimal sign) a centred dot.

Example: 2.3×3.4 or $2{,}3 \cdot 3{,}4$.

Division of one number by another may be indicated in any of the ways:

$\dfrac{123}{456}$ or $123/456$ or $123 \times (456)^{-1}$.

Certain of the units widely used do not conform to the SI system of units however they are sufficiently common to have special names.

Physical quantity	Name of unit	Symbol	Definition of unit
Length	angstrom	Å	10^{-10} m
Cross section	barn	b	10^{-28} m^2
Volume	litre	L	10^{-3} m^3
Mass	tonne	t	10^3 kg
Pressure	bar	bar	10^5 Pa

There are in addition a number of special names based on the cgs system; among these are the erg (10^{-7} J), the dyne (10^{-5} N), the pose (0.1 Pas), the stokes (10^{-4} m^2 s^{-1}), the gauss (10^{-4} T), the oersted ($100/4\pi$ Am^{-1}) and the maxwell (10^{-8} Wb). The micron (10^{-6} m) is not acceptable as an SI unit.

Fundamental constants

The Committee on Data for Science and Technology (CODATA) of the International Council of Scientific Unions recommended at its General Assembly in Stockholm on September 11, 1973 a revised set of consistent values of fundamental constants for international use.

References

M. L. McGlashan, M. A. Paul and D. H. Whiffen, Manual of symbols and terminology for physiochemical quantities and units, *Pure and Applied Chem*, **51**, 1 (1979).

The above document was adopted by the IUPAC Council at Cortina d'Ampezzo Italy on 7th July 1969.

International Union of Pure and Applied Chemistry. Manual of symbols and terminology for physiochemical quatities and units, Butterworths, London (1970).
International Union of Pure and Applied Chemistry, Symbols and terminology for physiochemical quantaties and units, *Pure Appl. Chem.*, **21**, 1 (1970).

Recommended values of physical quantities

Quantity	Symbol and equivalence	Value
Rest mass of electron	m_e	$9.109\ 534(47) \times 10^{-31}$ kg
Rest mass of proton	m_p	$1.672\ 648\ 5(86) \times 10^{-27}$ kg
Rest mass of neutron	m_n	$1.674\ 954\ 3(86) \times 10^{-27}$ kg
Atomic mass unit	$lu = 10^{-3}$ kg mol^{-1}/L	$1.660\ 565\ 5(86) \times 10^{-27}$ kg
Avogadro constant	N_a	$6.022\ 045(31) \times 10^{23}$ mol^{-1}
Planck constant	$\hbar\quad h = h/2\pi$	$6.626\ 176(36) \times 10^{-34}$ JHz^{-1}
Speed of light	c	$2.997\ 924\ 58(1) \times 10^8$ ms^{-1}

Quantity	Symbol and equivalence	Value
Permeability of vacuum	μ_0	$4\pi \times 10^{-7} \, \text{Hm}^{-1}$
Permittivity of	$\varepsilon_0 = (\mu_0 c^2)^{-1}$	$8.845\ 187\ 82(5) \times 10^{-12} \, \text{Fm}^{-1}$
Gas constant	R	$8.314\ 41(26) \, \text{J K}^{-1} \, \text{mol}^{-1}$
Boltzmann constant	$k = R/N_0$	$1.380\ 662(44) \times 10^{-23} \, \text{J K}^{-1}$
Elementary charge	e	$1.602\ 189\ 2(46) \times 10^{-19} \, \text{C}$
Rydberg constant	$R_\infty = \mu_0^2 m_e e^4 c^3 / 8h^3$	$1.097\ 373\ 177(83) \times 10^7 \, \text{m}^{-1}$
Fine structure constant	$a = \mu_0 ce^2 / 2h$	$7.297\ 350\ 6(60) \times 10^{-3}$
Bohr radius	$a_0 = a/4\pi R_\infty$	$5.291\ 770\ 6(44) \times 10^{-11} \, \text{m}$
Hartree energy	$E_h = 2hcR_\infty$	$4.359\ 814(24) \times 10^{-18} \, \text{J}$
Electron magnetic moment	μ_e	$9.284\ 832(36) \times 10^{-24} \, \text{J T}^{-1}$
Bohr magnetron moment	$\mu_B = e\lambda/2m_e$	$9.274\ 078(36) \times 10^{-24} \, \text{J T}^{-1}$
Nuclear magnetron	$\mu_N = eh/4\pi m_p$	$5.050\ 824(20) \times 10^{-27} \, \text{J T}^{-1}$
Lande g factor for free electron	$g_e = 2\mu_e/\mu_B$	$2.002.319\ 31(70)$
Proton gyromagnetic ratio	γ_p	$2.675\ 198\ 7(75) \times 10^8 \, \text{s}^{-1} \, \text{T T}^{-1}$
Proton resonance frequency (H_2O) per unit field	$\gamma_p'/2\pi$	$4.257\ 602(12) \times 10^7 \, \text{Hz T}^{-1}$
Normal atmospheric pressure	p_0	$1.013\ 25 \times 10^5 \, \text{Pa}$
Zero on the Celcius temperature scale	T_0	$273.16 \, \text{K}$
	RT_0	$2.271\ 081(71) \times 10^3 \, \text{J mol}^{-1}$
Standard molar volume of ideal gas	$V_0 = RT_0/p_0$	$2.241\ 383(70) \times 10^{-2} \, \text{m}^3 \, \text{mol}^{-1}$

The digit(s) in parentheses following a numerical value in the above table represents the standard deviation of that value in the decimal place(s) indicated for its digit(s).

Conversion factors and equivalence of SI units

The following table contains conversion factors widely used in polymer science. The table has been constructed from data obtained in various U.S. Government and IUPAC publications and from calculations based on values given in these publications.

Equivalence of energy	J mol^{-1}	cal mol^{-1}	erg mol^{-1}
Wavenumber of $1 \, \text{cm}^{-1}$	11.96	2.859	1.986×10^{-16}
Energy of 1 electron vole per molecule	9.649×10^4	2.306×10^4	1.602×10^{-12}

Quantity	SI unit		Conversion to						
			thou or mil	in	ft	yd			
Length	1 m	=	3.937×10^4	3.937×10^1	3.281	1.094			
Area	$1\ m^2$	=	sq in 1.550×10^3	sq ft 1.076×10^1	sq yd 1.196				
Volume	$1\ m^3$	=	cu in 6.102×10^4	cu ft 3.531×10^1	cu yd 1.308	gal(US) 2.642×10^2	gal(UK) 2.20×10^2	litre 10^3	
Mass	1 kg	=	grain 1.543×10^4	ounce 3.527×10^1	lb m 2.205	cwt(UK) 1.968×10^{-2}	ton(UK) 9.842×10^{-4}	ton(US) 1.102×10^{-3}	
Density	$1\ kg/m^3$	=	lb m/cu in 3.613×10^{-5}	lb m/cu ft 6.243×10^{-2}	lb m/gal(US) 8.345×10^{-3}	lb m/gal(UK) 1.002×10^{-2}			
Force	1 N	=	dyn 1.0×10^5	kgf 1.020×10^{-1}	lbf 2.248×10^{-1}	tonf(UK) 1.004×10^{-4}	ton(US) 1.124×10^{-4}		
Pressure	$1\ Pa = N/m^2 = kg/m\,s^2$		atm 9.869×10^{-6}	bar 1.0×10^{-5}	kgf/cm^2 1.020×10^{-5}	psi (lbf/sq in) 1.450×10^{-4}	torr 7.501×10^{-3}	mbar 1.0×10^{-2}	
Energy	$1\ J = (Nm = kg\,m^2/s^2)$		erg 1.0×10^7	lbf/ft 7.376×10^{-1}	kgf m 1.020×10^{-1}	kcal 2.388×10^{-4}	BTU 9.478×10^{-4}	kWh 2.778×10^{-7}	HPh 3.725×10^{-7}
Heat	$1\ J\,(N\,m = kg\,m^2/s^2)$		kcal 2.388×10^{-4}	BTU 9.478×10^{-4}					
Power	1 W $(J/s = kg\,m^2/s^3)$		erg/s 1.0×10^7	BTU/h 3.412	kcal/h 8.598×10^1	kgf m/s 1.020×10^1	HP 1.341×10^3		

Useful conversions of material properties

Quantity SI unit	psi	N/mm²	kgf/cm²	kgf/mm²
1 Pa	1.450×10^{-4}	1.0×10^{-6}	1.020×10^{-5}	1.020×10^{-7}
$(N/m^2 = kg/m\,s^2)$				
Viscosity (dynamic)	mPa s	cP	kp s/m²	
	$(mN\,s/m^2)$			
1 Pa s	10^3	1×10^3	1.020×10^{-1}	
$(N\,s/m^2 = kg/s\,m)$				
Viscosity (kinematic)	sq ft/s	sq ft/h	cSt	
1 m²/s	1.076×10^1	3.875×10^4	10^6	
Surface tension	mN/m	kgf/m	dyn/cm	
1 N/m	10^3	1.020×10^{-1}	10^3	

Equivalence of quantities

Electron volt (eV)	$1.602\ 1892 \times 10^{-19}$ J
Electric dipole moment (Debye)	$(10^{-21}/c)$ A m² $= 3.3356 \times 10^{-30}$ Cm.
British Thermal Unit	1055.0585262 J
Pressure torr	101325/760 Pa
mmHg	$13.5951 \times 980.665 \times 10^{-2}$ Pa
atm	101325 Pa
Radioactivity curie (Ci)	3.7×10^{10} Bq
Exposure to X or radiation (röntgen)	2.58×10^{-4} C kg^{-1}
Ionizing radiation absorbed (rad)	10^{-2} Gy
Atomic mass unit 1 u	$1.660\ 565\ 5 \times 10^{-27}$ kg

Additional useful conversion factors

Quantity	Unit	Symbol	Conversion factor
Lengths	Angstrom	1 Å	10^{-10} m $= 0.1$ nm
	millimicron	1 m	10^{-9} m $= 1$ nm
	micron	1	10^{-6} m $= 1$ m
	mil	1 mil	2.54×10^{-5} m
	inch	1 in	0.025 4 m
	foot	1 ft	0.304 8 m
	yard	1 yd	0.914 4 m
	mile (statute)	1 mile	1609.344 m
	mile (nautical)	1 nmile	1852
	light year	1 light year	9.461×10^{15} m
Areas	square inch	1 sq in	$6.451\ 6 \times 10^{-4}$ m²
	square foot	1 sq ft	$9.290\ 304 \times 10^{-2}$ m²
	square yard	1 sq yd	0.836 127 36 m²
	acre	1 acre	4.047×10^3 m²
	hectare	1 ha	10^4 m²
	square mile	1 sq mile	$2.589\ 988\ 11 \times 10^6$ m²
Volumes	cubic inch	1 cu in	$1.638\ 706\ 4 \times 10^{-5}$ m³
	U.S. liquid ounce	1 ounce	$2.957\ 4 \times 10^{-5}$ m³
	U.K. liquid ounce	1 ounce	$2.841\ 3 \times 10^{-5}$ m³
	U.S. quart	1 qt	$9.463\ 353 \times 10^{-4}$ m³

	liter (SI)	1 l	$1.0 \times 10^{-3}\,\text{m}^3$
	liter (cgs)	1 l	$1.000\,028 \times 10^{-3}\,\text{m}^3$
	U.S. liquid gallon	1 gal	$3.785\,412 \times 10^{-3}\,\text{m}^3$
	U.S. dry gallon	1 gal	$4.4 \times 10^{-3}\,\text{m}^3$
	Imperial (U.K.)		
	gallon	1 gal	$4.545\,9 \times 10^{-3}\,\text{m}^3$
	cubic foot	1 cu ft	$2.831\,684\,659\,2 \times 10^{-2}\,\text{m}^3$
	U.S. barrel	1 bbl	$0.119\,\text{m}^3$
	U.S. barrel		
	(petroleum)	1 bbl	$0.158\,97\,\text{m}^3$
	imperial barrel	1 bbl	$0.163\,6\,\text{m}^3$
	cubic yard	1 cu yd	$0.764\,544\,857\,\text{m}^3$
	store	1 st	$1\,\text{m}^3$
Masses	grain	1 gr	$6.480 \times 10^{-5}\,\text{kg}$
	carat (jewels)	1 ct	$2 \times 10^{-4}\,\text{kg}$
	ounce		
	(avoirdupois)	1 oz	$0.028\,349\,52\,\text{kg}$
	pound		
	(apothekers)	1 lb	$0.373\,242\,\text{kg}$
	pound		
	(avoidupois)	1 lb	$0.453\,592\,37\,\text{kg}$
	stone	1 stone	$6.350\,293\,18\,\text{kg}$
	slug	1 slug	$14.59\,\text{kg}$
	hundredweight		
	(U.S.)	1 sh cwt	$45.359\,2\,\text{kg}$
	hundredweight		
	(U.K.)	1 cwt	$50.802\,3\,\text{kg}$
	ton (SI)	1 t	$100\,\text{kg}$
	ton (U.K.)	1 ton	$1016.046\,909\,\text{kg}$
Time	min	1 min	$60\,\text{s}$
	hour	1 hour	$3600\,\text{s}$
	day	1 d	$86\,400\,\text{s}$
	month	1 m	$2.629\,8 \times 10^6\,\text{s}$
	year	1 yr	$3.155\,76 \times 10^7\,\text{s}$
Temperature	degree Rankine	$^\circ$R	$= (\tfrac{5}{9})\text{K}$
	degree Fahrenheit	$^\circ$F	$= (\tfrac{9}{5})^\circ\text{C} + 32$
	degree Celsius	$^\circ$C	$= \text{K} + 273.15$
Angles	degree	1°	$= (\pi/180)\,\text{rad} =$
			$= 1.745\,329\,2 \times 10^{-2}\,\text{rad}$
	min	$1'$	$= 2.908\,882 \times 10^{-4}\,\text{rad}$
	second	$1''$	$= 4.848\,136\,6 \times 10^{-6}\,\text{rad}$
	grad	1 gon	$= (\pi/200)\,\text{rad}$
Concentration	mole per liter	1 M	$10^3\,\text{mol}\,\text{m}^{-3}$
Densities	1 lb/cu in	$= 27.679\,904\,71\,\text{g}\,\text{cm}^{-3}$	
	1 oz/cu in	$= 1.729\,993\,853\,\text{g}\,\text{cm}^{-3}$	
	1 lb/cu ft	$= 1.601\,846\,337 \times 10^{-2}\,\text{g}\,\text{cm}^{-3}$	
	1 lb/U.S. gal	$= 7.489\,150\,454 \times 10^{-3}\,\text{g}\,\text{cm}^{-3}$	

Velocities	knots (nautical miles per hour)	1 knot	$= 0.514\,44\,\mathrm{m\,s^{-1}}$
	miles per hour	1 mph	$= 0.447\,04\,\mathrm{m\,s^{-1}}$
Forces	dyne	1 dyn	$10^{-5}\,\mathrm{N}$
	gram force	1 gf	$9.806\,65 \times 10^{-3}\,\mathrm{N}$
	pond	1 p	$9.806\,65 \times 10^{-3}\,\mathrm{N}$
	poundal	1 pdl	$0.138\,3\,\mathrm{N}$
	pound-force	1 lbf	$4.448\,22\,\mathrm{N}$
	kilogram-force	1 kgf	$9.806\,65\,\mathrm{N}$
	impact strength with notch	1 ft-lbf/in-notch	$53.378\,64\,\mathrm{N}$
Energies ($1\,\mathrm{J} = 1\,\mathrm{N\,m}$ $= 1\,\mathrm{W\,s}$)	foot-pound-force	1 ft-lbf	$1.355\,818\,\mathrm{J}$
	calorie	1 cal	$4.184\,\mathrm{J}$
	IT calorie	$1\,\mathrm{cal_{IT}}$	$4.186\,8\,\mathrm{J}$
	foor-poundal	1 ft-pdl	$4.215\,384\,\mathrm{J}$
	meter-kilogram-force	1 m-kgf	$9.806\,65\,\mathrm{J}$
	liter-atmosphere	1 l-atm	$1.013\,250 \times 10^2\,\mathrm{J}$
	British Thermal Unit	1 BTU	$1.055\,056 \times 10^3\,\mathrm{J}$
		$1\,\mathrm{ft^3\,lb/in^2}$	$1.952\,378 \times 10^2\,\mathrm{J}$
	cubic foot atmosphere	1 cb ft atm	$2.869\,205 \times 10^3\,\mathrm{J}$
	horse power hour (U.K.)	1 hph	$2.685 \times 10^6\,\mathrm{J}$
	kilowatt-hour	1 kWh	$3.6 \times 10^6\,\mathrm{J}$
Power ($1\,\mathrm{W} = 1\,\mathrm{J\,s^{-1}}$)	horsepower	1 hp	$9.810 \times 10^3\,\mathrm{W}$
		1 BTU/h	$0.293\,275\,\mathrm{W}$
		1 cal/h	$1.162\,222 \times 10^{-3}\,\mathrm{W}$
Length related quantities	impact strength	1 kp/cm	$= 980.665\,\mathrm{N\,m^{-1}}$
	impact strength	1 lbf/ft	$= 14.593\,898\,\mathrm{N\,m^{-1}}$
	surface tension	1 dyn/cm	$= 10^{-3}\,\mathrm{N\,m^{-1}}$
Pressure related quantities ($1\,\mathrm{Mpa}$ $= 1\,\mathrm{M\,N\,m^{-2}}$ $= 1\,\mathrm{N\,mm^{-2}}$ $= 10^{-6}\,\mathrm{N\,m^{-2}}$)	physical atmosphere	1 atm	$= 0.101\,325\,\mathrm{M\,Pa}$
		1 bar	$= 0.1\,\mathrm{M\,Pa}$
	technical atmosphere	1 at	$= 0.098\,0665\,\mathrm{M\,Pa}$
		$1\,\mathrm{kp/cm^2}$	$= 0.098\,0665\,\mathrm{M\,Pa}$
		$1\,\mathrm{kgf/cm^2}$	$= 0.098\,0665\,\mathrm{M\,Pa}$
		1 lbf/sq in	$6.894\,76 \times 10^{-3}\,\mathrm{M\,Pa}$
		1 psi	$6.894\,76 \times 10^{-3}\,\mathrm{M\,Pa}$
	inch mercury (32°F)	1 in Hg	$3.386\,388 \times 10^{-3}\,\mathrm{M\,Pa}$
	torr	1 torr	$(101\,325/760) \times 10^{-6}\,\mathrm{M\,Pa}$
	millimeter mercury	1 mm Hg	$13.591 \times 980.665 \times 10^{-6}\,\mathrm{M\,Pa}$
		$1\,\mathrm{dyn/cm^2}$	$10^{-5}\,\mathrm{M\,Pa}$
	millimeter water	$1\,\mathrm{mm\,H_2O}$	$9.806\,65 \times 10^{-6}\,\mathrm{M\,Pa}$
		1 pol/sq ft	$1.488\,649 \times 10^{-6}\,\mathrm{M\,Pa}$

Heat conductivities	1 cal/(cm s °C)	$418.6 \, W \, m^{-1} \, K^{-1}$
	1 BTU/(ft h °F)	$1.731\,956 \, W \, m^{-1} \, K^{-1}$
	1 kcal/(m h °C)	$1.162\,78 \, W \, m^{-1} \, K^{-1}$
	1 (BTU in)/(sq ft h °F)	$0.144\,42 \, W \, m^{-1} \, K^{-1}$

Heat transfer 'coefficients'	1 cal/(cm^2 s °C)	$4.186\,8 \times 10^4 \, W \, m^{-2} \, K^{-1}$
	1 BTU/(ft^2 h °F)	$5.682\,215 \, W \, m^{-2} \, K^{-1}$
	1 kcal/(m^2 h °C)	$1.163 \, W \, m^{-2} \, K^{-1}$

Permeabilities of gases and liquids	(24 h 1 g)/(sq in 1 mil)	$4.56 \times 10^{-11} \, g \, cm^{-1} \, s^{-1}$
	1 barrer	$7.50 \times 10^{-15} \, cm^3 \, s \, g^{-1}$
	1 cm^3(STP) cm	$7.50 \times 10^{-5} \, cm^3 \, s \, g^{-1}$

$$\frac{cm^2 \, s \, cm \, Hg}{}$$

$$\frac{1 \, cm^3 \, mil}{100 \, in^2 \, atm \, 24 \, h} \qquad 4.51 \times 10^{-17} \, cm^3 \, s \, g^{-1}$$

$$\frac{1 \, ml}{m^2 \, h \, atm} \qquad 2.75 \times 10^{-8} \, cm^2 \, s \, g^{-1}$$

Quantities used in radiation studies

radioactivity	1 Ci	$37 \, G \, Bq = 3.7 \times 10^{10} \, s^{-1}$
doses	1 R	$2.58 \times 10^{-4} \, C \, kg^{-1}$
equivalent doses	1 rem	$10^{-2} \, Gy = 10^{-2} \, J \, kg^{-1}$
	1 rd	$10^{-2} \, Gy = 10^{-2} \, J \, kg^{-1}$

Quantities used in the textile industry

fineness or titer	1 tex	$10^{-6} \, kg \, m^{-1}$
fineness or titer	1 den	$0.111 \times 10^{-6} \, kg \, m^{-1}$
tenacity	1 g/den	$9 \times 10^3 \, m$
specific tenacity	1 gf/den	$0.082\,599 \, N \, tex^{-1}$
specific tenacity	1 gf/den	$98.06 \times$ (density in g/cm^3) M Pa

Energy contents of various fossil fuels
In the following table it is assumed that substitute natural gas (SNG) and liquid natural gas (LNG) contain approximately the same energy as natural gas.

Name	Symbol	Energy in SI units
Natural gas	1 cu ft = 1 CF	1.055 J
	10^3 cu ft = 1 MCF	1.055 kJ
	10^6 cu ft = 1 MMCF	1.055 MJ
U.S. barrel crude oil	1 bbl	5.904 kJ
U.S. ton bituminous coal	1 T	26.368 kJ
British Thermal Unit	1 BTU	1.055 kJ
Steinkohleeinheit (German coal unit)	1 SKE	29.300 MJ

References

The data and constants presented in the preceding section have been obtained from a series of reference articles of the SI system of units and related publications. Additional information has been obtained from the following reference works:

SI System of Units

Manual of Physiochemical Symbols and Terminology, IUPAC, Butterworths, London (1959); also reprinted in *Journal of American Chemical Society, 82*, 551 (1960).

The International System of Units (SI), National Bureau of Standards Special Publication 330, Edition SD Catalog No. C13.10: 330/4, U.S. Government Printing Office, Washington D.C., 20402 (1977).

SI The International System of Units, National Physical Laboratory 3rd Ed., HMSO, London (1977).

Manual of Symbols and Terminology for Physiochemical Quantities and Units, reprinted in *Pure and Appl. Chem.*, **51**, 1 (1979).

Conversion units

M. L. McGlashan, *Physiochemical Quantities and Units*, Royal Institute of Chemistry, London (1968).

G. W. C. Kaye and T. H. Laby, *Tables of Physical and Chemical Constants*, 13th Ed., Longmans, London (1966).

R. C. Weast, Ed. 57th Edition, *Handbook of Chemistry and Physics*, Chemical Rubber Publishing Company 1977.

Synopsis of Elements of Polymer Science

POLYMERIZATIONS

The following section contains a summary of basic concepts, definitions and equations used in polymer science. The reader is referred to the references group at the end of each section for more details on each of the topics.

1.1 Classification of polymerization processes

Polymerization can be broadly subdivided into two types:

a) *Addition reactions*—involving double and triple bonds. These reactions may be initiated by thermal or photo decomposition of a molecule such as AIBN (azoisobutyronitrile) or benzoyl peroxide. Termination will occur by disproportionation, recombination or transfer to solvent. Addition reactions may also involve use of stereospecific alkyl metal reaction centres.

b) *Condensation reactions*—self initiating reactions. These usually proceed with the elimination of water, hydrogen chloride or other small molecule species.

1.2 Stereospecific polymerization

These can be divided into three groups depending upon whether they proceed from prochiral or chiral monomer molecules and whether they have chiral or achiral monomeric units. These are classified as follows:

a) Chirality producing polymerizations leading to an excess of one stereoisomeric form—asymmetric synthesis or asymmetric induced polymerization.

b) Chirality transfer polymerization leading to an excess of one stereoisomeric form—asymmetric selective, asymmetric transforming or stereoselective polymerization.

c) Chirality transfer polymerizations leading to the same amount of stereoisomeric structures—stereoselective polymerization.

Atactic polymer   A strictly random array of segments will contain 50% syndiotactic diads and 25% isotactic, 25% syndiotactic and 50% heterotactic triads.

2.1 *Free radical unipolymerization*

Polymerizations are initiated by the introduction of a free radical species, propagate by addition to a growing macro radical and are terminated either by recombination, disproportionation or transfer to solvent.

Initiation   Thermal or photodecomposition of benzoyl peroxide, azoisobutyronitrile (AIBN) or similar material.

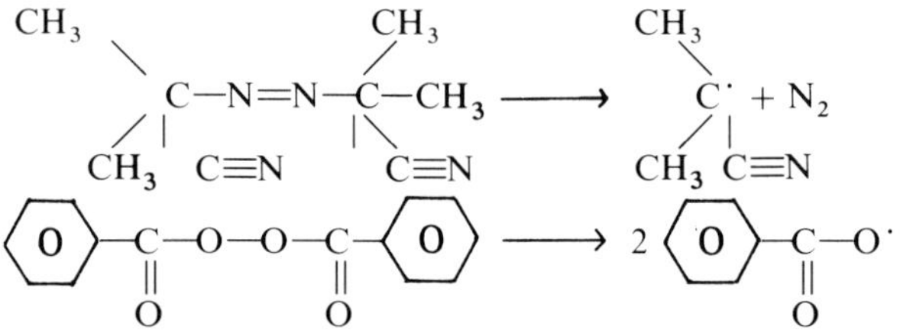

Redox systems   consist of a reducing and oxidizing agent and generate intermediate radicals which can initiate polymerization. When the concentration of reducing agent becomes comparable with the oxygen content, then the rate of polymerization is increased.

Four types of *redox initiators* can be designated:

a) The peroxide amine systems.

b) Systems composed of a reducing agent, a metal ion (Fe, Cu, Co) and a peroxide compound. The reaction can be summarized

$$H_2O_2 + M^{2+} \rightarrow HO^- + HO^{\cdot} + M^{3+}$$

which in the presence of a vinyl monomer will initiate polymerization and generate of polymer with hydroxyl end groups.

c) A combination of a metal in its zeroth oxidation state and an organo halide e.g. CCl_4

$$M^{\cdot} + RX \rightarrow M^+X^- + R^{\cdot}$$

d) Systems consisting of a reducing and an oxidizing agent. The subsequent reaction will usually involve the formation of two radicals:

$$K_2S_2O_8 + RSH \rightarrow RS^{\cdot} + KSO_4^{\cdot} + KHSO_4$$

Termination   There are three possible processes for a free radical polymerization:

a) Recombination—two radicals combine to form a stable entity. Since the radical is an energetically activated state this process will usually involve three molecule interactions and hence is identified by a low propagation constant. The velocity of reaction is given by

$$v_{t(rc)} = k_{t(rc)}[P^{\cdot}]^2$$

b) Disproportionation—two radicals interact however the net result is a proton transfer reaction and the generation of two neutral species. The velocity constant is defined as:

$$v_{t(dp)} = k_{t(dp)}[P^{\cdot}]^2$$

c) Transfer to solvent—a radical abstracts a proton from a solvent generating another initiator centre for reaction.

2.2 Molecular weight distribution

By definition every radical generated will produce a polymer. Thus the total number of polymer molecules will be given by the Schulz–Flory distribution:

$$X_x = \beta(1 + \beta)^{-x}$$

where

$$\beta = \frac{[f(I)k_{t(pp)}]^{0.5}}{k_p[M]}$$

where $k_{t(pp)}$ is the termination rate constant and is usually that for disproportionation, $f(I)$ is the initiator generation rate which in the case of photo-initiation is proportional to the amount of light absorbed, k_p is the propagation rate constant.

2.3 Transfer reactions

A free radical can undergo a number of different types of transfer reaction: reaction with another monomer, the polymer, the initiator, the solvent and other foreign substances dissolved in the polymerization mixture. The velocity constant is defined as:

$$v_{tr,x} = k_{tr,x}[P^{\cdot}][RX]$$

for the process

$$CH_2CHR^{\cdot} + R'X \rightarrow CH_2CHRX + R''$$

2.3.1 *Transfer to monomer* This process leads to the termination of the growth of one polymer and the initiation of the growth of another. The degree of polymerization can then be defined by:

$$\frac{1}{X_n} = C_m + f(I)\frac{[I]}{v_{tot}}$$

where v_{tot} is the velocity of propagation for the overall reaction. If the rate of transfer is much greater than that for termination then the above reduces to

$$\frac{1}{X_n} = \frac{k_{tr,m}}{k_p}$$

where $k_{tr,m}$ and k_p are respectivel the transfer rates of radical to monomer and the propagation rate for the radical growth.

2.3.2 *Transfer to solvent* If the solvent molecules are capable of denoting atoms, usually protons to the growing radical then the possibility of transfer to solvent arises. The degree of polymerization in this case can be defined by

$$\frac{1}{X_n} = C_m + C_s \frac{[S]}{[M]} \quad \text{where} \quad C_s = \frac{k_{tr,s}}{k_p}$$

where $k_{tr,s}$ is the transfer to solvent and this can be investigated by monitoring the degree of polymerization as a function of the solvent concentration.

2.8.3 *Transfer to initiator* At high concentrations and in viscous media transfer to initiator can occur. In such media the fragment generated during the initial break down of the radical generator can be trapped close to the initial site and this leads to the so-called *cage effect*. The degree of polymerization will be similarly given by:

$$\frac{1}{X_n} = C_m + C_1 \frac{[I]}{[M]} \quad \text{where} \quad C_1 = \frac{k_{tr,i}}{k_p}$$

where $k_{tr,i}$ is the rate of transfer to initiator and I is the concentration of initiator. Azobisisobutyronitrile has practically no transfer ability.

2.3.4 *Transfer to polymer* This process is best studied by the addition of low molecular weight polymer to the reaction mixture. The degree of polymerization can be determined with and without added low molecular weight material and leads to:

$$\frac{1}{X_n} - \frac{1}{X_n^0} = C_{poly} \frac{[Poly]}{[M]}$$

where C_{poly} is the ratio of the rate of transfer to polymer to that for normal propagation.

References

C. H. Bamford, W. G. Barb, A. D. Jenkins and P. F. Onyon, *Kinetics of Vinyl Polymerization by Radical Mechanism*, Butterworths, London (1958).

A. M. North, *The Kinetics of Free Radical Polymerization*, Pergamon Press, Oxford (1965).

C. Starks, *Free Radical Telomerization*, Academic Press, New York (1974).

J. R. Ebdon, Thermal polymerization of styrene-critical review, *Brit Polym. J.*, **3**, 9 (1971).

G. E. Scott and E. Senogles, Kinetic relationships in radical polymerization, *J. Macromol Sci. C. ((Rev. Macromol Chem.)*, **9**, 49 (1973).

3.1 *Condensation reactions—elimination reactions yielding small molecules*

Conversion and degree of polymerization   Condensation reaction.

In polycondensation reactions, the number average degree of polymerization is defined by

$$X_n = \frac{\text{amount of monomeric units in the system}}{\text{amount of molecules in the system}}$$

$$= \frac{n_{\text{mer}}}{n_{\text{mol}}}$$

Where n_{mer} for the polymerization of A groups with B groups is defined by

$$n_{\text{mer}} = \frac{n_A + n_B}{2} = \frac{n_A(1 + (1/r))}{2}$$

where $r = n_A/n_B$ which leads to

$$X_n = \frac{r_0 + 1}{2r_0(1 - p_A) + 1 - r_0}$$

where p_A is the extent of reaction and r_0 is the initial mole ratio of monomers.

3.2 *Molecular weight distribution and conversion*

Let the probability of forming an ester group in a condensation reaction by p. Then the probability of forming a given x-mer is given by: p^{x-1} where x is the degree of polymerization of the x-mer. The mass fraction w_x of molecule with the degree of polymerization X is given by

$$w_x = \frac{N_x X}{N_{\text{mer}}}$$

or

$$w_x = Xp^{x-1}(1 - p)^2$$

where p is therefore also equal to the extent of reaction. Number average degree of polymerization defined by

$$X_w = w_x = \frac{1 + p}{1 - p}$$

where w_x is the weight fraction of x-mer.

3.3 *Equilibria for multifunctional polycondensation*

The probability that N linear chain sections add on further linear chain sections through polyfunctional molecules is determined by the branching coefficient α and the functionality f:

$$\alpha_{crit} = \frac{1}{f - 1}$$

Gel point—the transition of the viscous reactive mixture into an elastic gel. At the gel point

$$\frac{1}{r - 1} = \frac{r_0 P_{A,crit}^2 x_A^\lambda}{1 - r_0 P_{A,crit}^2 (1 - x_A^\lambda)}$$

where $r_0 = (N_A/N_B)$ and $P_{A,crit}$ is the extent or reaction at the gel point and

$$x_A^\lambda = \frac{\text{number of branched monomer molecules}}{\text{total number of monomer molecules of type A}}$$

Number average degree of polymerization will be equal to

$$X_n = \frac{f(1 - x_A^\lambda + (1/r_0) + 2x_A^\lambda}{f(1 - x_A^\lambda + (1/r_0) - 2p_A) + 2x_A^\lambda}$$

3.4 *Functionality*

Number of potentially reactive sites per monomer unit. A bifunctional monomer will form a linear chain structure, whereas trifunctional monomer can form a network and in certain instances tetrafunctional monomers will give ladder polymers.

3.5 *Kinetics of ring formation*

The ratio of ring to chain formation is determined by the relative rates of formation. Chain formation v_c is a bimolecular process and has the form

$$v_c = 2k_c P^{\cdot} M$$

where $P^{\cdot}$ and M are respectively the radical and monomer concentrations. The factor of 2 arises because the reaction involves two double bonds.

In contrast, ring formation is mono molecular, since it is intramolecular and has the form

$$v_r = k_r \mathbf{P}^{\cdot}$$

The number fraction of rings f_r in the product of reaction is defined by

$$f_r = \frac{v_r}{v_r + v_c}$$

This equation becomes

$$\frac{1}{f_r} = 1 + \frac{2k_c}{-k_r}[\bar{M}] = 1 + \frac{1}{r_c}[M]$$

where $r_c = k_r/k_c$. Hence the higher the monomer concentration, the smaller will be f_r—Ruggli–Zeigler dilution principle.

Reference

G. B. Butler, G. C. Corfield and C. Aso, *Poly Polymer Science*, **4**, 71 (1975).

4.1 *Polyinsertion*

Polyinsertions are defined as polymerizations in which the monomer is inserted in between the growing chain and the initiator entity. These reactions are also sometimes called 'coordinative polymerization.' *Ziegler* catalysts are formed from transition metals of groups IV–VIII with hydrides, alkyl or aryl compounds of metals of the nontransition groups I–III. Ziegler catalysts can initiate various polymerization mechanisms. An anionic polymerization will occur when the metal alkyl component can itself induce polymerization in the monomer e.g. $C_4H_9Li/TiCl_4$/isoprene. Polymerization takes place cationically when one of the components of the Ziegler catalyst is a strong electron acceptor e.g. $TiCl_4$, VCl_4, $C_2H_5AlCl_2$, $(C_2H_5)_2AlCl$ and the monomer is an electron donor.

4.2 *Mechanism of action*

The monometallic site is approached by an olefin with its bond at the vacant ligand site of the transition metal and is complexed by it:

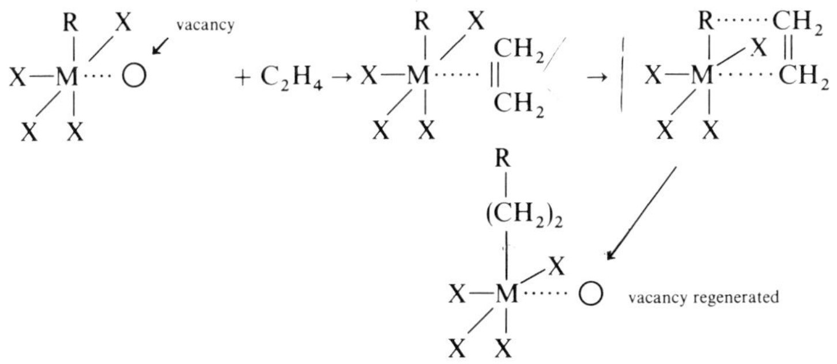

The M—R bond between the alkyl group R and the transition metal is destabilized by this coordination and as a consequence reacts with the double bond of the coordinated olefin. As a consequence the monomer is inserted in between the growing end (R) and the active centre, hence an insertion reaction. The number average degree of polymerization is determined by the propagation reaction as well as by both the concentration of true active centre $[C^*]$ and transfer reactions involving both absorbed monomer and absorbed metal alkyl.

$$\frac{1}{X_n} = \frac{k_{tr,mon}}{k_p} + \frac{k_{tr,A} K_A [A]}{k_p K_{mon} [M]}$$

where $k_{tr,mon}$ and k_p are respectively the rate constants for transfer to monomer reactions and transfer to polymer reactions, K_A and K_{mon} are the absorption equilibrium constants and A and M are the molar concentrations of metal alkyl and monomer.

References

N. C. Billingham, The polymerization of olefins at transition metal-carbon bonds, *Brit Polym, J.*, **6**, 299 (1974).

N. G. Gaylord and H. F. Mark, Linear and stereoregular addition polymers; polymerization with controlled propagation, Wiley Interscience, New York (1958).

T. Keii, Kinetics of Zeigler-Natta polymerization, Kodansha Sci. Books, Tokyo 1972).

5.1 Ionic polymerization

Ionic initiators are rarely completely associated or dissociated in organic media, but prefer to adopt partial degrees of association.

$$R-X \rightleftharpoons \underset{\substack{\text{Contact} \\ \text{Ion Pair}}}{R^+ X^-} \rightleftharpoons \underset{\substack{\text{Solvent Separated} \\ \text{Ion Pair}}}{R^+ S_n X^-} \rightleftharpoons \underset{\text{Free Ions}}{R^+ + X^-}$$

$\underbrace{}_{\text{Associated}}$

Two cases can be identified: transfer of two electrons with the formation of a bond between initiator and monomer and transfer of one electron without formation of a bond.

Initiation occurring via a two electron mechanism. Three sub groups can be identified:

a) An anion adds on to a monomer molecule thereby forming a monomer anion by electrophilic reaction.

$$R^- + CH_2=CH\emptyset \rightarrow R^- —CH_2—CH\emptyset^-$$

b) A cation adds to a monomer molecule and forms a monomer cation by nucleophic reaction

$$R^+ + CH_2\!\!=\!\!C(CH_3)_2 \rightarrow R\!\!-\!\!CH_2\!\!-\!\!C^+(CH_3)_2$$

c) A neutral molecule adds on to a monomer and forms a zwitterion

$$R_3N + \underset{\underset{\displaystyle CH_2\!\!-\!\!O}{|}}{CH_2}\!\!-\!\!\underset{\underset{\displaystyle}{|}}{C}\!\!=\!\!O \rightarrow R_3N^+\!\!-\!\!CH_2CH_2COO$$

Initiation occurring via a one electron mechanism. In the first step, a radical ion is always produced, where according to electron spin resonance measurements, the charges are not separated. In the second step, a dimerization to a diions occurs. Once more three subgroups can be identified:

i) Electron transfer from an electron donor to the monomer molecule. Example of this would be the formation of the lithium naphthalide.

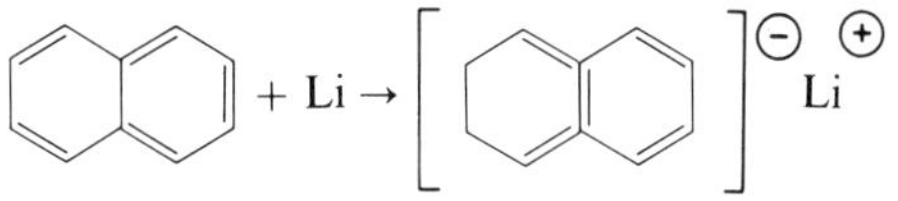

ii) Electron transfer from a monomer to an electron acceptor.

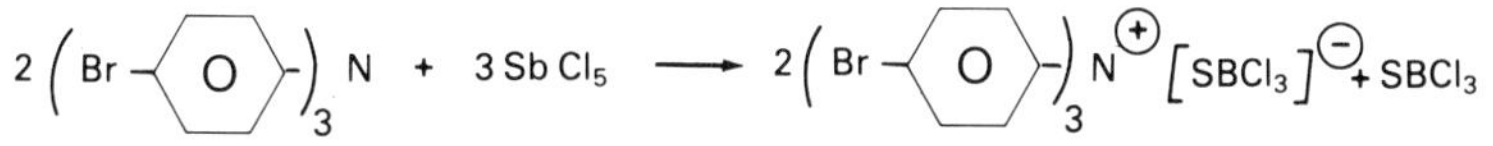

iii) A monomer molecule with donor or acceptor properties forms a charge transfer complex (CT complex) with another molecule. Thermal electron transfer leads to a radical cation with a radical anion:

$$D + A \rightleftharpoons [D\!\!-\!\!A] \rightleftharpoons \dot{D}^+/\dot{A}^- \rightleftharpoons \dot{D}^+S + \dot{A}^-S$$

The radical cations and radical anions may homopolymerize to give dications and dianions or heterodimerize to give diradicals.

Number average degree of polymerization—$\bar{X}_n$ is equal to:

$$\frac{1}{\bar{X}_n} = C_m + \frac{C}{1 + C}\frac{[I]_0}{[M]_0}$$

where $[I]_0$ and $[M]_0$ are respectively the initial concentrations of initiator and monomer, $C_m = -k_{tr,m}/k_p$ where $k_{tr,m}$ and k_p are respectively the rates for the monomer transfer reaction and propagation rate constant. The distribution about this average will be given by $1 + 1/\bar{X}$. In other words for a high molecular weight polymer the distribution will approximate to monodispersed.

References

M. Szwarc Ed., *Ions and Ion Pairs in Organic Reactions*, Wiley Interscience, New York, **1** (1972; **2** (1974).

M. Szwarc, *Carbanions Living Polymers and Electron Transfer Processes*, Wiley Interscience (1968).

J. P. Kennedy, Cationic polymerization, *Macromolecular Science*, **8**, Physical Chemistry Series 1, C. E. H. Bawn, *MTP Inter. Rev. of Sci.*, **49** (1972).

6.1 Copolymerization processes

Copolymers from two monomers are termed bipolymers, whereas those from three are called terpolymers and from four quaterpolymers. The product obtained from copolymerizations can be classified according to the sequence structure of the backbone:

i) random—no identifiable sequence of units—A—A—B—A—B—B—A—A—B— etc.

ii) alternating—sequence structure is strictly alternating—A—B—A

iii) a block copolymer—distinct blocks of monomer units chemically bonded together—A—A—A—A—A—B—B—B—B—B

iv) graft copolymers—a backbone of one polymer type with pendant chains of another type—

$$
\begin{array}{c}
A{-}A{-}A{-}A{-}A{-}A{-}A{-}A{-}A{-}A{-}A \\
\quad\ \, | \qquad\qquad\quad | \\
\quad\ \, B \qquad\qquad\quad B \\
\quad\ \, B \qquad\qquad\quad B \\
\quad\ \, B \qquad\qquad\quad B \\
\quad\ \, B \qquad\qquad\quad B \\
\quad\ \, B \qquad\qquad\quad B
\end{array}
$$

6.2 Principle of kinetics

The following assumptions are made in constructing the kinetic relationships:

a) It is usually assumed that both monomers are polymerized by the same type of bimolecular mechanism. For instance for a random copolymer this may correspond to a free radical addition process

$$\sim\!\!A^{\cdot} + A \rightarrow \, \sim\!\!A{-}A^{\cdot} \qquad v_{11} = k_{11}[A^{\cdot}][A]$$

$$\sim\!\!A^{\cdot} + B \rightarrow \, \sim\!\!A{-}B^{\cdot} \qquad v_{12} = k_{12}[A^{\cdot}][B]$$

$$\sim\!\!B^{\cdot} + B \rightarrow \, \sim\!\!B{-}B^{\cdot} \qquad v_{22} = k_{22}[B^{\cdot}][B]$$

$$\sim\!\!B^{\cdot} + A \rightarrow \, \sim\!\!B{-}A^{\cdot} \qquad v_{21} = k_{21}[B^{\cdot}][A]$$

Implicit in the above scheme is the assumption that the penultimate unit in the growing polymer chain does not affect the reactivity of the reaction centre. The copolymerization parameters (reactivity ratios) are defined as follows:

$$r_1 = (k_{11}/k_{12}) \qquad r_2 = (k_{22}/k_{21})$$

b) The overall concentrations A and B of the monomer are the effective concentrations at the reaction sites. This assumption excludes the possibility of complex formation between the monomer and the growing end. This assumption may not be strictly correct however it is a good working approximation.

c) Monomers are consumed by the propagation process and the loss in concentration due to other reactions is negligible for a high degree of polymerization.

The probability of formation of constitutive diads —A—A— will be defined by:

$$p_{11} = \frac{v_{11}}{v_{11} + v_{12}} = \frac{r_1}{r_1 + [B]/[A]}$$

correspondingly the probability of forming the —B—B— diad is given by

$$p_{22} = \frac{r_2[B]}{r_2[B] + [A]}$$

The mixed diads will be obtained from the relationships:

$$p_{12} = \frac{[A]}{r_2[B] + [A]} \qquad p_{21} = \frac{[B]}{r_1[A] + [B]}$$

The sum of the probabilities of addition to a given growing end must be equal to unity:

$$p_{11} + p_{12} = 1 \qquad p_{22} + p_{21} = 1$$

The number average sequence length $(M_A)_n$ of a monomer sequence of type A will be given by:

$$(M_A)_n = \frac{p_{12}}{(1 - p_{11})^2} = \frac{r_1[A] + [B]}{[B]}$$

likewise

$$(M_B)_n = \frac{p_{21}}{(1 - p_{22})^2} = \frac{r_2[B] + [A]}{[A]}$$

The mole ratio of monomeric A and B in the copolymer are given by

$$\frac{(M_A)_n}{(M_B)_n} = \frac{p_{21}}{p_{12}} = \frac{1 + r_1[A]/[B]}{1 + r_2[B]/[A]}$$

The copolymer ratio is usually determined either by nmr analysis or in suitable cases by UV or infrared spectroscopy.

6.3 Several special cases can be identified

The following cases can be identified:

a) $r_1 = 0$, the rate constant k_{11} is zero. The growing end only adds on unlike monomer units. The result of this polymerization will be a strictly alternating copolymer.

b) $r_1 = 1$, the rate constants of both propagation reactions are equal. The growing end adds on with equal facility both like and unlike monomer units and a random copolymeric species arises.

c) $r_1 = \infty$, only unipolymerization takes place; no copolymerization.

These limiting conditions are rarely observed in practice and the sequence structure observed will relate to the milder constraints of:

$r_1 < 1$ where the growing end adds monomers of both type but prefers unlike monomer addition

$r_1 > 1$ where the growing end adds like monomers but not exclusively.

6.4 Sequence distribution

The probability of finding a sequence of N units of monomer A in the copolymer chain will be given by $(p_{11})^{N-1}$. The probability of addition of this unit to a growing chain is then $p_{12} = 1 - p_{11}$ and thus the probability that a monomeric unit will occur in a sequence of N units is given by

$$(p_A) = (p_{11})^{N-1}(1 - p_{11})$$

Correspondingly the weight average value is

$$(p_A) = (p_{11})^{N-1}(1 - p_{11})^2 N$$

TABLE I

Reactivity ratios products $r_1 r_2$ in free radical copolymerization at 333 K

	Butadiene	Styrene	Vinyl acetate	2,5-di-chloro-styrene	Methyl metha-crylate	Vinyl-idene chloride	Acrylo-nitrile
Styrene	1.08						
Vinyl acetate	—	0.55					
Vinyl chloride	0.31	0.34	0.39				
2,5-dichlorostyrene	0.21	0.16	—				
Methyl methacrylate	0.19	0.24	0.30	1.0	0.61		
Vinylidene chloride	0.10	0.16	0.1	—	0.24		
Acrylonitrile	0.08	0.02	0.25	0.015	0.12	0.34	
Maleic anhydride	—	—	0.004	—	—	—	—

6.5 The Q–e scheme

Monomer pairs used in free radical copolymerization can be arranged according to the product of the copolymerization parameters, Table I. On the left hand side of the table are monomers with electron donating capability and those on the left electron attracting ability. The product r_1r_2 decreases from one to zero in the vertical series, whereas in the horizontal series it increases from zero to unity. The reason for this correlation is the assumption that the reactivity as defined by the activation energy E_{12}^+ is a consequence of polarity and resonance stabilization effects.

$$k_{12} = A_{12} \exp(-E_{12}^+/RT)$$

where E_{12}^+ is the preexponential factor and can be assumed to be approximately independent of the reaction mechanism. The activation energy can be split into two parts. Firstly a contribution from the electrostatic interactions between the charges in the radical $(e_1^{\cdot})$ and the monomer (e_2):

$$k_{12} = A_{12} \exp(-(p_1 + q_2 + e_1^{\cdot}e_2))$$

where p_1 and q_2 are respectively contributions from polarization and resonance stabilization of the radicals by the reaction centre. The above can be rewritten as

$$k_{12} = P_1 Q_2 \exp(-e_1^{\cdot}e_2)$$

Analogous equations can be written down for the other reaction rates. If it is further assumed that $e_1^{\cdot}$ and e_1 are equal, then the reaction rate for unipolymerization will be

$$k_{11} = P_1 Q_1 \exp(-e_1^2).$$

TABLE II

Values of e and Q for various monomers used in free radical copolymerization

Monomer	e	Q
N-vinyl urethane	−1.62	0.12
Isopropylvinylether	−1.70	0.03
Butadiene	−1.05	2.39
Styrene (reference monomer)	−0.80	1.00
Vinyl acetate	−0.22	0.026
Ethylene	−0.20	0.015
Vinyl chloride	+0.20	0.044
2,5-dichlorostyrene	+0.09	1.60
Methyl methacrylate	+0.40	0.74
Vinylidene chloride	+0.36	0.22
Acrylonitrile	+1.20	0.60
Vinylidene cyanide	+2.58	20.1

The values of Q are assigned relative to styrene which is taken as the standard. A very low value of Q are indicative of a system in which there is no resonance stabilization. A value of unity is taken for the standard styrene in which it was assumed there was the largest possible resonance stabilization. Values of the parameters are presented in the accompanying tables.

In practice, the values obtained are unable to predict the observed experimental variations however they are a very useful guide to values.

REFERENCES

P. W. Tidswell and G. A. Mortimer, Science of determining copolymerization reactivity ratios, *J. Macromol Sci. C* (Rev. Macromol Chem.), **4**, 281 (1970).
H. Sawada, Thermodynamic of polymerization, V. Thermodynamics of copolymerization, Pt 1, *J. Macromol Sci. C* (Rev. Macromol Chem), **11**, 293 (1974).
T. J. Alfrey, J. J. Bohrer and H. Mark, *Copolymerization,* Wiley Interscience, New York (1952).
G. E. Ham, Ed., *Copolymerization,* Wiley Interscience, New York (1964).

MOLECULAR WEIGHTS AND MOLECULAR WEIGHT DISTRIBUTION

The molecular weight of a polymer because of the nature of the processes whereby the molecules are synthesized will *not* have a single value which is a simple multiple of the monomer molecular weight and is best described by a statistical mean value and a distribution.

The number average molecular weight M_n or P_n. Let w_i be the weight of that portion of a macromolecular substance having molecular weight M_i. The $w_i = n_i M_i$, where n_i is the number of moles of substance with molecular weight M_i. The *number average molecular weight M_n* of the macromolecular substance is evaluated from weights w_i and the numbers of moles n_i of the individual portions according to the equation

$$\bar{M}_n = \sum_i \frac{w_i}{n_i} = \sum_i \frac{n_i M_i}{n_i}$$

Similarly the *weight average molecular weight* is defined as

$$\bar{M}_w = \sum_i \frac{w_i M_i}{w_i} = \sum_i \frac{n_i M_i^2}{n_i M_i}$$

Types of distribution function

The molecular weight distribution may be described by one of the following mathematical functions:

1.1 *Gaussian distribution*   It represents the error law about the arithmetic mean. The mole fraction differential distribution of a property R is given by the Gaussian function

$$x(R) = \frac{1}{\sigma_n(2\pi)^{\frac{1}{2}}} \exp\left[\frac{(R - R_m)^2}{2\sigma_n^2}\right]$$

where R_m is the mean value of the property and fulfils the condition

$$\int_{-\infty}^{R_m} dx = 0.5$$

The mean of the Gaussian distribution will equal the number average since the distribution is symmetric. Gaussian distributions are rarely observed in polymer science since negative values are included, which is a nonsense when one considers a molecular weight distribution.

1.2 *Logarithmic normal distribution* The form of the distribution function is essentially the same as that for the Gaussian distribution except that in this case the property variation is assumed to be logarithmic

$$x(R) = \frac{1}{\sigma_n(2\pi)^{\frac{1}{2}}} \exp\left[-\frac{(\ln R - \ln R_m)^2}{2(\sigma_n)^2}\right]$$

The distribution is symmetric about $\ln R_m$. The median curve is not identical with the number average R_m. The function corresponds to the error distribution about the geometric mean.

1.3 *Poisson distribution* This distribution occurs when a constant number of polymer chains begin to grow simultaneously and when the addition of monomeric units is random and occurs independently of the previous addition of other monomeric units. This type of distribution is to be expected for 'living polymers'.

For the differential molar degree of polymerization distribution, the Poisson distribution gives

$$x = \frac{v^{R-1} \exp(-v)}{\Gamma(R)}$$

where $v = \bar{X}_n - 1$ and $\Gamma(R)$ is the gamma function. Further, the number average degree of polymerization is related to the weight average degree by

$$\frac{\bar{X}_w}{\bar{X}_n} = 1 + \left(\frac{1}{X_n}\right) - \left(\frac{1}{X_n}\right)^2$$

Consequently, the ratio X_w/X_n in the Poisson distribution depends on the number average degree of polymerization and on no other parameter.

1.4 *Schulz–Flory distribution*   The distribution models the growth of chains whos number is constant in time randomly adding monomer until the growth centre of individual chains is destroyed by termination. The differential molar distribution is given by

$$x = \frac{\beta^{k+1} X^{k-1} X_n \exp(-\beta X)}{\Gamma(k+1)}$$

and the differential mass fraction distribution is given by

$$w = \frac{\beta^{k+1} X^k \exp(-\beta X)}{\Gamma(k+1)}$$

where

$$\beta = \frac{k}{X_n}$$

The simple one moment degree of polymerization averages are related to each other by

$$\frac{X_n}{k} = \frac{X_w}{k+1} = \frac{X_z}{k+2}$$

This type of distribution is expected theoretically for most normal free radical polymerizations and also for condensation polymerization processes.

Polydispersity for various types of polymerization

Polymers	M_w/M_n
Living polymers	1.01–1.05
Polymer formed by the combination of two radicals	1.5
Polymer formed by the disproportionation, condensation or addition of two radicals	2.0
Vinyl polymers obtained from polymerization to a high degree of conversion	2–5
Polymers, synthesized by the 'Trommsdorf effect'	5–10
Coordination polymers	8–30
Branched polymers	20–50

Methods for the determination of the molecular weight of polymers, types of average values and approximate limits of measurement

Method	Type of average	Upper limit of average	
Ultracentrifuge	M_w	$\sim 10^7$	
Light scattering	M_w	$\sim 10^7$	
Gel chromatography	M_n, M_w	$\sim 10^7$	
Solution viscometry	M_v	$\sim 10^7$	
Melt viscosity	M_w	$\sim 10^6$	
Osmometry	M_n	$\sim 10^6$	
Vapour pressure osmometry	M_n	$\sim 10^5$	
	M_n	$\sim 5 \times 10^4$	
Cryoscopic measurements	M_n	$\sim 10^4$	Depends on
Ebullioscopic measurements	M_n	$\sim 10^4$	solvent system.

2 *Summary of equations used in molecular weight determinations*

Many experiments provide the observer with a particular type of molecular weight average.

2.1 *Measurement of osmotic pressure* For solutions of nonassociating nonelectrolytes at finite concentration, the osmotic pressure π can be described by a power series of the concentration

$$\lim_{c_2 \to 0} \frac{\pi}{RTc_2} = (M_n)^{-1} = A_1 + A_2 c + A_3 c^2 + \cdots$$

where A_1, A_2 and A_3 are virial coefficients. The intercept is equal to the reciprocal of the number average molecular weight of polymer.

2.2 *Ebulliometry and cryoscopy* The elevation of the boiling point or depression of the freezing point of a solvent is a direct function of the added solute. The elevation of the boiling point T_b divided by the concentration of solute at infinite dilution can be described by the following relationship

$$\text{for } c_2 \to 0 \qquad \frac{\Delta T_{bp}}{c_2} \left(\frac{p_1 \Delta H_{bp}}{T_{bp} M_1} \right) = \frac{R T_{bp}}{M_2}$$

where M_2 is the molecular weight of the solute, ΔH_{bp} is the latent heat of vapourization of the solvent, p_1 is its density and M_1 its molecular weight. In order to obtain the largest possible boiling point elevation ΔT_{bp} for a solute should be as high as possible and a low heat of vapourization. The corresponding relationship for cryoscopy is obtained by replacing ΔT_{bp} and ΔH_{bp} by the melting point depression and the enthalpy of fusion respectively.

2.3 *Light scattering measurements* The intensity of light scattered by passing through a medium containing scattering centres is given by

$$I_0 - I_s = I_0 \exp(-\tau r)$$

where I_0 is the incident intensity and I_s is the scattered intensity, r is the path length through the medium and τ is the extinction coefficient of the scattered light. The scattering intensity is in turn defined by

$$\tau = HMc$$

where $H = (32\pi^3/3)(n_0^2/N\lambda^4)((n - n_0)/c)^2$ where $(n - n_0)/c$ is the refractive index increment with n the refractive index of solution, n_0 the refractive index of solvent and c the concentration of polymer in grams per cc and λ the wavelength of light. It has been shown that the scattering function $P(\theta)$

D

which is defined by

$$P(\theta) = 1 - \frac{1}{3}\left(\frac{4\pi}{\lambda}\right)^2 \langle R_G^2 \rangle \sin^2\left(\frac{\theta}{2}\right) + \cdots$$

has a concentration dependence of

$$\frac{Kc}{R_\theta} = \frac{1}{M_w P(\theta)} + \frac{2A_2}{Q(\theta)}c + \cdots = \frac{4\pi^2 n_1^2 (\partial n/\partial c)^2 c}{N_L \lambda^4 R_\theta}$$

According to the Zimm Kc/R_θ plotted against $\sin^2(\theta/2) + kc$ where k is an arbitary constant should give a grid plot which extrapolates to $(M_w)^{-1}$ at zero concentration and zero angle. The optical constant k is defined by

$$k = 4\pi^2 n_1^2 \left(\frac{\partial n}{\partial c}\right)^2 N_L^{-1} \lambda^4 \frac{(1 + \cos^2\theta)}{2}$$

and A_2 is the second virial coefficient characteristic of the strength of polymer solvent interactions.

2.4 *Viscosity measurements*   The specific viscosity η_{sp} is the increase in viscosity which results from the dissolution of a polymer in a solvent.

$$\eta_{sp} = \eta_{rel} - 1 = 2.5\phi_2$$

The value of the limit of η_{sp}/c as the concentration tends to infinity is called the Staudinger index or limiting-intrinsic viscosity number. The term ϕ_2 is the volume fraction of polymer treated as spheres. The concentration dependence of the viscosity would be given by

$$\eta_{sp} = 2.5\phi_2 + A_2\phi_2^2 + A_3\phi_2^3 + \cdots$$

where A_2 and A_3 are virial coefficients, $\phi = V_2/V$, the volume of all solute molecules (V_2) in a solution of volume V. According to Schulz-Blaschke the above can be written as

$$\frac{\eta_{sp}}{c} = [\eta] + k[\eta]\eta_{sp} \quad \text{where} \quad \lim_{c \to 0} \frac{\eta_{sp}}{c} = [\eta]$$

this can be simplified into

$$\frac{\eta_{sp}}{c} = [\eta] + k[\eta]^2 c$$

according to Huggins. For higher concentrations the Martin–Bungenberg–

de Jong is used in the form

$$\log \frac{\eta_{sp}}{c} = \text{long}\,[\eta] + kc$$

The Staudinger index can be empirically determined via the Fuoss equation

$$\frac{c}{\eta_{sp}} = \frac{1}{[\eta]} + Bc^{\frac{1}{2}} - \cdots$$

where the values of c/η_{sp} for $c > c_{max}$ are plotted against $c^{\frac{1}{2}}$ and extrapolate to $c^{\frac{1}{2}} \to 0$.

A series of values can be identified:

2.4.1 Unsolvated spheres   The intrinsic viscosity $[\eta]$ will depend upon both the molecular weight and the hydrodynamic volume. The mass m_{mol} of an individual molecule is related to its hydrodynamic volume via $m_{mol} = V_n p_2$. For a molecule of molecular weight M_2 the equations become $M_2 = N_L V_n p_2$ where

$$[\eta] = 2.5/p_2$$

2.4.2 Solvated spheres—If the hydrodynamic volume V_h is replaced in the above equation one obtains:

$$[\eta] = 2.5(\tilde{v}_2 + \Gamma\tilde{v}_1)$$

The intrinsic viscosity depends on the partial specific volume $\tilde{v}_2$ of the solute, the specific volume $\tilde{v}_1$ of solvent and the mass ratio $\Gamma = m_1^0/m_2$ degree of solvation of both components in the interior of the polymer. The quantity m_1^0 is the mass of solvent attached—bound to the polymer molecule. It is clear that as a consequence it is impossible to determine the molecular weight of the solvated polymer species.

2.4.3 Unsolvated rods   The intrinsic viscosity can be defined by

$$[\eta] = \frac{\Phi^*\langle R_G^2\rangle^{3/2}}{M_2}$$

where Φ^* is a general proportionality constant, $\langle R_G\rangle$ is the radius of gyration and is related to end-to-end distance $\langle L^2\rangle = 12\langle R_G^2\rangle_{rod}$.

2.4.4.Coil-like molecules—There are two conditions which we can consider:

2.4.4.1 Non-draining coil—defined as the condition where the solvent molecules move within the coil with the same velocity as the nearest segments of the coil itself during the transport processes. As a consequence the hydrodynamic volume of the coil is greater in a good solvent than in a theta solvent. The hydrodynamic volume can then be rewritten including an expansion coefficient term α_R^q in the form

$$V_h = (V_h)_\theta \alpha_R^q$$

For an ideal random coiled chain q would have a value of 3, however the non-Gaussian distribution of segments causes q to possess a lower value. For a sphere-like coil $q = 2.43$ and for an ellipsoid-like coil it will have a value of 2.18. Hence

$$[\eta] = 2.5 N_L (V_n)_\theta \frac{\alpha_R^q}{M_2}$$

In the theta condition, the hydrodynamic volume becomes proportional to the third power of the radius of gyration and the above equation has the form

$$[\eta] = 2.5 N_L \Phi' \frac{\langle R_G^2 \rangle^{\frac{3}{2}} \alpha_R^q}{M_2} = 2.5 N_L \Phi' \left(\frac{\langle R_G^2 \rangle}{M_2} \right)^{\frac{3}{2}} M_2^{\frac{1}{2}} \alpha_R^q$$

$$= \Phi \left(\frac{\langle R_G^2 \rangle}{M_2} \right)^{\frac{3}{2}} M_2^{\frac{1}{2}} = \Phi \frac{(\langle R_G^{2C} \rangle)^{\frac{3}{2}}}{M_2}$$

where $\langle R_G^2 \rangle_0 = \langle R_G^2 \rangle \alpha_R^q$ and $\Phi = 2.5 N_L \Phi'(\alpha_R/\alpha_R^3)$.

Two methods can be used to obtain the variation of the intrinsic viscosity as a function of molecular weight:

a) The radius of gyration $\langle R_G^2 \rangle$ can be expressed as follows:

$$\langle R_G^2 \rangle = \text{constant} \times M_2^{1+\varepsilon}$$

In which case the intrinsic viscosity has the form

$$[\eta] = \Phi(\text{constant})^{\frac{3}{2}} M_2^{\frac{1}{2}(1+3\varepsilon)}$$

or with $K = \Phi(\text{constant})^{\frac{3}{2}}$ and $a_\eta = \frac{1}{2}(1 + 3\varepsilon)$ we have

$$[\eta] = K M_2^{a_\eta}$$

the so called Kuhn–Mark–Houwink–Sakurada equation. Ideally should have a value of 0.5 however in practice values of between 0.55–0.85 are obtained.

b) According to the Burchard–Stockmayer–Fixman equation

$$\left(\frac{6 \langle R_G^2 \rangle}{M_2} \right) = A^3 + 0.632 B M_2^{\frac{1}{2}}$$

Hence

$$\frac{[\eta]}{M^{\frac{1}{2}}} = K_\theta + \frac{0.632}{6^{\frac{3}{2}}} \Phi B M^{\frac{1}{2}}$$

where it is assumed that

$$K_\theta = \frac{\Phi}{6^{\frac{3}{2}}} A^3$$

By plotting $[\eta]/M^{\frac{1}{2}} = f(M^{\frac{1}{2}})$ the quantity K_θ can be obtained. The A factor is associated with steric hinderance to free internal rotation of chain elements and defined by

$$A = \left[\left(\frac{1 - \cos v}{1 + \cos v} \right) \frac{\sigma^2 l^2}{M_u} \right]^{\frac{1}{2}}$$

where σ is the steric hinderance parameter, l is the bond length of the backbone chain elements, M_u is the formula molecular weight and v is the valence angle.

2.4.4.2 Free draining coils—in this case the relative velocity of the solvent both without and within the polymer coil are identical. This situation will be encountered in the case of a rigid rod in a good solvent, whereas non draining coils are associated with the limit of a flexible chain in a poor solvent.

Two approaches to this problem can be identified. In the first, Kirkwood and Riseman considered the perturbation of the rate of flow of the solvent by $N - 1$ chain elements averaged over all possible conformations. They obtain two parameters from the theory: an effective bond length (which need not be identical with the real bond length)—b and the frictional coefficient per segment ζ. The second approach is due to Debye–Bueche, they consider the partially draining soil as a more or less permeable sphere within which are homogeneously distributed the segments. The viscosity is due to the drag produced by one segment on another. The drag is defined in terms of a length L, which corresponds to the distance from the surface of the sphere to a point at which the rate of solvent flow is reduced to $1/e$ that of the value at the surface. The ratio of the radius of the sphere R_s to this effective length L is called the 'shielding factor' and expressed as

$$\zeta = R_s/L$$

The shielding function has the form

$$F(\zeta) = 2.5 \frac{1 + (3/\zeta^2) - (3/\zeta)\cos\zeta}{1 + (10/\zeta^2)[1 + (3/\zeta^2) - (3/\zeta)\cot\zeta]}$$

and the intrinsic viscosity has a value of

$$[\eta] = F(\zeta)N_L\left(\frac{4\pi R_s^3}{3}\right)M_2^{-1}$$

Molecules in reality rarely have a completely spherical contour and in general a more ellipsoidal profile is observed. For such coils, it is found experimentally that the shielding factor and the quantity ε obey the relation $\zeta\varepsilon = 3$. Hence

$$L = R_s\varepsilon/3$$

Once more we can obtain

$$[\eta] = KM^{a_\eta}$$

and find that $a_\eta = 0.8$ with a value of $\varepsilon = 0.2$, this gives L a value of $0.067R_s$. This illustrates the fact that for an expanded coil the shielding depth is 6.7% of the major rotational axis and solvent only partially penetrates the coil
A number of theoretical values of the coefficient a_η are worth noting.

Theoretical exponent a_η of the viscosity molecular weight relationship

Shape of polymer	Description of assumptions in model	a_η
Coils	Unbranched; free draining; no excluded volume	1
Coils	Unbranched; non draining; excluded volume	0.51–0.9
Spheres	Constant density, unsolvated or uniformly solvated	0
Disks	Diameter proportional to M, height constant	0.5
Disks	Diameter constant, height proportional to M	-1
Rods	Constant diameter; height proportional to M; no rotational diffusion	2
Rods	Same as above but with rorational diffusion	1.7
Rods	Diameter proportional to $M^{\frac{1}{2}}$ height constant	-1
Rods	Diameter proportional to M, height constant	-2

References

Molecular Weight and Molecular Weight Distribution
P. W. Allen, ed., *Techniques of Polymer Characterization*, Butterworths, London (1959).
Characterization of Macromolecular Structure, Natl. Acad. Sci. U.S.A. Pub. (1973); Washington D.C. (1968).
Membrane Osmometry
H.-G. Elias, *Dynamic Osmometry in Characterization of Macromolecular Structure*, Natl. Acad. Sci. U.S. Publ. 1573 Washington D.C. (1968).
M. P. Tombs and A. R. Peacock, *The Osmotic Pressure of Biological Macromolecules*, Clarendon Press Oxford (1975).

Ebulliometry and Cryoscopy
R. S. Lehrle, 'Ebulliometry applied to polymer solutions', *Prog. High Polymers*, **1**, 37 (1961).
Vapour Phase Osmometry
J. van Dam, *Vapour phase osmometry in Characterization of Macromolecular Structure*, Natl. Acad. Sci. U.S. Publ., 1573 Washington D.C. (1968).
Light Scattering
M. B. Huglin, *Light Scattering from Polymer Solutions*, Academic Press, London (1972).
Viscometry
M. Kurata and W. H. Stockmayer, 'Intrinsic viscosities and unperturbed dimensions of long chain molecules'. *Adv. Polym. Sci.*, **3**, 196 (1961–1964).
H. Yamakawa, *Modern Theory of Polymer Solutions*, Harper & Row, New York (1971).
General References
H.-G. Elias, *Macromolecules Vol 1 Structure and Properties*, John Wiley & Sons (1977).
H. Batzer and F. Lohse, *Introduction to Macromolecular Chemistry*, 2nd Ed., Hohn Wiley & Sons (1979).

Gel Permeation Chromatography Equations

L. H. TUNG

Dow Chemical, Central Research Plastics Laboratory, Midland, Michigan 48640, USA

INTRODUCTION

Gel permeation chromatography (GPC), developed in the early sixties by Moore, is now recognized as the most versatile and useful method for the determination of the molecular weight and the molecular weight distribution of polymers. Separation in GPC is achieved principally by the ability of the polymer molecules to penetrate into the pores of the beads packed in columns. Low molecular weight materials are thus retained longer in the column and high molecular weight materials are eluted first. A match between the pore-size of the packing to the molecular weight range of the polymer sample is needed for optimum separation. If the pores are too small separation amongst molecules that are totally excluded from the pores will not be possible. Packings of a broad range of pore sizes are available commercially.

Section I deals with column efficiency equations. Besides the matching of pore-size with sample molecular weights, column efficiency is another factor to be considered in selecting the proper columns to use. Column efficiency is expressed conventionally in plate height or plate count.

To relate elution volume to molecular weight, a GPC column set must be calibrated with samples of known molecular weights. Narrow distribution polystyrene standards are often used for this purpose and the logarithm of molecular weight is usually plotted vs the elution volume to establish a calibration curve as shown in Fig. 1. This curve may be represented by a polynomial of various degrees. When only the center portion of the curve is used the calibration sometimes can be approximated by a linear relation which simplifies the routine data treatment calculations. Universal calibration is a conventional technique to transfer the calibration relation determined for one polymer, usually polystyrene, to the calibration relation for another polymer. Key equations for calibration are given in Section II.

To convert the chromatogram to a molecular weight distribution curve

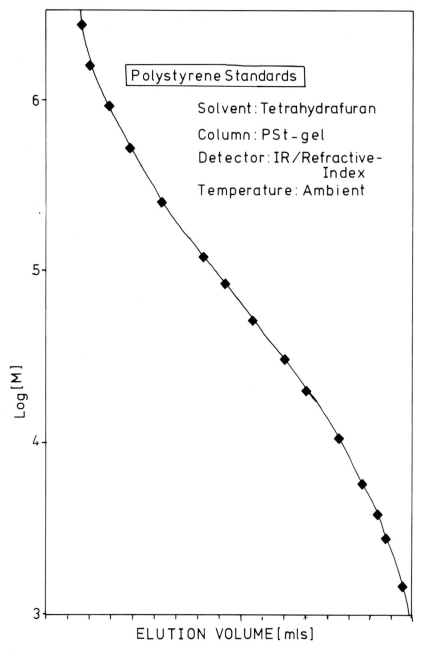

Figure 1.   Gel Permeation Chromatograph Calibration for Polystyrene—Narrow Molecular Weight Distribution Polystyrene Standards—Solvent Tetrahydrafuran.

the slopes of the calibration curve are used in the calculation. Often the molecular weight distribution is better represented with the molecular weight axis expressed on a logarithmic scale. The equations relating these variables and the equations for calculating the average molecular weights are given in Section III.

The GPC chromatogram of a polymer is a sum of the chromatograms of thousands of the individual species in the sample. For precise treatment of data the overlap of the chromatograms of the individual species due to the zone spreading characteristics of the chromatographic process must be corrected. The correction is more important if the column efficiency is low and if the distribution of the sample is narrow. Key equations for zone spreading correction are shown in Section IV. A number of numerical methods are available for solving the zone spreading equations and original articles should be consulted for details of these methods.

GENERAL REFERENCES

1. J. C. Moore, *J. Polym. Sci.*, **A**2, 835 (1964).
2. L. H. Tung and J. C. Moore, chapter 6 in *Fractionation of Synthetic Polymers*, Ed., L. H. Tung, Marcel Dekker, New York 1977.
3. W. W. Yau, J. J. Kirkland, and D. D. Bla, *Modern Size Exclusion Liquid Chromatography*, Wiley, New York 1979.

APPENDIX

I. Column efficiency

a) Plate height

$$H = L\left(\frac{\sigma}{v_0}\right)^2 \tag{1}$$

where H = height of a theoretical plate

L = column length

σ = standard deviation of the chromatogram of a monodisperse compound

v_0 = retention volume of the compound, usually the peak elution volume.

$$H = L\left(\frac{d}{4v_0}\right)^2 \tag{2}$$

where d = base-line width of the chromatogram—defined as the base line

width intercepted by the tangents drawn from the inflection points of the bell-shaped (Gaussian) chromatogram.
Other symbols are the same as in Equation (1).

b) *Plate count*

$$P = \frac{1}{H} = \left(\frac{16}{L}\right)\left(\frac{v_0}{d}\right)^2 \tag{3}$$

where P = number of plates per unit length of the column.
Other symbols are the same as in Equations (1) and (2).

c) *Plate height relationship with flow rate—the Van Deemter Equation* [1]

$$H = A + \left(\frac{B}{u}\right) + Cu \tag{4}$$

where A, B, C = constants
    u = flow rate.
Other symbols are the same as in Equation (1).
Note: The value of H at high u is found to be always lower than that described by Equation (4).

II. Calibration equations

a) *General calibration*

$$\log M = \phi(v) \tag{5}$$

where M = molecular weight
    v = elution volume
    $\phi(v)$ = polynomial of degree 3 or higher.

b) *Linear calibration*

$$v = C_1 - C_2 \log M \tag{6}$$

where C_1, C_2 = constants.
Other symbols are the same as in Equation (5).
Note: Equation (6) applies to specially packed columns, or is used in limited elution volume range.

c) *Universal calibration* [2]

$$J = [\eta]M \sim \text{a function of } (v) \tag{7}$$

where J = the product of $[\eta]M$ for any polymer
    $[\eta]$ = the intrinsic viscosity.

Other symbols are the same as in Equation (5).
Note: J relates directly to the hydrodynamic volume of the polymer in solution and regardless whether the polymer is of a different chemical structure; is branched; or is a copolymer.

$$\log J = \Phi(v) \tag{8}$$

where $\Phi(v)$ = polynomial of degree 3 or higher.
Other symbols are the same as in Equation (7).
Note: If calibration is determined for polymer 1, calibration for polymer 2 may be derived from (7) provided that the Mark–Houwink relation.

$$[\eta] = K M^\alpha \tag{9}$$

for both polymers are known. In Equation (9) K and α are constants.

d) *Derivation of Mark–Houwink constants from known broad distribution samples* [3]

$$[\eta]_{av} = K^{1/(\alpha+1)} \int w(v) J^{\alpha/(\alpha+1)} \, dv \tag{10}$$

$$M_w = K^{-1/(\alpha+1)} \int w(v) J^{1/(\alpha+1)} \, dv \tag{11}$$

$$M_n = \frac{K^{-1/(\alpha+1)}}{\int \left[\dfrac{w(v)}{J^{1/(\alpha+1)}} \right] dv} \tag{12}$$

where $[\eta]_{av}$ = known average intrinsic viscosity of the sample
M_w = known weight average molecular weight of the sample
M_n = known number average molecular weight of the sample
$w(v)$ = the experimental chromatogram.
The other symbols are the same as in Equations (7) and (9).
Note: By using Equations (10) and (12) or in Equation (11) and (12) the constants K and α may be solved.

III. Conversion of chromatogram to molecular weight distribution

a) *General conversion*

$$W(M) = -\frac{w(v)}{(2.303 s M)} \tag{13}$$

where $W(M)$ = normalized molecular weight distribution function
M = molecular weight
$w(v)$ = chromatogram as a normalized function of elution volume
s = slope, $d(\log M)/dv$, of calibration relation (5).

b) *Conversion when calibration is linear*

$$W(M) = \frac{w(v)C_2}{(2.303\,M)} \tag{14}$$

where C_2 = constant in (6).
The other symbols are the same as in Equation (13).

c) *Conversion to distribution of log molecular weight*

$$W_L(\log M) = 2.303\,M\,W(M) \tag{15}$$

where $W_L(\log M)$ = normalized distribution of log molecular weight.
The other symbols are the same as in Equation (13).
Note: When plotting distribution on semilogarithmic paper $W_L(\log M)$ should be used.

$$W_L(\log M) = -\frac{w(v)}{s} \tag{16}$$

where the symbols are the same as in Equation (13) and (15).

$$W_L(\log M) = C_2 w(v) \tag{17}$$

where the symbols are the same as in Equations (13), (14) and (15).
Note: Equation (17) is for the case where calibration is linear.

d) *Average molecular weights*

$$M_n = \frac{1}{\left\{ \int \left[\frac{W(M)}{M} \right] dM \right\}} \tag{18}$$

$$M_w = \int W(M)M\,dM \tag{19}$$

$$M_z = \frac{\left\{ \int W(M)M^2\,dM \right\}}{\left\{ \int W(M)M\,dM \right\}} \tag{20}$$

where M_n = the number average molecular weight
   M_w = the weight average molecular weight
   M_z = the z-average molecular weight.
Other symbols are the same as in Equation (13).
Note: $w(v)$ and $W_L(\log M)$ may be substituted for $W(M)$ in Equations (18), (19) and (20), thus

$$M_w = \int w(v)M \, dv = \int W_L(\log M)M \, d \log M \tag{21}$$

IV. Correction of zone spreading (axial dispersion)

a) *Zone spreading integral equation* [4]

$$f(v) = \int w(y)g(v - y) \, dy \tag{22}$$

where $f(v)$ = experimental chromatogram
   $w(y)$ = spreading corrected chromatogram as a function of retention volume
   $g(v - y)$ = spreading characteristic function
   v = elution volume
   y = elution volume under integration, may be identified as the retention volume of each specie in the distribution.

b) *Zone spreading equation with Gaussian spreading function*

$$f(v) = \int w(y)\left(\frac{h}{\sqrt{\pi}}\right)\exp\left[-h^2(v - y)^2\right]dy \tag{23}$$

where h = parameter relating to the broadness of Gaussian distribution.
The other symbols are the same as in Equation (22).
Note: The parameter h is related to the standard deviation of the Gaussian spreading function by

$$h = \frac{1}{\sigma\sqrt{2}} \tag{24}$$

where σ = standard deviation of the Gaussian spreading function. Experimental calibration of h may be found in references [5, 6]. Correction of spreading involves the solution for $w(y)$ from experimental $f(v)$ with known $g(v - y)$ or h from Equations (22) or (23). Typical solutions may be found in references [7–10].

c) Corrected average molecular weights when calibration is linear[11]

$$M_k = M'_k \exp\left\{(3 - 2k)\left(\frac{2.303}{2hC_2}\right)^2\right\} \qquad (25)$$

where $M_k = M_n$, corrected number average molecular weight if $k = 1$
$M_k = M_w$, corrected weight average molecular weight if $k = 2$
$M_k = M_z$, corrected z-average, molecular weight if $k = 3$
M'_k = corresponding uncorrected molecular weights
C_2 = constant in linear calibration Equation (6).
Other symbols are the same as in Equation (23).

REFERENCES

1. J. J. Van Deemter, F. J. Zuiderweg and A. Klinkenberg, *Chem. Eng. Sci.*, **5**, 271 (1956).
2. Z. Grubisic, P. Rempp and H. Benoit, *J. Polym. Sci., Part B*, **5**, 753 (1967).
3. A. R. Weiss and E. Cohn-Ginsberg, *J. Polym. Sci., Part A-2*, **8**, 148 (1970).
4. L. H. Tung, *J. Appl. Polym. Sci.*, **10**, 375 (1966).
5. L. H. Tung, J. C. Moore and G. W. Knight, *J. Appl. Polym. Sci.*, **10**, 1261 (1966).
6. L. H. Tung and J. R. Runyon, *J. Appl. Polym. Sci.*, **13**, 2397 (1969).
7. P. E. Piece and J. E. Armonas, *J. Polym. Sci., Part C*, **21**, 23 (1968).
8. L. H. Tung, *J. Appl. Polym. Sci.*, **13**, 775 (1969).
9. T. Ishige, S.-I. Lee and A. E. Hamielec, *J. Appl. Polym. Sci.*, **15**, 1607 (1971).
10. K. S. Chang and R. Y. Huans, *J. Appl. Polym. Sci.*, **16**, 329 (1972).
11. A. E. Hamielec, *J. Appl. Polym. Sci.*, **14**, 1519 (1970).

Polymer Networks and their Characterization

W. FUNKE

Institute for Technical Chemistry, University of Stuttgart, D-7000 Stuttgart 60, West Germany.

Crosslinked polymers are three-dimensional chain structures consisting of network chains interconnected by crosslinks. Apart from very small crosslinked particles (microgels), polymer networks are insoluble in all solvents and in some cases unswellable.[1,2] Due to the insolubility conventional methods for characterizing the structure of polymers are not applicable to polymer networks. As a matter of fact, the complex nature of polymer networks makes the determination of the structure of most practical polymer networks even still more difficult. Besides the chemical nature, for a complete characterization the following invariable network parameters should be known:

1. Number, functionality and distribution of crosslinks.
2. Length and length distribution of network chains.
3. Number and kind of terminal-end groups.
4. Length and length distribution of terminal polymer chains bearing terminal groups.
5. Spatial distribution of structural density and crosslinked concentration.

More recently it has been recognized, that many polymer networks are not homogeneous in structure nor in the spatial distribution and concentration of crosslinks.[3-6] Accordingly one may classify:

Homogeneous polymer networks, having an uniform or statistical distribution of crosslinks.

Heterogeneous polymer networks, exhibiting a regular or irregular spatial fluctuation of crosslink concentration throughout the network volume.

The large majority of fundamental work to characterize network structures is based on rubber-elastic polymers, which after crosslinking (vulcanization) of linear macromolecules (primary molecules) exhibit an approximately uniform or statistical distribution of crosslinks throughout the bulk material with comparatively little order in the spatial arrangement of the network chains. Accordingly all physico-chemical methods to determine the crosslink concentration of polymer networks have been developed and initially tested experimentally on rubber-elastic polymers.

To characterize the heterogeneity of polymer networks mostly physico-optical methods e.g. light scattering[7,8] or electron microscopy[9-11] are used. By these methods spatial fluctuations in density of the polymer material are detected. These fluctuations are frequently but not always related to the spatial variation of crosslink concentration (crosslink density).

Among all parameters for characterizing polymer networks the gross concentration of crosslinks is most important.

A crosslink (network junction) is a point connecting three or more network chains.

A network chain is the portion of a macromolecule contained between two adjacent crosslinks.

There are several kinds of crosslinks[2]:

(a) Permanent crosslinks (chemical crosslinks).
(b) Temporary crosslinks (secondary valency interactions).
(c) Chain entanglements (permanent or transitory).

DEFINITIONS FOR THE CONCENTRATION OF CROSSLINKS IN A POLYMER NETWORK

In analogy to the molecular weight of linear macromolecules the crosslink concentration of a network may be expressed by the mean molecular weight (molar mass) of a network chain, $\bar{M}_c$. which is defined as the average mass of a polymer chain which connects two adjacent crosslinks.

Other parameters for the description of the crosslink concentration are:

Crosslink density, v = moles of crosslinked basic units per weight (or volume) unit of the crosslinked polymer material.

Degree of crosslinking, ρ = moles of crosslinked basic units per total moles of basic units.

Crosslinking index, γ = crosslinked basic units per primary (linear) macromolecule.

The simplest way to describe the crosslink concentration of a polymer network would be to give the mass of crosslinked basic units per total mass of basic units. However, because the total mass of the crosslinkable system may change during network formation, e.g. by condensation reactions, it is more appropriate to express this ratio in moles.

The degree of crosslinking and the crosslink density are related by

$$\rho = v \cdot N_0^{-1} \qquad (1)$$

where N_0 is the total number of moles of basic units per unit weight (or volume).

The relation between $\bar{M}_c$ and v is given by

$$M_c = M_0 \cdot N_0 \cdot v^{-1} = \bar{M}_n \cdot N \cdot v^{-1} = d_2 \cdot v^{-1} = (\bar{v}_2 \cdot v)^{-1} \, \text{g mole}^{-1} \quad (2)$$

where M_0 is the molecular weight of a basic unit, $\bar{M}_n$ the number average molecular weight of a primary polymer molecule, N the number of moles of primary molecules per weight (or volume) unit, d_2 the specific weight of the crosslinked polymer and $\bar{v}_2$ its specific volume.

M_c and ρ are connected by

$$\bar{M}_c = \bar{M}_n \cdot N \cdot \rho^{-1} \cdot N_0^{-1} = \bar{M}_n \cdot \rho^{-1} \cdot \bar{P}_n^{-1} = M_0 \cdot \rho^{-1} \quad (3)$$

where $\bar{P}_n$ is the number average degree of polymerization of a primary (linear) macromolecule.

Contrary to the degree of crosslinking, $\bar{M}_c$ not only is determined by the number of crosslinks formed, but is also influenced by weight changes involved in the crosslinking reaction.

Finally, the crosslinking index relates to the crosslink density resp. degree of crosslinking by

$$\gamma = v \cdot N^{-1} = \rho \cdot N_0 \cdot N^{-1} = \rho \cdot \bar{P}_n = \bar{M}_n \cdot \bar{M}_c^{-1} \quad (4)$$

When networks are formed from low-molecular monomers like in vinyl–divinyl copolymerization, above definitions may be used correspondingly. In this case the kinetic chains of the copolymer may be formally considered as the primary macromolecules. A divinyl basic unit with both double bonds polymerized then corresponds to two crosslinked trifunctional basic units.

For a more rigorous description of a polymer network, besides $\bar{M}_c$ also the molecular weight distribution of the network chains should be given. Anomalous molecular weight distributions of network chains are frequently encountered in heterogeneous polymer networks. No allowance is made in the definitions above for freely ending polymer chains. Such chains, which are attached to the network only at one of their ends, do not contribute to the crosslink concentration and are considered as network defects.

Which of the above definitions for the crosslink concentration is most suitable, depends on the mechanism of crosslinking, the kind and functionality of the monomers and the special structure of the network.

When the networks are formed by crosslinking of primary macromolecules (Figure 1a), as in vulcanization of rubber, all definitions are useful. In crosslinking copolymerization reactions (Figure 1b), crosslinking density or degree of crosslinking is suitable.

When networks are formed via primary polymer particles (Figure 1c)[12–15]

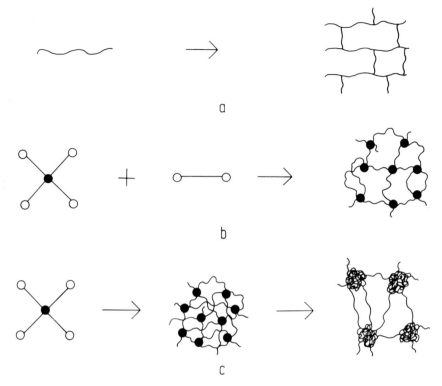

Figure 1   Reaction mechanism for formation of crosslinked polymers.
a) Crosslinking of primary macromolecules.
b) Crosslinking polyreaction with di- and tetrafunctional monomers.
c) Crosslinking polyreaction of a tetrafunctional monomer.

all definitions above are gross quantities, which exhibit a very complex relationship to mechanical or physical properties of the network.

NETWORK DEFECTS

An ideal polymer network contains no freely ending polymer chains. In real networks, however, a crosslinking unit, which functionality has been completely used up during the crosslinking reaction, is not necessarily also an effective crosslink. Primary macromolecules or copolymer chains have always a finite length and besides crosslinked basic units always contain crosslinkable units, part of which functionality is used for formation of branches (Figure 2a,b,c). Because of these defects, network chains may differ

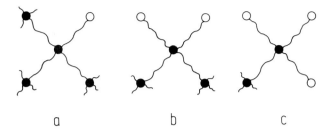

Figure 2.   Network defects by terminating chains.

in effectivity or are not effective at all as far as their influence on mechanical or physico-chemical properties is regarded.

Other network defects are closed loops, which are formed by intra-molecular reaction (cyclization) (Figure 3a,b).

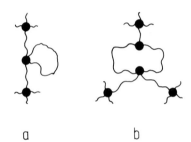

Figure 3   Network defects by intramolecular reactions (loop formation).

Such loops may be considered as precursors of tightly crosslinked domains of which heterogeneously crosslinked polymers are composed. The crosslinks within such domains may be ineffective at all as individuals, and the whole domain acts as a multifunctional junction which should be classified as a special type of a network and not as a defect (Figure 4).

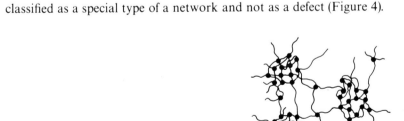

Figure 4   Heterogeneously crosslinked poly-mers with domain structure.

Some authors[4] also consider chain entanglements (Figure 5a,b) as network defects. However, as these entanglements may be also effective

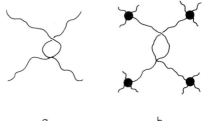

Fig. 5   Chain entanglements.
a) Temporary.
b) Permanent.   a   b

mechanically or physico-chemically, it seems more appropriate not to rank them among network defects.

In relations between crosslink concentration and mechanical resp. physico-chemical properties of polymer networks, ineffective crosslinks are usually allowed for by various corrections.[16-19]

EXPERIMENTAL DETERMINATION OF CROSSLINK CONCENTRATION

The response of polymer networks to mechanical or osmotic forces strongly depends on the crosslink concentration. Accordingly, physico-chemical methods, like stress-strain or equilibrium swelling measurements are widely used to determine the effective crosslink concentration.

Contrary to physico-chemical methods by chemical or spectroscopic methods, which have a more limited application, the total number of chemically crosslinked units may be determined, irrespective of whether they are effective or not. By physico-chemical methods besides chemical crosslinks, also temporary crosslinks and chain entanglements are measured. Therefore these values are usually larger than those from chemical or spectroscopic methods. Nevertheless the values obtained by physico-chemical methods may also be smaller if many of the chemical crosslinks are immobilized in densely crosslinked domains and therefore can't be effective as individuals.

MECHANICAL MEASUREMENTS

These methods are based on the kinetic theory of rubber elasticity of ideally elastic materials with Gaussian distribution of the distances between the ends of the network chains.[16] The deformational response of elastic polymer networks to uniaxial stresses depends on the crosslink concentration according to

$$\sigma = \frac{F}{A} = v \cdot k \cdot T \cdot \frac{\langle \bar{r}^2 \rangle_i}{\langle \bar{r}^2 \rangle_0} \cdot (\lambda - \lambda^{-2})$$ (5)

where σ is the force F exerted on the cross-sectional area, A, of the deformed (by stretching or compression) sample, $\lambda = l/l_0^{-1}$ is the relative deformation, $\langle \bar{r}^2 \rangle_0$ is the mean square end-to-end distance of the network chains and $\langle \bar{r}^2 \rangle_i$ is the corresponding value for free chains in an isotropic undisturbed condition, v is the number of elastic effective crosslinks per unit volume (crosslink density). The ratio $\langle \bar{r}^2 \rangle_i / \langle \bar{r}^2 \rangle_0$ is also called dilation-, memory- or front-factor. According to the above equation the crosslink, v, may be calculated if the deformation resulting from the applied stress is measured. Unfortunately experimental values frequently deviate considerably from this equation, even in the range of small deformations, and also extended theories, e.g. the Langevin approximation for large deformations, are not satisfactory. The same holds true for an empirical 2-parameter equation of Mooney and Rivlin[20,21]

$$\sigma = (2 \cdot C_1 \cdot \lambda + 2 \cdot C_2) \cdot (\lambda - \lambda^{-2})$$ (6)

where C_1 corresponds to the ideal deformation law. The meaning of C_2 is still somewhat controversial. It is probably related to specific features of the network, like chemical structure, entanglements, length distribution and close order arrangement of network chains and free chain ends.

Measurements of deformation of polymers in the swollen state usually correspond a little better to theory because deviations from deformation equation decrease with increasing degree of swelling. The relation between the force per unit dry unstrained cross-section, σ_d, and the relative deformation λ is given by

$$\sigma_d = \frac{A \cdot v \cdot R \cdot T}{V_d \cdot N_A} \cdot q_i^{1/3} \cdot q_0^{-2/3} \cdot \left(\lambda - \frac{q}{q_i} \cdot \lambda^{-2} \right)$$ (7)

where q is the volume degree of swelling of the deformed, q_i the volume degree of swelling of the isotropic gel, N_A the Avogadro number and V_d the volume of the dry gel.

$q_0^{-2/3}$ is the degree of dilution of the system when the network is formed and is given by

$$q_0^{2/3} = \frac{\langle \bar{r}^2 \rangle_{0.5}}{\langle \bar{r}^2 \rangle_c} \cdot q_c^{2/3} = \frac{\langle \bar{r}^2 \rangle_{0.5}}{\langle \bar{r}^2 \rangle_d}$$ (8)

where $\langle \bar{r}^2 \rangle_{0.5}$ is the mean square end-to-end distance of the network chains of the swollen gel in the undisturbed free state, $\langle \bar{r}^2 \rangle_c$ the corresponding value after crosslinking and q_c is the degree of dilution by the swelling agent. For the determination of the crosslink, v, besides the experimentally accessible parameters σ, λ and q_i and q_0 must be available, for which usually a combination of deformation and equilibrium swelling measurements is necessary. Of course q_0 may be also calculated by equation (8) via $\langle \bar{r} \rangle$ from the Kuhn–Mark–Houwink equation and $\langle \bar{r}^2 \rangle_d$ from the specific volume of the polymer with given functionality and distribution of network chains.

PHOTOELASTIC MEASUREMENTS

On the deformation of a rubber-elastic polymer birefrigence occurs,[22] which is related to the deformation by the equation (9)

$$\Delta n = \frac{2 \cdot \pi}{45} \cdot \frac{(\bar{n}^2 + 2)^2}{\bar{n}} \cdot v \cdot \Delta\alpha \frac{\langle \bar{r}^2 \rangle}{\langle \bar{r}_0^2 \rangle} \cdot (\lambda^2 - \lambda^{-1}) \tag{9}$$

where Δn is the difference of the refractive index in direction and perpendicular to the deformation. $\bar{n}$ the mean refractive index of the sample, $\Delta\alpha$ the difference of the main polarizabilities of the statistical chain segment in direction of this element and perpendicular to it.

Experimental values for determining v are Δn, λ and the stress-optical coefficient C, which is obtained by combining equation (9) and (5):

$$C = \frac{\Delta n}{\sigma} = \frac{2 \cdot \pi}{45kT} \cdot \frac{(\bar{n}^2 - 2)^2}{\bar{n}} \cdot \Delta\alpha \tag{10}$$

Because according to equation (5) and (9) σ and Δn should change on swelling of the elastomer network likewise, C should be independent of the degree of swelling. However, C usually decreases strongly on swelling with isotropic solvents. This decrease is related to a close range order in completely amorphous networks.

MEASUREMENT OF EQUILIBRIUM SWELLING

A sample of a crosslinked polymer is placed in a solvent until the chemical potential inside the gel is equal to that of the outside phase. The degree of equilibrium swelling is related to the crosslink density, v, by the modified Flory-Rehner equation:[16]

$$v = \frac{\ln(1 - \phi_2) - \phi_2 - \chi\phi_2^2}{V_1(\phi_2^{1/3} \cdot q_0^{2/3} - \phi_2/2)} \tag{11}$$

where $\phi_2 = q^{-1}$ the volume fraction of the polymer in the swollen network, V_1 the molar volume of the solvent and χ the Huggins's interaction parameter.

Equation (11) implies the validity of rubber elastic theory of an ideal network, no network defects and tetrafunctional crosslinks, applicability of the Flory-Huggins theory, i.e. nonpolar, athermic systems of small swellability and crosslink concentration, and independency of χ from the degree of swelling. As usually one or more of these requirements are not met, the quantitative determination of the crosslink concentration by this method has to be considered with reservation. However, equilibrium swelling measurements are very useful for qualitative classification of crosslink densities of networks which have the same chemical structure but have been prepared in different ways or under different conditions.

SOL-GEL-ANALYSIS

Chemical crosslinking of macromolecules is accompanied by a rapid increase of the insoluble fraction (gel fraction w_g) of the reaction mixture, especially after the gel point has been passed. The crosslinking index γ is related to the fraction $w_g = 1 - w_s$ by the equation[23-25]

$$w_s = \frac{\sum H(P)P \exp\left[-\gamma(1 - w_s)\right]}{\sum H(P)P} \tag{12}$$

where $H(P)P$ is the mass distribution function for the degree of polymerization of the primary macromolecules before crosslinking takes place. The application of this equation requires the same probability of crosslinking of all primary macromolecules, small degrees of crosslinking and no intramolecular reactions.

For the crosslinking of monodisperse primary macromolecules equation (12) simplifies to

$$w_s = \exp\left[-\gamma(1 - w_s)\right] \quad \text{or} \quad \ln w_s = -\gamma(1 - w_s) \tag{13}$$

For a random probability distribution of the primary macromolecules the crosslinking index γ becomes

$$\gamma = (w_s + w_s^{1/2})^{-1} \tag{14}$$

It should be noted that with increase of the degree of polymerization of the primary macromolecules, increasing number of crosslinkable basic units and increasing degree of crosslinking the sol-gel analysis for determining the crosslink concentration becomes progressively less accurate.

CONCLUSION

In addition to the methods for the determination of the crosslink concentration of a polymer network discussed above, a few additional methods may be mentioned, as the depression of the freezing point of solvents in a swollen network,[26] which is related to the crosslink concentration and light scattering measurements with swollen networks,[27] which are sensitive for density differences due to crosslinking.

In conclusion it may be said that the determination of the crosslink concentration is still a problematic, not satisfactorily treated field of polymer characterization. More or less all methods discussed above are submitted to serious limitations and need critical considerations before applied to optional polymer networks. The critical statement[29] that 'a quantitative description of the network structure is only possible in case of ideal networks of low crosslink concentration and known conditions of synthesis' is fully justified. This statement becomes even more stringent as many polymer networks have a heterogeneous structure[5] which is not allowed for in any method for the determination of the crosslink concentration. On the other hand it is sufficient in many practical cases to characterize polymer networks on a relative scale. For this purpose any polymer property may be chosen which is related to the crosslink concentration.

References

1. H. G. Elias, *Chimia*, **3**, 101 (1968).
2. A. Ziabicki, R. Taksermann, Krozer, *J. Polym. Sci.* Pt. A2, **7**, 2005 (1969).
3. W. Funke, *Kolloid Zeitschrift und Zeitschrift f. Polymere*, **197**, 71 (1964).
4. K. Dusek, W. Prins, *Adv. Polym. Sci., Fortschrittsberichte der Hochpolymerenforsch.*, **6**, 1 (1969).
5. W. Funke, *Chimia*, **22**, 111 (1968).
6. W. Funke, *J. Polym. Sci., Pt. C*, **16**, 1497 (1967).
7. L. Gallacher and F. A. Bettelheim, *J. Polym. Sci.*, **58**, 697 (1962).
8. F. Beuche, *J. Colloid and Interface Sci.*, **33**, 61 (1970).
9. H. P. Wohnsiedler, *J. Polym. Sci. C*, **3**, 77 (1963).
10. V. A. Kargin, I. V. Prismenko and Ye. P. Cherneva, *Vyskomol. soyed.*, **A10**, 846 (1968).
11. E. H. Erath and M. Robinson, *J. Polym. Sci., Pt. C*, **16**, 1497 (1967).
12. H. Kast and W. Funke, *Makromol. Chem.*, **180**, 1334 (1979).
13. K. Dusek, H. Galina and J. Mikes, *Polym. Bull.*, **3**, 35 (1980).
14. K. Dusek, *Makromol. Chem., Supp.*, **2**, 35 (1979).
15. H. Gallina and K. Rupicz, *Polym. Bull.*, **3**, 473 (1980).
16. P. J. Flory, *Principles of Polymer Chemistry*, Cornell University Press (1953).
17. L. Mullins and A. G. Thomas, *J. Polym. Sci.*, **43**, 13 (1960).
18. J. Scanlan, *J. Polym. Sci.*, **43**, 501 (1960).
19. L. C. Case, *J. Polym. Sci.*, **45**, 397 (1960).
20. M. Mooney, *J. Appl. Phys.*, **11**, 582 (1940).
21. R. S. Rivlin, *Phil. Trans. Roy. Soc.*, **A240**, 459, 509 (1948); **A241**, 379 (1948).
22. G. Gebhard, G. Rehage and J. Schwarz, *Progr. Colloid and Polym. Sci.*, **66**, 51 (1979).

23. P. J. Flory, *J. Am. Chem. Soc.*, **65**, 30 (1947).
24. A. Charlesby, *J. Polym. Sci.*, **11**, 513 (1953).
25. A. Charlesby, *Proc. Roy. Soc. (London)*, **A222**, 542 (1954).
26. W. Kuhn and H. Majer, *Angew. Chem.*, **68**, 345 (1956).
27. K. L. Wun and W. Prins, *J. Polym. Sci.*, **12**, 533 (1974).
28. E. Schröder, *Plaste und Kautschuk*, **26**, 1 (1979).

Compositions, Trade Names and Producers of Polymers

Name of polymer	Trade names	Abbre-viation	Producers
Polyethylene	Alathon	PE	Du Pont/USA
	Alkathene	PE (high pressure)	ICI/UK
	Courlene	PE (fibre)	Courtaulds/UK
	Ethron	PE	Dow/USA
	Fertene	PE (high pressure)	Montecatini/Italy
	Hi-flex	PE (low pressure)	Hercules/USA
	Hostalen G	PE	Hoechst/Germany
	Lacqtene	PE	Aquitaine/France
	Lupolen	PE	BASF/Germany
	Marlex	PE	Philips/USA
	Natene	PE	Pechiney/France
	Royalene	PE	US-Rubber/USA
	Supralen	PE (pipes)	Mannesmann/Germany
	Vestolen A	PE (low pressure)	Hüls/Germany
Polypropylene	Chevron	PP	Chevron/USA
	Courlene PY	PP (fibre)	Courtaulds/UK
	Hostalen PP	PP	Hoechst/Germany
	Luparen	PP	BASF/Germany
	Meraklon	PP	Montecatini/Italy
	Napril	PP	Pechiney/France
	Pro-fax	PP	Hercules/USA
	Propathene	PP	ICI/UK
	Royalene	PP	US-rubber/USA
	Vestolen P	PP	Hüls/Germany
Polystyrene	Bextrene	PS	British Xylonite/UK
	Carinex	PS	Shell/Neth.
	Celatron	PS	Celanese/USA
	Gedex	PS	Charbonnages/France
	Hostyren	PS	Hoechst/Germany
	Styrocell	PS plastic foam	Shell/UK
	Styroflex	PS fibre	Ndd. Seekabelwerke/Germany
	Styrofoam	PS plastic foam	Dow/USA
	Styron	PS	Dow/USA

Name of polymer	Trade names	Abbreviation	Producers
	Styropor	PS plastic foam	BASF/Germany
	Trolitul	PS	Dynamit Nobel/Germany
	Vestyron	PS	Hüls/Germany
Polyvinylchloride	Carina	PVC	Shell/Neth.
	Corvic	PVC	ICI/UK
	Ecavyl	PVC	Kuhlmann/France
	Elvic	PVC	Solvay/Belgium
	Geon	PVC	Goodrich/USA
	Hostalit	PVC	Hoechst/Germany
	Igelit	PVC	Bitterfeld/Germany
	Lacqvyl	PVC	Aquitaine/France
	Lucoflex	PVC	Pechiney/France
	Luvitherm	PVC foil	BASF/Germany
	Mipolam	PVC	Dynamit Nobel/Germany
	Rhovyl	PVC	Soc. Rhovyl/France
	Trovidur	PVC	Dynamit Nobel/Germany
	Trovitherm	PVC sheet	Dynamit Nobel/Germany
	Vestolit	PVC	Hüls/Germany
	Vinidur	PVC	BASF/Germany
	Vinnol	PVC	Wacker/Germany
Polymethylmethacrylate	Bonoplex	PMMA	AB Bofors/Sweden
	Degalan	PMMA	Degussa/Germany
	Elvacite	PMMA	Du Pont/USA
	Lucite	PMMA and co-polymers	Du Pont/USA
	Perspex	PMMA	ICI/UK
	Plexiglas	PMMA	Rohm & Haas/Germany
Polyacrylonitrile	Acrilan	PAN	Chemstrand Corp./USA
	Courtelle	PAN	Courtaulds/UK
	Dolan	PAN	Südd Zellwolle/Germany
	Dralon	PAN fibre	Bayer/Germany
	Leacril	PAN	ACSA/Italy
	Orlan	PAN	Du Pont/Germany
Polycarbonate	Baylon	PC	Bayer/Germany
	Jupilon	PC	Mitsubishi/Jap.
	Lexan	PC	Gen. Electric/USA
	Makrolon	PC	Bayer/D
	Merlon	PC	Mobay/USA
Polytetrafluoroethylene	Algoflon	PTFE	Montecatini/Italy
	Fluon	PTFE	ICI/UK
	Halon	PTFE	Allied Chem./USA
	Hostaflon	PTFE	Hoechst/Germany
	Soreflon	PTFE	Rhone-Poulenc/France
	Teflon	PTFE	Du Pont/USA
	Toyoflon	PTFE	Toyo/Jap.
Polyethylene terephthalate)	Arnite	PETP	AKU/Neth.
	Crastin	PETP	Ciba-Geigy/Switz.
	Dacron	PETP	Du Pont/USA

Name of polymer	Trade names	Abbre- viation	Producers
	Diolen	PETP	ENKA-Glazstoff/ Germany
	Hostadur	PETP	Hoechst/Germany
	Hostaphan	PETP	Kalle/Germany
	Hylar	PETP	Du Pont/USA
	Kodel-10	PETP	Kodak-Eastmann/USA
	Mylar	PETP foil	Du Pont/USA
	Terital	PETP	Montefibre/Italy
	Terlenka	PETP	AKU/Neth.
	Terylem	PETP	ICI/UK
	Trevira	PETP fibre	Hoechst/Germany
Poly(vinyl acetate)	Elvacet	PVAC	Du Pont/USA
	Flovic	PVAC	ICI/UK
	Mowicoll	PVAC- dispersions	Hoechst/Germany
	Mowilith	PVAC	Hoechst/Germany
	Vinavil	PVAC	Montecatini/Italy
	Vinnipas	PVAC	Wacker/Germany
Poly(vinyl alcohol)	Elvanol	PVAL	Du Pont/USA
	Mowiol	PVAL	Hoechst/Germany
	Vinaviol	PVAL	Montecatini/Italy
Poly(vinyl fluoride)	Tedlar	PVF	Du Pont/USA
Cellulose acetate	Bexoid	CA	British Xylonite/UK
	Cellit	CA	Bayer/Germany
	Cellon	CA	Dynamit Nobel/Germany
	Hercocel	CA	Hercules/USA
	Rhodester	CA	Rhone-Poulenc/France
	Rhodia	CA-$2\frac{1}{2}$	Soc Rhodiaceta/France
	Tricel	CA	Bayer/Germany
Cellulose nitrate	Herculoid	CN	Hercules/USA
	Nitron	CN	Monsanto/USA
	Trolit	CN	Dynamit Nobel/Germany
Cellulose propionate	Forticel	CP	Celanese/USA
Epoxy resins	Araldit	EP	Ciba Geigy/Switz.
	Beckopox	EP	Reichhold/Germany
	Epikote	EP	Shell/Neth.
	Epilox	EP	Leuna/E. Germany
	Epi-rez	EP	Devoe–Raynolds/USA
	Epocast	EP	Elastomer Chem./USA Furane Plastics/USA
	Epon	EP	Shell/Neth.
	Grilonit	EP	Emser–Werke/Switz.
	Lekutherm	EP	Bayer/Germany
	Scotchcast	EP	Minnesota Min. Mfg/USA
Nylon	Akulon	Nylon 6	AKU/Neth.
	Decaplast	Nylon 6, 12	Montefibre/Italy
	Perlenka	Nylon 6	AKU/Neth.
	Rhodiaceta- nylon	Nylon 6, 6	Soc. Rhodiaceta/France

Name of polymer	Trade names	Abbre- viation	Producers
	Rilsan	Nylon 11	Acquitaine/France
	Tynex	Nylon 6, 6	Du Pont/USA
	Ultramid A	Nylon 6, 6	BASF/Germany
	Ultramid B	Nylon 6	BASF/Germany
	Ultramid S	Nylon 6, 10	BASF/Germany
	Zytel	Nylon 6, 10	Du Pont/USA
	Zytel 101	Nylon 6, 6	Du Pont/USA
Copolymer of acrylonitrile butadiene and styrene	Absafil	ABS	Fiberfil/USA
	Abson	ABS	Goodrich/USA
	Abstrene	ABS	Distillers/USA
	Novodur	ABS	Bayer/Germany
	Toyarac	ABS	Toyo/Jap.
Poly(vinyl formal)	Bexone F	PVFM	British Xylonite/UK
Polyoxymethylene	Celcon	POM	Celanese/USA
	Formaldefil	POM	Fiberfil/USA
	Hostaform	POM	Hoechst/Germany
	Kematal	POM	ICI/UK
Urea-formaldehyde resins	Cibamin	UF	Ciba Geigy/Switz.
	Cibanoid	UF	Ciba Geigy/Switz.
	Ipoka	UF	BASF/Germany
	Kaurit-Leim	UF	BASF/Germany
	Melan	UF	Henkal/Germany
	Pollopas	UF	BASF/Germany
	Resolite	UF	Ciba Geigy/UK
	Roskydal	UF	Bayer/Germany
	Stratyl	UF	Pechiney/France
	Stypol	UF	Freeman/USA
	Urecoll	UF	BASF/Germany
Melamin-formaldehyde resin	Cibamin	MF	Ciba Geigy/Switz.
	Melacoll	MF	Piesteritz/E. Germany
	Melan	MF	Henkel/Germany
	Melolam	MF	Ciba Geigy/Switz.
	Melopas	MF	Ciba Geigy/Switz.
	Ultrapas	MF	Dynamit Nobel/Germany
Polyvinylpyrrolidone	Kollidon	PVP	BASF/Germany
Polyvinylcarbazole	Luvican	PNVC	BASF/Germany
Polyphenyleneoxide	Noryl	PPO	Gen. Electric/USA
	PPO	PPO	Gen. Electric/USA

Trade name	Polymer	Producers
Acronal	Dispersion of homo- and co-polymers of acrylic acid esters	BASF/Germany
Acrylan-Rubber	Butyl acrylate-acrylonitrile copolymer	Monomer Corp./USA
Adiprene	Diisocyanates	Du Pont/USA
Alloprene	Chlorinated rubber	ICI/UK
Amberlite	Synthetic ion exchanger	Rohm & Haas/USA
Ameripol	Polyisoprene	Firestone/USA
Ameripol SM	cis-1,4-polyisoprene	Firestone/USA
Astyr	Butyl rubber	Montecatini/Italy
Baygal Baymidur	Polyester or polyether with hydroxyl end groups for iso-cyanate cross-linking.	Bayer/Germany
Budene	cis-1,4-polybutadiene	Goodyear/USA
Buna N	Copolymer of butadiene and acrylonitrile	Chem. Werke Hüls/Germany
Buna S	Butadiene-styrene copolymer	Chem. Werke Hüls/Germany
Buna Hüls butacryl	Butadiene-acrylonitrile copolymer	Plastugil/France
Buton	Cross-linking butadiene-styrene copolymerizate	ESSO/UK
Butylkautschuk	Polyisobutylene with 5% isoprene	Bayer/Germany
BXL	Polysulphone	Union Carbide/USA
Caradol	Polyhydroxy compound for isocyanate crosslinking	Shell/UK
Carbowax	Poly(ethylene glycol)	Union Carbide/USA
Cariflex	Styrene-butadiene block	Shell/Nath.
Cellophan	Cellulose hydrate from pulp	Kalle/Germany
Celluloid	Cellulose nitrate plasticized with camphor	Dynamit Nobel/Germany
Chlorkautschuk	Chlorinated natural rubber	Bayer/Germany
Coral rubber	Cis-1,4-polisoprene	Goodrich/USA
Dalvor	Tetrafluoroethylene-perfluoro propylene copolymerized	Dow/USA
Desmodur	Isocyanate types	Bayer/Germany
Desmophen	Polyesters containing hydroxyl end groups, crosslinking compound for isocyanates	Bayer/Germany
Diofan	Poly(vinylidene chloride) copolymers	BASF/Germany
Durethan	Polyamides and polyurethanes	Bayer/Germany

E

Trade name	Polymer	Producers
Dutral	Ethylene-propylene copolymer	Montecatini/Italy
Dynel	Vinyl chloride-acrylonitrile copolymers	Union Carbide/USA
Econol	Poly(p-hydroxybenzoic acid ester)	Carborundum/USA
Enjay-butyl	Isobutylene-isoprene copolymer	Enjay/USA
Ethocel	Cellulose ether	Dow/USA
Fenilon	Polyamide from m-phenylenediamine and isophthalic acid	USSR
Filabond	Unsaturated polyester	Reichhold/Germany
Galalith	Plastics from milk protein	Int. Galalith-Ges/Germany
Gedamine	Unsaturated polyester	Charbonnages/France
Gemon	Maleimide	Gen. Electric/USA
Grafoil	Foils of pure graphite	Union Carbide/USA
Hetron	Polyester difficult to ignite	Hooker Chem/USA
H-Film	Polyamide from pyromellitic anhydride and 4,4-diaminodiphenyl ether	Du Pont/USA
Hydrin	Epichlorohydrin rubber	Goodrich/Hercules/USA
Hypalon	Chlorosulphonated PE	Du Pont/USA
Irrathene	PE, crosslinked by radiation	Gen. Electric/USA
Kapton H	Polyimide from pyromellitic anhydride and 4,4-diaminodiphenyl ether	Du Pont/USA
Kel F Elastomer	Copolymer from vinylidene fluoride and chlorotrifluoroethylene	Kellog/USA
Kinel	Maleimide	Rhone-Poulenc/France
Kodel-2	Polyester from tetephthalic acid and 1,4-bishydroxymethyl-cyclohexane	Easterman/USA
Lanital	Fibre from milk protein	Snia Viscosa/Italy
Larodur	Acrylic resins	BASF/Germany
Leguval	Unsaturated polyester	Bayer/Germany
Lewatit	Ion-exchange resins	Bayer/Germany
Lucite	PMMA and copolymers	Du Pont/USA
Luprenal	Acrylic resin	BASF/Germany
Lutonal	Poly(vinyl ether)	BASF/Germany
Lycra	Elastomer from diisocyanates and polyethers	Du Pont/USA

Trade name	Polymer	Producers
Moltopren	Foam material based on polyurethane	Bayer/Germany
Niax	Polyether from glycerin or hexane-1,2,6-triol	Union Carbide/USA
Nomex	Polyamide from isophthalic acid and m-phenylenediamine	Du Pont/USA
Oppanol B	Polyisobutylene	BASF/Germany
Oppanol C	Poly(vinyl isobutyl ether)	BASF/Germany
Oppanol O	Copolymer from 90% isobutylene and 10% styrene	BASF/USA
Paralac	Polyester resin	ICI/UK
Parlon	Chlorinated rubber	Hercules/USA
Parylen N	Poly-p-xylene	Union Carbide/USA
Parylen C	Polymonochloro-p-xylene	Union Carbide/USA
Penton	Poly(2,2-dichloromethyl-trimethylene oxide)	Hercules/USA
Phenoxy	Copolymer from Bisphenol A epichlorohydrin	Union Carbide/USA
Plexidur	Copolymer from acrylonitrile and methyl methacrylate	Rohm & Haas/Germany
Plexigum	Acrylate and methacrylate resins	Rohm & Haas/Germany
Pluronics	Ethylene oxide-propylene oxide copolymer containing hydroxyl end groups for isocyanate crosslinking	Wyandotte Che./USA
Polymin	Polyethyleneimine	BASF/Germany
Polysulfon	Copolymer from Bisphenol A and p,p-dichlorodiphenyl sulphone	Shell/Neth.
Propiofan	Poly(vinyl propionate)	BASF/Germany
PYR-ML	Polyimide	Du Pont/USA
Qiana	Fibre from trans,trans- diamino-dicyclohexylmethane and dodecane-dicarboxylic acid	Du Pont/USA
Q2	Polyamide from 1,4-bis-aminomethylcyclohexane and suberic acid	Eastman/USA
Rayon	Generic name for fibres from regenerated cellulose or cellulose derivatives.	
R-700	Cellulose (Viscose)	Glazstoff/Germany
Ryton	Poly(p-phenylene sulphide)	Phillips Petroleum/USA
Saflex	Polyvinylacetal	Monsanto/USA

Trade name	Polymer	Producers
Silocet	Silicone rubber	ICI/UK
Silopren	Silicone rubber	Bayer/Germany
Surlyn A	Ionomer, copolymer from ethylene and a small amount of acrylic or maleic anhydride	Du Pont/USA
Teflon FEP	Copolymer from tetrafluoroethylene and hexafluoropropylene	Du Pont/USA
Terluran	Impact-resistant PS	BASF/Germany
Thiokol	Polymers from ethylene chloride and sodium tetrasulphide	Du Pont/USA
Thornel	Graphite yarn	Union Carbide
Trogamid T	Polyimide from terephthalic acid and trimethylhexamethylenediamine	Dynamit Nobel/Germany
Versamide	Vegetable oil condensed with di and triamines, thermoplastic polyamides and reactive resins	Cray Valley/UK
Vestolen BT	Poly but-1-ene	Hüls/Germany
Vinoflex	Vinyl chloride-vinyl ether copolymer	BASF/Germany
Viscose	Generic name for fibres from regenerated cellulose after xanthogenate process	BASF/Germany
Vistanex	Polyisobutylene	Standard Oil/USA
Viton A	Vinylidene fluoride-hexafluoro-propylene copolymer	Du Pont/USA
Zetafin	Ethylene-vinyl acetate copolymer	Dow/USA

Abbreviations for Thermoplastics, Thermosets, Fibers, Elastomers and Additives

Nomenclature Committee, Division of Polymer Chemistry, Inc., American Chemical Society. Members: Hans-G. Elias (Chairman), Norbert Bikales, B. Peter Block, L. Guy Donaruma, Robert B. Fox, Josip Kratohvil, Kurt L. Loening, Raphael M. Ottenbrite, Calvin Schildknecht, Leslie H. Sperling, Bernhard Wunderlich.

Many international and national organizations have proposed 2-, 3-, and 4-letter abbreviations for the names of thermoplastics, thermosets, fibers, elastomers, and additives. Some of these abbreviations are even introduced by law, others are trademarks in certain countries. Sometimes the same abbreviation is used for different compounds, sometimes the same compound is characterized by different abbreviations. The identification of these abbreviations is therefore not always easy, especially in foreign texts or when the paper does not define the abbreviation.

The Polymer Nomenclature Committee has prepared a master list of all the abbreviations known to its members in the hope that the tabulation will be helpful to polymer scientists and technologists. The Committee does not recommend any particular set of abbreviations. It would appreciate receiving information on other national and international systems of abbreviations presently in use so that they may be included in future updates of this list.

The published lists of the following organizations were used:

ANSI	American National Standards Institute (ASTM D 1600-75; D 1418-77)
ASTM	American Society for Testing Materials (ASTM D-1600-64 T; D-1418-67)
BS	British Standards (Standard 3502-1862)
DDR	Standards of the German Democratic Republic (Deutsche Demokratische Republik; East Germany)
DIN	German Industrial Standards (DIN 7723, 7728, 60001) (Deutsche Industrie Normen; Federal Republic of Germany; West Germany)

EDV	Data processing key of the European Textile Characterization Law
EWG	European Community
ISO	International Organization for Standardization (ISO 1043-1978)
IUPAC	International Union of Pure and Applied Chemistry (*Pure Appl. Chem.* 18, 583 (1969); 40, 473 (1974))
NS	Norwegian Standards NS-4012.

The following list contains the abbreviations introduced by these organizations together with other abbreviations frequently used in the scientific and technical literature. The name of the organization is given in square brackets.

Alphabetical list of abbreviations

ABR	Elastomeric copolymer from an acrylate (ester) and butadiene; acrylate-butadiene rubber [ANSI/ASTM, IUPAC]; see also AR
ABS	Thermoplastic terpolymer from acrylonitrile, butadiene, and styrene; acrylonitrile/butadiene/styrene rubber [ASTM, DIN, ISO, IUPAC, NS]
AC	Cellulose acetate; acetate fiber [EDV]; see also CA
ACM	Elastomeric copolymer from an acrylate (ester) and a small amount of a vulcanizable monomer, e.g., 2-chlorovinyl ether [ANSI/ASTM]
ACS	Thermoplastic blend of a copolymer from acrylonitrile and styrene with chlorinated poly(ethylene)
ACSP	Acetyl cyclohexylsulfonyl peroxide
ADA	Adipic acid
AE	Amioethyl- (with cellulose)
AES	Thermoplastic quaterpolymer from acrylonitrile, ethylene, propylene, and styrene
AFK	Asbestos fiber reinforced plastic
AFMU	Elastomeric terpolymer from tetrafluoroethylene, trifluoronitrosomethane, and a small amount of nitrosoperfluorobutyric acid; nitroso rubber [ANSI/ASTM]; see also CNR

AI	Amide-imide polymers
AIBN	Azobisisobutyronitrile
AMMA	Thermoplastic copolymer from acrylonitrile and methyl methacrylate [ASTM]; see also A/MMA
A/MMA	Thermoplastic copolymer from acrylonitrile and methyl methacrylate [DIN, ISO, NS]; see also AMMA
AN	Acrylonitrile
AN-AE	Acrylonitrile-acrylate (ester) copolymer; see also ANM
ANM	Elastomeric copolymer from acrylonitrile and an acrylate (ester) [ANSI/ASTM]; see also AN-AE
AO	Antioxidant
AP	Elastomeric copolymer from ethylene and propylene; see also APR, E/P, EPM, and EPR [older German technical literature]
APR	Elastomeric copolymer from ethylene and propylene; see also AP, E/P, EPM, EPR [older German technical literature]
APT	Elastomeric copolymer from ethylene, propylene, and an unsaturated termonomer; see also EPD, EPDM, EPT, EPTR
AQ	Anthraquinone
AR	Elastomeric copolymer from acrylates (esters) and olefins; see also ABR
ASA	Thermoplastic copolymer from acrylonitrile, sytrene, and acrylates (esters) [ASTM]; see also A/S/A
A/S/A	Thermoplastic copolymer from acrylonitrile, styrene, and acrylates (esters) [DIN, ISO, NS]; see also ASA
ASE	Alkylsulfonic acid ester [ISO]
ATE	Triethylaluminum
ATH	Alumina trihydrate
ATO	Antimony trioxide
AU	Elastomeric polyurethane with polyester segments [ANSI/ASTM]
BBP	Benzyl butyl phthalate [DIN, ISO]
BD	1,4-Butanediol
BFK	Boron fiber reinforced plastic [German technical literature]

BHA	Butylated hydroxyanisole; *tert*-Butyl-4-methoxyphenol
BHET	Bis(2-hydroxyethyl) terephthalate
BHT	Butylated hydroxytoluene; 2,6-Di-*tert*-butyl-*p*-cresol
BIIR	Brominated elastomer from isobutene and isoprene; bromo-butyl rubber [ANSI/ASTM]
BMC	Bulk moulding compound
BOA	Benzyl octyl adipate (i.e., benzyl 2-ethylhexyl adipate) [ISO]
BOP	Benzyl octyl phthalate
BOPP	Balanced oriented poly(propylene)
BP	Polybutadiene rubber
BPO	Dibenzoyl peroxide
BR	Elastomeric polybutadiene; butadiene rubber [ANSI/ASTM, IUPAC]; see also CBR
BT	Thermoplastic poly(l-butene); see also PB, PBT
BTDA	3,3',4,4'-Benzophenonetetracarboxylic dianhydride
BTX	Benzene, toluene, and xylenes
Butyl	Elastomeric copolymer from isobutene and isoprene; butyl rubber [BS]; see also GR-I, IIR, PIBI
CA	Cellulose acetate [ASTM, DIN, ISO, IUPAC, NS]; see also AC
CAB	Cellulose acetate-butyrate [ASTM, DIN, ISO, IUPAC, NS]
CAP	Cellulose acetate-propionate [ASTM, DIN, ISO, IUPAC, NS]
CAR	Carbon fiber
CBA	Chemical blowing agent
CBR	Butadiene rubber; see also BR
CBS	*N*-Cyclohexyl-2-benzothiazolesulfenamide
CDB	Conjugated diene butyl elastomer
CDP	Cresyl diphenyl phosphate
CE	Cellulose plastics in general [ASTM]
CEM	Poly(chlorotrifluoroethylene) [ASTM]; see also CFM, CTFE, PCTFE
CF	Thermoset from cresol and formaldehyde; cresol-formaldehyde resin [ASTM, DIN, ISO, IUPAC, NS]

CFK	1. Synthetic fiber reinforced plastic [German technical literature]
	2. Carbon fiber reinforced plastic [German technical literature]
CFM	1. Poly(chlorotrifluoroethylene) [ANSI/ASTM]; see also CEM, CTFE, PCTFE
	2. Copolymer from chlorotrifluoroethylene and vinylidene fluoride
CFRP	Carbon fiber reinforced plastics
CHC	Elastomeric copolymer from epichlorohydrin and ethylene oxide; see also ECO
CHDM	1,4-Cyclohexanedimethanol
CHR	Elastomeric poly(epichlorohydrin); see also CO
CIIR	Postchlorinated elastomeric copolymer from isobutene and isoprene [ANSI/ASTM]
CL	Poly(vinyl chloride) fiber [EWG, EDV]; see also PVC
CM	Postchlorinated poly(ethylene) [ANSI/ASTM]; see also CPE
CMC	1. Carboxymethylcellulose [ASTM, DIN, ISO, IUPAC, NS]
	2. Critical micelle concentration
CMHEC	(Carboxymethyl) (hydroxyethyl) cellulose
CMF	Classified mineral filler
CN	Cellulose nitrate [ASTM, DIN, ISO, IUPAC, NS]; see also NC
C/N	Cellulose nitrate
CNR	Elastomeric terpolymer from tetrafluoroethylene, trifluoronitrosomethane, and a small amount of an unsaturated termonomer, e.g., nitrosoperfluorobutyric acid; nitroso rubber; carboxynitroso rubber; see also AFMU
CO	Elastomeric poly[(chloromethyl)oxirane]; poly(epichlorohydrin) [ANSI/ASTM]; see also CHR
COX	Carboxylic rubber
CP	Cellulose propionate [ASTM, DIN, ISO, IUPAC, NS]
C.P.	Chemically pure
CPE	Chlorinate poly(ethylene) [ASTM]; see also CM

CPI	*cis*-Poly(isoprene); see also IR, PIP
CPVC	1. Postchlorinate poly(vinyl chloride) [ASTM]; see also PC, PeCe, PVCC
	2. Critical pigment volume concentration
CR	Elastomeric poly(chloroprene) [ANSI/ASTM, BS, IUPAC]
CRP	Carbon fiber reinforced plastics
CS	Casein [ASTM, DIN, ISO, IUPAC, NS]
CSM	Chlorosulfonated poly(ethylene) [ANSI/ASTM]; see also CSPR, CSR
CSPR	Chlorosulfonated poly(ethylene) [BS]; see also CSM, CSR
CSR	Chlorosulfonated poly(ethylene); see also CSM, CSPR
CT	1. Cellulose triacetate (DIN); see also CTA, TA
	2. Continuous thread
CTA	Cellulose triacetate [ASTM]; see also CT, TA
CTFE	Poly(chlorotrifluoroethylene) [ASTM]; see also CEM, CFM, PCTFE
CTS	Compact tension specimen
Cullp	Copper laminated hard paper [German technical literature]
CV	Viscose [DIN]; see also VI
DABCO	Triethylenediamine
DAC	Diallyl chlorendate; diallyl 1,4,5,6,7,7-hexachlorobicyclo-[2.2.2]hept-5-ene-2,3-dicarboxylate [ASTM]
DAF	Diallyl fumarate [ASTM]
DAIP	Diallyl isophthalate resin [ASTM]
DAM	Diallyl maleate [ASTM]
DAP	Diallyl phthalate resin [ASTM, DIN]; see also FDAP
DATBP	Diallyl tetrabromophthalate
DBDPO	Decabromodiphenyl oxide
DBEP	Bis(2-butoxyethyl)phthalate
DBF	Dibutyl fumarate
DBNPG	Dibromoneopentyl glycol
DBP	Dibutyl phthalate [ASTM, DIN, ISO, IUPAC]

DBS	Dibutyl sebacate
DBTDL	Dibutyltin dilaurate
DCHP	Dicyclohexyl phthalate
DCP	Dicapryl phthalate [ASTM, DIN, ISO, IUPAC]
DDP	Didecyl phthalate
DE	Diatomaceous earth
DEA	N,N-Diethylaniline
DEAE	(Diethylamino)ethyl- (with cellulose)
DEG	Diethylene glycol
DEHP	Bis(2-ethylhexyl) phthalate; see also DOP
DEP	Diethyl phthalate [ISO]
DGEBA	Diglycidyl ethers of bisphenol A
DHA	Dehydroacetic acid
DHP	Diheptyl phthalate [ISO]
DHXP	Dihexyl phthalate [ISO]
DIBA	Diisobutyl adipate
DIBP	Diisobutyl phthalate [DIN, ISO]
DIDA	Diisodecyl adipate [ASTM, DIN, ISO, IUPAC]
DIDG	Diisodecyl glutarate
DIDP	Diisodecyl phthalate [ASTM, DIN, ISO, IUPAC]
DINA	Diisononyl adipate [ISO]
DINP	Diisononyl phthalate [DIN, ISO]
DIOA	Diisooctyl adipate [ASTM, DIN, ISO, IUPAC]
DIOF	Diisooctyl fumarate
DIOP	Diisooctyl phthalate [ASTM, DIN, ISO, IUPAC]
DIOS	Diisooctyl sebacate
DIOZ	Diisooctyl azelate
DIPP	Diisopentyl phthalate
DITDP	Diisotridecyl phthalate [DIN, ISO]; see also DITP
DITP	Diisotridecyl phthalate [DIN]; see also DITDP
DLTDP	Dilauryl 3,3'-thiodipropionate
DMA	N,N-Dimethylacetamide

DMMA	*N*,*N*-Dimethylacetoacetamide
DMC	Dough moulding compound
DME	1. Dimethylethanolamine; 2-(Dimethylamino)ethanol
	2. Dimethyl ether
DMEP	Bis(2-methoxyethyl) phthalate
DMF	*N*,*N*-Dimethylformamide
DMG	Dimethyl glutarate
DMP	Dimethyl phthalate [ISO]
DMSO	Dimethyl sulfoxide
DMT	Dimethyl terephthalate
DMTDP	Dimyristyl 3,3'-thiodipropionate
DNA	Dinonyl adipate
DNHZ	Di-*n*-hexyl azelate
DNODA	Di-*n*-octyl *n*-decyl adipate
DNODP	Di (octyl-decyl) phthalate [ASTM]
DNP	Dinonyl phthalate [ASTM, ISO, IUPAC]
DOA	Dioctyl adipate (= bis(2-ethylhexyl) adipate) [ASTM, DIN, ISO, IUPAC]
DODP	Dioctyl decyl phthalate [ISO]
DOF	Dioctyl fumarate
DOIP	Dioctyl isophthalate (= bis(2-ethylhexyl) isophthalate) [DIN, ISO]
DOP	Dioctyl phthalate (= bis(2-ethylhexyl) phthalate) [ASTM, DIN, ISO, IUPAC]; see also DEHP
DOS	Dioctyl sebacate (= bis(2-ethylhexyl) sebacate) [ASTM, DIN, ISO, IUPAC]
DOTP	Dioctyl terephthalate (= bis(2-ethylhexyl) terephthalate) [DIN, ISO]
DOZ	Dioctyl azelate (= bis(2-ethylhexyl) azelate) [ASTM, DIN, ISO, IUPAC]
DP	Degree of polymerization
DPA	Diphenylamine
DPCF	Diphenyl cresyl phosphate [ASTM, ISO]; see also DPCP

DPCP	Diphenyl cresyl phosphate; see also DPCF
DPOF	Diphenyl 2-ethylhexyl phosphate [ASTM, ISO]; see also EHDPP
DPOP	Diphenyl octyl phosphate
DPP	Diphenyl phthalate
DRC	Dry rubber content
DS	Degree of substitution
DSTDP	Distearyl 3,3'-dithiopropionate
DTDP	Ditridecyl phthalate
DUP	Diundecyl phthalate
DVB	Divinylbenzene
EA	Segmented polyurethane fiber [EDV], see also PUE
EAA	Ethylene-acrylic acid copolymers
EAM	Elastomeric copolymer of ethylene and vinyl acetate [ANSI/ASTM]
EBM	Extrusion blow moulding
EC	Ethyl cellulose [ASTM, DIN, ISO, IUPAC, NS]
ECB	Blends from ethylene copolymers with bitumen
ECO	Elastomeric copolymer from ethylene oxide and epichlorohydrin (=(chloromethyl)oxirane) [ANSI/ASTM]; see also CHC
ECTFE	Poly(ethylene-co-chlorotrifluoroethylene)
EDC	Ethylene dichloride; 1,2-Dichloroethane
EDTA	Ethylenediaminetetraacetic acid
EEA	Elastomeric copolymer from ethylene and ethyl acrylate [ASTM]; see also E/EA
E/EA	Elastomeric copolymer from ethylene and ethyl acrylate [ISO, NS]; see also EEA
EG	Ethylene glycol
EHDPP	2-Ethylhexyl diphenyl phosphate; see also DPOF
ELO	Epoxidized linseed oil [DIN, ISO]
EMA	1. Copolymer from ethylene and methacrylic acid [ASTM]
	2. Copolymer from ethylene and maleic anhydride

EOT	Ethylene ether polysulfide
EP	Epoxy resin [ASTM, DIN, ISO, IUPAC, NS]
E/P	Copolymer from ethylene and propylene [ISO, NS]; see also AP, APR, EPM, EPR
EPC	Easy processing channel black
EPD	Elastomeric terpolymer from ethylene, propylene, and a non-conjugated diene [ASTM]; see also APT, EPDM, EPT, EPTR
EPDM	Elastomeric terpolymer from ethylene, propylene, and a non-conjugated diene [ASTM]; see also APT, EPD, EPT, EPTR
EPE	Ester of an epoxy resin [DIN]
EP-G-G	Prepreg from an epoxy resin and a textile glass fabric [German technical literature]
EP-K-L	Prepreg from an epoxy resin and a carbon fiber fabric [German technical literature]
EPM	Elastomeric copolymer of ethylene and propylene [ANSI/ASTM, ISO]; see also AP, APR, E/P, EPR
EPR	Elastomeric copolymer of ethylene and propylene [BS]; see also AP, APR, E/P, EPM
EPS	Poly(styrene) foam expanded poly(styrene))
EPT	Elastomeric copolymer from ethylene, propylene, and a diene; see also APT, EPD, EPDM, EPTR
EPTR	Elastomeric copolymer from ethylene, propylene, and a diene [BS]; see also APT, EPD, EPDM, EPT
E-PVC	Poly(vinyl chloride), polymerized in emulsion
ERM	Elastic reservoir molding
E-SBR	Copolymer from styrene and butadiene, polymerized in emulsion
ESC	Environmental stress cracking
ESCR	Environmental stress cracking resistance
ESO	Epoxidized soya bean oil [DIN, ISO]
ET	Ethylene polysulfide
ETCH	Ethylene chlorohydrin; 2-Chloroethanol
ETFE	Copolymer of ethylene and tetrafluoroethylene

ETG	Ethylene glycol
EU	Elastomeric polyurethane with polyether segments [ANSI/ASTM]
EVA	Copolymer from ethylene and vinyl acetate [ASTM, DIN, ISO]; see also E/VAC, VAE
E/VAC	Copolymer from ethylene and vinyl acetate [ASTM, ISO, NS]; see also EVA, VAE
EVE	Ethylene vinyl ether
FC	Fluorocarbon
FDAP	Diallyl phthalate or resins therefrom; see also DAP
FE	Fluorine containing elastomer
FEF	Fast extruding furnace black
FEP	Elastomeric copolymer from tetrafluoroethylene and hexafluoropropylene [ASTM, DIN, ISO, NS]; see also PFEP
FF	1. Resin from furan and formaldehyde [ASTM]
	2. Fine furnace black
FK	Fiber reinforced plastic [German technical literature]
FKM	Elastomer with saturated main chain and fluoro; per-fluoroalkyl, or perfluoroalkoxy substituents [ANSI/ASTM]
FPM	1. Elastomeric copolymer from vinylidene fluoride and hexafluoropropylene [ASTM]
	2. Fluoroelastomers in general
FPVC	Flexible PVC film
FQ	Elastomeric silicone with fluorine containing substituents [ASTM]; see also FSI
FR	Flame retardant
FRP	Glass fiber reinforced polyester
FSI	Fluorinated silicone [ASTM]; see also FQ
FT	Fine thermal black
FVMQ	Silicone rubber with fluorine, vinyl, and methyl substituents [ANSI/ASTM]
FVSI	Fluorosilicone
GC	Glass coupled
GEP	Glass fiber reinforced epoxy resin

GF	Glass fiber reinforced plastic; see also GFK
GF-EP	Glass fiber reinforced epoxy resin
GFK	Glass fiber reinforced plastic [German technical literature]; see also GF
GF-PF	Glass fiber reinforced phenolic resin
GF-UP	Glass fiber reinforced unsaturated polyester resin [German technical literature]
GP	1. Gutta-percha
	2. General purpose [ASTM]
GPF	General purpose furnace black
GPO	Elastomeric copolymer from propylene oxide and allyl glycidyl ether [ANSI/ASTM]
GR	Glass fiber reinforced
GRAS	General recognized as safe
GR-A	Former US symbol for nitrile rubber; see also GR-N
GR-I	Former US symbol for butyl rubber
GR-N	Former US symbol for nitrile rubber; see also GR-A
GR-S	Former US symbol for styrene/butadiene rubber
GUP	Glass fiber reinforced unsaturated polyester resins [German technical literature]
GV	Glass fiber reinforced thermoplastics
HAF	High abrasion furnace black
HDPE	Poly(ethylene) with high density; in older German technical literature sometimes used for high pressure poly(ethylene), i.e., for PE's with mostly low densities
HEC	Hydroxyethylcellulose
HEMA	2-Hydroxyethyl methacrylate
HFIP	Hexafluoroisopropanol; 1,1,1,3,3,3-Hexafluoro-2-propanol; see also HFP
HFP	Hexafluoroisopropanol; 1,1,1,3,3,3-Hexafluoro-2-propanol; see also HFIP
Hgw	Hard fabric [German technical literature]
HIPS	High impact poly(styrene)
Hm	Hard mat [German technical literature]

HM	Hot melt adhesive
HMDI	Hydrogenated methylene diisocyanate
HMF	High-modulus furnace black
HMPT	Hexamethylphosphoric acid triamide
HMWPE	Poly(ethylene) with very high molar mass (molecular weight)
HP	Hard paper [German technical literature]
HPA	Hydroxypropyl acrylate
HPC	1. Hydroxypropylcellulose
	2. Hard processing channel black
HPMC	(Hydroxypropyl)methylcellulose
HR	High resiliency foams
HS	1. Heat stabilized
	2. High percent solids
HTBA	High temperature blowing agent
HTE	Hydroxyl terminated polyether
HTST	High temperature, short time processing
ICR	Initial concentration rubber (a type of natural rubber)
IEN	Interpenetrating elastomeric network
IFR	Intumescent flame retardent
IHPN	Interpenetrating homopolymer network
IIR	Elastomeric copolymer from isobutene and isoprene [ANSI/ASTM, IUPAC]; see also Butyl, GR-I, PIBI
IM	Poly(isobutene), see also PIB
IMW	Intermediate molecular weight
IPA	Isophthalic acid
IPP	Isopropyl percarbonate; diisopropyl peroxydicarbonate
IPN	Interprenetrating polymer network
IR	Synthetic *cis*-1,4-poly(isoprene) [ANSI/ASTM, BS, IUPAC]; see also CPI, PIP
IRT	Insect resistant treated
IT	Interrupted or split thread
IVE	Isobutyl vinyl ether

KEK	Carbon fiber reinforced plastic [DIN]
KFA	Potassium fatty acid soap
LDPE	Poly(ethylene) with low density; in older German technical literature sometimes used for low pressure poly(ethylene), i.e., for PE's with mostly high densities; see also NDPE
LIM	Liquid impingement molding (now RIM)
LIPN	Latex interpenetrating polymer network
LLDPE	Linear low density poly(ethylene)
LRM	Liquid reaction molding (now RIM)
LRMR	Reinforced liquid reaction molding
L-SBR	Solution polymerized SBR
MA	1. Modacryl fiber [EDV]; see also PAM
	2. Maleic anhydride
MABS	Copolymer from methyl methacrylate, acrylonitrile, butadiene, and styrene
MAN	Methacrylonitrile
MBI	Mercaptobenzimidazole; 2-Benzimidazolethiol
MBS	Copolymer from methacrylate, butadiene, and styrene
MBT	Mercaptobenzothiazole; 2-Benzothiazolethiol
MBTS	2-Benzothiazolyl disulfide
MC	Methylcellulose
MCB	Monochlorobenzene
MD	Modal fiber [EDV]
MDA	Methylenediamine
MDAC	4-Methyl-7-(diethylamino) coumarin
MDI	4,4'-Diphenylmethane diisocyanate; methylenedi-p-phenylene diisocyanate
MDPE	Poly(ethylene) of medium density (ca. 0,93—0,94 g/cm^3)
MEK	Methyl ethyl ketone; 2-Butanone
MEKP	Methyl ethyl ketone peroxide
MF	Melamine/formaldehyde resin [ASTM, DIN, ISO, IUPAC, NS]
MKF	Metal fiber reinforced plastic [German technical literature]

MMA	Methyl methacrylate [ASTM]
MOCA	4,4′-Methylenebis (2-chloroaniline)
MOD	Modacryl fiber [EWG]
MP	Melamine/phenol/formaldehyde resin
MPE	Metallized polyester film
MPF	Melamine/phenol formaldehyde resin [ISO, NS]
M-PVC	PVC, polymerized in bulk [German technical literature]
MQ	Elastomeric silicones with methyl substituents [ANSI/ASTM]
MR	Mold resistant
MTM	Mixed tertiary mercaptans
MWR	Molding with rotation
NBR	Elastomeric copolymer from butadiene and acrylonitrile; nitrile rubber [ANSI/ASTM, BS, IUPAC]; see also GR-N, PBAN
NC	Cellulose nitrate; see also CN
NCR	Elastomeric copolymer from acrylonitrile and chloroprene [ANSI/ASTM, IUPAC]
NDGA	Nordihydroguaiaretic acid
NDOP	n-Decyl n-octyl phthalate
NDPE	Poly(ethylene) of low density; see also LDPE
NEM	N-Ethylmorpholine
NHDP	n-Hexyl n-decyl phthalate
NIR	Elastomeric copolymer from acrylonitrile and isoprene [ANSI/ASTM]
NK	Natural rubber; see also NR
NODA	n-Octyl n-decyl adipate
NODP	n-Octyl n-decyl phthalate; see also ODP
NPG	Neopentyl glycol
NPP	Propyl percarbonate; dipropyl peroxydicarbonate
NR	Natural rubber [ANSI/ASTM, IUPAC]; see also NK
OBSH	p,p′-Oxybis(benzenesulfonyl hydrazide)
ODP	Octyl decyl phthalate [ISO]; see also NODP

ODPP	Octyl diphenyl phthalate
OER	Oil-extended rubber
OPP	Oriented poly(propylene), film or bottles
OPR	Elastomeric polymer from propylene oxide
OPS	Oriented poly(styrene) films
OPVC	Oriented poly(vinyl chloride)
PA	1. Polyamide [ASTM, DIN, ISO, IUPAC, NS]; the abbreviation PA is normally followed by a number, a combination of numbers, a letter or a combination of letters and numbers. A single number refers to the polyamide from an α,ω-amino acid or its lactam. A combination of two numbers is often separated by a comma. The first number following the symbol PA indicates the number of methylene groups of aliphatic diamines, the second number the number of carbon atoms of aliphatic dicarboxylic acids. An I stands for isophthalic acid, a T for terephthalic acid.
	2. Phthalic anhydride
PAA	Poly(acrylic acid) [ASTM]
PAB	p-Aminobiphenyl
PABM	Polyaminobismaleimide
PAC	Poly(acrylonitrile) fiber [IUPAC, DIN]; see also PAN, PC
PAE	Phthalate ester plasticizer
PAI	Polyamide-imide [ASTM]
PAM	Modacryl fiver [DIN]; see also MA
PAMS	Poly(α-methylstyrene)
PAN	Poly(acrylonitrile) fiber; see also PAC, PC [ASTM; IUPAC] (PAN is also a German trademark)
PAPA	Polyazelaic polyanhydride
PAPI	Polymethylenepolyphenylene isocyanate; see also PMPPI
PARA	Polyarylamide [ASTM]
PAS	Polyaryl sulfone
PAT	Polyaminotriazole
PB	Poly(l-butene) [DIN, ISO, NS, ASTM]; see also BT, PBT
PBA	Physical blowing agent

PBAN	Copolymer from butadiene and acrylonitrile [ASTM]; see also GR-N, NBR
PBB	Polybrominated biphenyls
PBI	Poly(benzimidazoles)
PBMA	Poly(*n*-butyl methacrylate)
PBNA	Phenyl-β-naphthylamine; *N*-Phenyl-2-naphthylamine
PBR	Copolymer from butadiene and vinylpyridine [ANSI/ASTM, IUPAC]
PBS	Copolymer from butadiene and styrene [ASTM]: see also GR-S, SBR
PBT	1. Poly(l-butene); see also BT, PB
	2. Poly(butylene terephthalate) [ASTM]; see also PBTP
PBTP	Poly(butylene terephthalate) [DIN, ISO, NS]; see also PTMT
PC	1. Polycarbonate [ASTM, DIN, ISO, IUPAC, NS]; see also PCO
	2. Poly(acrylonitrile) [EWG, EDV]; see also PAC, PAN
	3. formerly: chlorinated PVC; see also CPVC, PcCe, PVCC
PCB	Polychlorinated biphenyl
PCD	Poly(carbodiimide)
PCDT	Poly(1,4-cyclohexylenedimethylene terephthalate)
PCF	Poly(chlorotrifluoroethylene) fiber
PCO	Polycarbonate
PCTFE	Poly(chlorotrifluoroethylene) [ASTM, DIN, ISO, IUPAC, NS]; see also CEM, CFM, CTFE
PCU	Poly(vinyl chloride) [old German technical literature]
PDAP	Poly(diallyl phthalate) [ASTM, DIN, ISO, IUPAC, NS]; see also DAP, FDAP
PDMS	Poly(dimethylsiloxane)
PE	1. Poly(ethylene) [ASTM, DIN, ISO, IUPAC, NS]; see also PL
	2. Polyester fiber [DDR; EWG]
	3. Pentaerythritol
PEA	Poly(ethyl acrylate)

PEC	Chlorinated poly(ethylene) [DIN, ISO]; also see CPE
PeCe	Chlorinated PVC; see also CPVC, PC, PVCC
PEG	Poly(ethylene glycol)
PEH	Poly(ethylene) with high density [DIN, NS]
PEL	Poly(ethylene) with low density [DIN, NS]
PEM	Poly(ethylene) with medium density [DIN, NS]
PEO	Poly(ethylene oxide) [ASTM, ISO, IUPAC]; see also PEOX
PEOX	Poly(ethylene oxide) [DIN, ISO, NS]; see also PEO
PES	1. Polyester fiber [DIN]
	2. Polyethersulfone [American technical literature]; see also PESU
PESU	Polyethersulfone [ASTM]; see also PES
PET	Poly(ethylene terephthalate) [ASTM]; see also PETP
PETP	Poly(ethylene terephthalate) [ASTM, DIN, ISO, IUPAC, NS]; see also PET
PF	Phenol-formaldehyde resin [ASTM, DIN, ISO, IUPAC, NS]
PFA	Perfluoroalkoxy resins
PFEP	Copolymer from tetrafluoroethylene and hexafluoropropylene; see also FEP
PF-P-B	Phenolic resin/paper prepreg
PG	Propylene glycol
PHEMA	Poly(2-hydroxyethyl methacrylate)
PHP	Physiological hydrophilic polymers
PI	1. *trans*-1,4-Poly(isoprene), gutta-percha [BS]
	2. Polyimide [ASTM]
PIB	Poly(isobutene) [ASTM, BS, DIN, ISO, IUPAC, NS]
PIBI	Elastomeric copolymer from isobutene and isoprene, butyl rubber; see also Butyl, GR-I, IIR
PIBO	Poly(isobutene oxide)
PIP	*cis*-1,4-Poly(isoprene), synthetic; see also CPI, IR
PIR	Poly(isocyanurate) [DIN, NS]
PIS	Poly(isobutylene)

PL	1. Poly(ethylene) [EWG]; see also PE
	2. Polyester fiber [EDV]; see also PES
PMA	1. Pyromellitic acid
	2. Pyromellitic dianhydride
	3. Poly(methyl acrylate)
PMAC	Polymethoxy acetal
PMCA	Poly(methyl α-chloromethacrylate) [ASTM]
PMF	Processed mineral fiber
PMI	Poly(methacrylimide)
PMMA	Poly(methyl methacrylate) [ASTM, DIN, ISO, IUPAC, NS]
PMMI	Polypyromellitimide
PMP	Poly(4-methyl-l-pentene) [ASTM, DIN, ISO, NS]; see also TPX
PMPPI	Polymethylenepolyphenylene isocyanate; see also PAPI
PMQ	Silicone rubbers with methyl and phenyl substituents [ANSI/ASTM]
PNA	Polynuclear aromatics
PNAH	Polynuclear aromatic hydrocarbons
PO	1. Elastomeric poly(propylene oxide) [ASTM]
	2. Poly(olefin)
	3. Phenoxy resin
POB	Poly(p-hydroxybenzoate) [ASTM]
POM	Poly(oxymethylene), polyformaldehyde [ASTM, DIN, ISO, IUPAC, NS]
POP	Poly(phenylene oxide) [ASTM]; see also PPO
POR	Elastomeric copolymer from propylene oxide and allyl glycidyl ether
PP	Poly(propylene) [ASTM, DIN, EDV, ISO, IUPAC, NS]
PPA	Polyparabanic acid
PPI	Polymeric polyisocyanate
PPO	Poly(phenylene oxide); (a trademark); see also POP
PPOX	Poly(propylene oxide) [ASTM, DIN, ISO, NS]
PPS	Poly(phenyl sulfide) [ASTM]

PPSU	Poly(phenylene sulfone) [ASTM, DIN, ISO, NS]; see also PSU
PPT	Poly(propylene terephthalate)
PQ	Elastomeric silicone with phenyl substituents [ASTM]
PS	Poly(styrene) [ASTM, DIN, ISO, IUPAC, NS]
P-S	Pressure sensitive adhesive
PSA	Pressure sensitive adhesive
PSAB	Copolymer from styrene and butadiene [DIN]; see also SB, S/B
PSAN	Thermoplastic copolymer from styrene and acrylonitrile [DIN]; see also SAN
PSBR	Elastomeric terpolymer from vinylpyridine, styrene, and butadiene [ANSI/ASTM]
PSF	Polysulfone; see also PSO
PSI	Poly(methylphenylsiloxane) [ASTM]
PSO	Polysulfone; see also PSF
PST	Poly(styrene) fiber
PS-TSG	Poly(styrene) foam, processed by thermoplastic injection [German technical literature]
PSU	Poly(phenylene sulfone); see also PPSU
PTA	Purified terephthalic acid
PTF	Poly(tetrafluorethylene) fiber
PTFE	Poly(tetrafluorethylene) [ASTM, DIN, ISO, IUPAC, NS]
P3FE	Poly(trifluorethylene) [IUPAC]
PTMEG	Poly(tetramethylene ether glycol)
PTMT	Poly(tetramethylene terephthalate) = Poly(butylene terephthalate); see also PBTP
PU	1. Polyurethane elastomer [BS]
	2. Polyurethane fiber [EDV]
PUA	Polyurea fiber
PUE	Segmented polyurethane fiber [DIN]; see also EA
PUR	Polyurethane [ASTM, DIN, ISO, IUPAC, NS]: see also PU
PVA	1. Poly(vinyl acetate); see also PVAC

2. Poly(vinyl alcohol); see also PVAL, PVOH

3. Poly(vinyl ether) [old German technical literature]

PVAC   Poly(vinyl acetate) [ASTM, DIN, ISO, IUPAC, NS]; see also PVA

PVAL   Poly(vinyl alcohol) [ASTM, DIN, ISO, IUPAC, NS]; see also PVA, PVOH

PVB   Poly(vinyl butyral) [ASTM, DIN, ISO, IUPAC, NS]

PVC   1. Poly(vinyl chloride) [ASTM, DIN, ISO, IUPAC, NS]; see also CL

2. Pigment volume concentration

PVCA   Copolymer from vinyl chloride and vinyl acetate [DIN, IUPAC]; see also PVCAC

PVCAC   Copolymer from vinyl chloride and vinyl acetate [ASTM]; see also PVCA

PVCC   Chlorinated PVC [DIN, ISO]; see also CPVC, PC, PeCe

PVDC   Poly(vinylidene chloride) [ASTM, DIN, ISO, IUPAC, NS]

PVDF   Poly(vinylidene fluoride) [ASTM, DIN, ISO, IUPAC, NS]; see also PVF_2

PVF   Poly(vinyl fluoride) [ASTM, DIN, ISO, IUPAC, NS]

PVF_2 Poly(vinylidene fluoride); see also PVDF

PVFM   Poly(vinyl formal) [ASTM, DIN, ISO, IUPAC, NS]; see also PVFO

PVFO   Poly(vinyl formal) [DIN]; see also PVFM

PVI   Poly(vinyl isobutyl ether)

PVID   Poly (vinylidene cyanide)

PVK   Poly(N-vinylcarbazole) [ASTM, DIN, ISO, NS]

PVM   Copolymer from vinyl chloride and vinyl methyl ether

PVMQ   Silicone rubber with methyl, phenyl, and vinyl substituents [ANSI/ASTM]

PVOH   Poly(vinyl alcohol); see also PVA, PVAL

PVP   Poly(N-vinylpyrrolidone) [ASTM, DIN, ISO, NS]

PVSI   Poly(dimethylsiloxane) with phenyl and vinyl substituents [ASTM]

PY   Unsaturated polyester resins [BS]; see also UP

RAM	Restricted area molding
RAN	Residual acrylonitrile (monomer in polymer)
RF	Resorcine/formaldehyde resin
RHB	Reheat blow molding
RHC	Rubber hydrocarbon content
RIM	Reaction injection molding
RP	Reinforced plastics
RP/C	Reinforced plastics/composite
RP/M	Reinforced polyester/mortar
RPVC	Rigid PVC film
RRIM	Reinforced reaction injection molding
RTP	Reinforced thermoplastic
RTV	Room temperature vulcanization
RVCM	Residual vinyl chloride monomer
S	Styrene
SAF	Super abrasion furnace black
SAF-HS	High structure SAF
SAF-LS	Low structure SAF
SAIB	Sucrose acetate isobutyrate
SAN	Thermoplastic copolymer from styrene and acrylonitrile [ASTM, DIN, ISO, IUPAC, NS]; see also PSAN
SB	Thermoplastic copolymer from styrene and butadiene [ASTM]; see also PASB, S/B
S/B	Copolymer from styrene and butadiene [ANSI, ASTM, ISO, IUPAC, NS]; see also PASB, SB
SBP	*sec*-Butyl percarbonate; *di-sec*-butyl peroxydicarbonate
SBR	Elastomeric copolymer from styrene and butadiene [ASTM, BS, IUPAC]; see also GR-S
SBS	Solid bleached sulfate
SCR	Elastomeric copolymer from styrene and chloroprene [ANSI, ASTM, IUPAC]
SDD	Sodium dimethyldithiocarbamate
SE	Sulfoethyl- (with cellulose)

S-EPDM	Sulfonated ethylene-proplylene-diene terpolymers
SF	Structural foam
SFK	Synthetic fiber reinforced plastic [German technical literature]
SFP	Scrapless forming process
SHIPS	Super-high impact poly(styrene)
SI	1. Silicones in general [ASTM, DIN, ISO, NS]
	2. Poly(dimethylsiloxane) [ASTM]
	3. Thermoplastic silicone [IUPAC]
SIN	Simultaneous interpenetrating network
SIPN	Sequential interpenetrating polymer network
SIR	1. Elastomeric copolymer from styrene and isoprene [ANSI, ASTM, IUPAC]
	2. Standardized Indonesian rubber
SM	Styrene monomer
SMA	Copolymer from styrene and maleic anhydride [ASTM]
SMC	Sheet molding compound
SMR	Standardized Malaysian rubber
SMS	Copolymer from styrene and α-methylstyrene [ASTM, DIN]; see also S/MS
S/MS	Copolymer from styrene and α-methylstyrene [DIN, ISO, NS]; see also SMS
SP	Superior processing rubber (a type of natural rubber)
SPF	Super processing furnace black
SPPF	Solid-phase pressure forming
SPSF	Solid-phase stretch forming
S-PVC	Poly(vinyl chloride), polymerized in suspension
SRF	Semi-reinforcing furnace black
SRF-HM	High modulus SRF
SRF-HM-NS	Non-staining SRF-HM
SRF-LM	Low modulus SRF
SRF-LM-NS	Non-staining SRF-LM

SRP	Rubber reinforced poly(styrene) [ASTM]
SS	1. Single stage [ASTM]
	2. Steam-stripped (to remove residual monomer)
SWP	1. Solvent welded plastics pipe [ASTM]
	2. Synthetic wood pulp
T	Polysulfide rubber
TA	Cellulose triacetate [EDV]; see also CT, CTA
TAC	Triallyl cyanurate [ASTM]
TBL	Tootal Broadhurst Lee process
TBP	*tert*-Butyl perbenzoate; see also TBPB
TBPA	Tetrabromophthalic anhydride
TBPB	*tert*-Butyl perbenzoate; see also TBP
TBS	*tert*-Butylstyrene
TBT	Tetrabutyl titanate
TC	Technically classified natural rubber
TCBO	Trichlorobutylene oxide
TCE	Trichloroethylene
TCEF	Trichloroethyl phosphate [ASTM, ISO]; see also TCEP
TCEP	Trichloroethyl phosphate; see also TCEF
TCF	Tricresyl phosphate, tritolyl phosphate [ASTM, DIN, ISO]; see also TCP, TKP, TTP
TCP	Tricresyl phosphate [IUPAC] (a registered trademark in Great Britain); see also TCF, TKP, TTP
TDI	Tolylene diisocyanate; tolylene isocyanate; methylphenylene diisocyanate
TEAE	Triethylaminoethyl- (with cellulose)
TEP	Triethyl phosphate
TETA	Triethylenetetramine
TFE	Tetrafluoroethylene [ASTM]
THBP	2,4,5-Trihydroxybutyrophenone
THF	Tetrahydrofuran
TIOTM	Triisooctyl trimellitate [DIN, ISO]

TKP	Tricresyl phosphate; see also TCF, TCP, TTP
TMA	Trimellitic anhydride
TMC	Thick molding compound
TMP	Thermal mechanical pulp
TMS	Tetramethylsilane
TOF	Trioctyl phosphate, i.e. tris(2-ethylhexyl) phosphate [ASTM, DIN, ISO]; see also TOP
TOP	Trioctyl phosphate, tris(2-ethylhexyl) phosphate [IUPAC]; see also TOF
TOPM	Tetraoctyl pyromellitate, i.e. tetrakis(2-ethylhexyl) pyromellitate [DIN, ISO]
TOTM	Trioctyl mellitate, i.e. tris(2-ethylhexyl) pyromellitate [DIN, ISO]
TPA	1. 1,5-*trans*-Poly(pentenamer); see also TRP
	2. Terephthalic acid
TPE	Thermoplastic elastomer
TPEL	Thermoplastic rubber or elastomer [ASTM]
TPES	Thermoplastic polyester in general [ASTM]
TPF	Triphenyl phosphate [DIN, ISO]; see also TPP
TPO	Thermoplastic olefinic elastomers
TPP	Triphenyl phosphate [ASTM, IUPAC]; see also TPF
TPR	1. 1,5-*trans*-Poly(pentenamer); see also TPA
	2. Thermoplastic elastomer; see also TR
TPS	Toughened poly(styrene); [in the U.K. for HIPS]
TPU	Thermoplastic urethanes
TPUR	Thermoset polyurethane [ASTM]
TR	1. Thermoplastic elastomer
	2. Thio rubber (U.K. term for polysulfides)
TSUR	Thermoset polyurethane [ASTM]
TTP	Tricresyl phosphate, tritolyl phosphate [ISO]; see also TCF, TCP, TKP
UE	Polyurethane elastomer [ASTM]; see also UR

UF	Urea-formladehyde resin [ASTM, DIN, ISO, IUPAC, NS]
UHMPE	Poly(ethylene) with ultrahigh molar mass
UHMW-PE	Poly(ethylene) with ultrahigh molar mass (molar mass over 3.1 $\times 10^6$ g/mol) [ASTM D-4020-81]
UP	Unsaturated polyester [ASTM, DIN, ISO, IUPAC, NS]
UP-G-G	Prepreg from unsaturated polyesters and textile glass fibers
UP-G-M	Prepreg from unsaturated polyesters and textile glass mats
UP-G-R	Prepreg from unsaturated polyesters and textile glass rovings
UR	Polyurethane elastomers [BS]; see also UP
VA	Vinyl acetate
VAC	Vinyl acetate
VAE	Vinyl acetate/ethylene copolymers; see also EVA, E/VAC
VC	Vinyl chloride: see also VCM
VC/E	Copolymer from ethylene and vinyl chloride [ISO, NS]
VC/E/MA	Copolymer from ethylene, vinyl chloride, and maleic anyhdride [ISO, NS]
VC/E/VAC	Copolymer from ethylene, vinyl chloride, and vinyl acetate [ISO, NS]
VCI	Volatile corrosion inhibitor
VCM	Vinyl chloride (monomer); see also VC
VC/MA	Copolymer from vinyl chloride and methyl acrylate [ISO, NS]
VC/MMA	Copolymer from vinyl chloride and methyl methacrylate [ISO, NS]
VC/OA	Copolymer from vinyl chloride and octyl acrylate [ISO, NS]
VC/VAC	Copolymer from vinyl chloride and vinyl acetate [ISO, NS]
VC/VDC	Copolymer from vinyl chloride and vinylidene chloride [ISO, NS]
VDC	Vinylidene chloride
VF	Vulcan fiber
VF_2/HFP	Copolymer from vinylidene fluoride and hexafluoropropylene
VI	Viscose [EDV]; see also CV
VMQ	Silicone rubber with methyl and vinyl substituents [ANSI/ASTM]

VPE	Crosslinked poly(ethylene) [DIN]
VQ	Elastomeric silicone with vinyl substituents [ASTM]
VRI	Volatile rust inhibitor
VSI	Poly(dimethylsiloxane) with vinyl groups [ASTM]
VT	Vinyl toluene
WFK	Whisker reinforced plastic [German technical literature]
WM	Plasticizer [German technical literature]
XABS	Elastomeric copolymer from acrylonitrile, butadiene, styrene, and a carboxylic group containing monomer [ASTM]
XLPE	Crosslinked poly(ethylene)
XNBR	Elastomeric copolymer from acrylonitrile, butadiene, and a carboxylic group containing monomer [ANSI, ASTM]
XPS	Expandable or expanded poly(styrene)
XSBR	Elastomeric copolymer from styrene, butadiene, and a carboxylic group containing monomer [ANSI, ASTM]
YSBR	Thermoplastic block copolymer from styrene and butadiene [ANSI, ASTM]
YXSBR	Thermoplastic block copolymer from styrene and butadiene, containing carboxylic groups [ANSI/ASTM]

MMI-4673, 3536, 3648, 1665, 9867, 9861

Cohesion Parameters

A. F. M. BARTON

School of Mathematical and Physical Sciences, Murdoch University, Western Australia 6150.

INTRODUCTION

The use of solubility parameters and other cohesion parameters has just one main purpose: to provide a simple method of correlating and predicting the cohesive and adhesive properties of materials from a knowledge of the properties of the individual components only. There are numerous more sophisticated theories and techniques which can be employed if precise descriptions are required, but there is none which can be used more easily to predict the properties of a mixture from those of its constituents, and manufacturers of polymers and solvents are now including these parameters in their specifications and information sheets. A selection of the many reviews on the subject is given in Table I, these include an extensive compilation of data in the 'Handbook of Solubility Parameters and Other Cohesion Parameters'.[1]

DEFINITIONS AND UNITS

Cohesion parameters are concerned with the extent of cohesion within condensed materials and of adhesion between condensed phases. Molecular materials exist in the form of liquids or solids over certain ranges of temperature and pressure because in some circumstances these condensed states are more stable than the corresponding gaseous state: there are energetic advantages in the molecules being packed together. In these condensed phases strong attractive forces exist between the molecules, each molecule being said to have considerable negative potential energy relative to a vapor phase molecule which has a negligible potential energy originating in this way. The internal energy of a condensed material is its potential energy relative to that of the ideal vapor at the same temperature, and is negative. The *cohesive energy*, defined as the internal energy with

F

TABLE I

A selection of reviews on cohesion parameters

Author	Topic	Language
Ahmad, Yaseen[61,119]	Group contributions	English
Allen et al.[120,121]	General	English
ASTM[26,27]	Standard polymer solubility test	English
Bagda[28,122,123]	Solution properties	German
Bagley, Scigliano[124]	Polymer solutions	English
Barton[1,125]	General	English
Beerbower, Dickey[90]	Elastomer/fluid	English
Beerbower et al.[106]	Elastomer/fluid	English
Bernardo, Burrell[126]	Plasticization	English
Blanks[127,128]	Polymer solution thermo-dynamics	English
Blanks, Prausnitz[17]	Polymer solution thermo-dynamics	English
Bristow, Watson[34,35]	Polymer, swelling, viscosity	English
Burrell[20–24,48,129]	Polymers	English
Corish[130,131]	Polymer blends	English
Cosaert[109]	General	French, Flemish
Crowley et al.[104,105]	Three components	English
Dayantis[132]	Polymer solubility	French
Dernini, Vargiu[133]	Paint resins	Italian
	Polymer swelling	English
Eastman Kodak[111]	General	English
Flory[11]	χ Interaction parameter	English
Fowkes[134,135]	Polymer surfaces	English
Gardon[99,100,136–138]	Polymer adhesion	English
Garden, Teas[98]	General	English
Geczy[139]	General	Hungarian
Guzman[140]	General	Spanish
Hadert[141]	Polymer solvent	German
Hansen[81–85,93,142]	Hansen parameters	English
Hansen, Beerbower[86]	General	English
Hildebrand et al.[4]	General	English
Hildebrand and Scott[2,3]	General	English
Hoy[54,55]	Data tables	English
Imoto[143,144]	General	Japanese
Johannsen[145]	General	German
Kaelble[146]	Adhesion	English
Karger et al.[75–77]	Chromatography	English
Keller et al.[74]	Chromatography	English
Klein, Smith[147]	Membranes	English
Koenhen, Smolders[148]	Correlation with polymer physical properties	English
Kumar, Prausnitz[149]	General	English
Lucas[150]	General	French
Mandik[151,152]	General	Czechoslovaki
Mandik, Kaspar[153]	Film forming	Czechoslovaki
Mandik, Stanek[154]	Coatings	Czechoslovaki
Marti[155]	General	German
Mellan[156]	General	English
Meyer zu Bexten[91]	General	German
Mitomo, Teshirogi[157]	Calculations	Japanese

TABLE I—*continued*

Author	Topic	Language
Moiseev, Pilyagin[158]	Spraying solvents	Russian
Moore[159]	Fibers	English
Nelson et al.[107]	*General; coatings*	*English*
Nunn[160]	*General*	*English*
Olabisi et al.[161]	Polymer-polymer miscibility	English
Orwoll[16]	Polymer-solvent inter-action parameter	English
Patterson[162–164]	Polymer solution thermo-dynamics	English
Paul[165]	General	English
Paul, Barlow[166]	Polymer blends	English
Ramsbotham[115–118]	Solubility maps	English
Rebenfeld et al.[167]	Polymer solvent inter-actions	English
Reichardt[168]	General	English
Rheineck, Lin[169]	Group calculations	English
Robu[170]	General	Romanian
Sandholm[171]	Evaluation for polymers	English
Sattelmeyer[172]	General	German
Schoenmakers et al.[173,174]	Chromatography	English, Dutch
Sevestre[175]	General	English
Sevestre[176]	General	French
Seymour[67,68,177]	Polymers, general	English
Seymour[178]	Polymers, general	Spanish
Seymour, Sosa[179]	Calculations	English
Shaw[180]	Polymer-polymer solubility	English
Sheehan, Bisio[181]	Polymer-solvent inter-action parameters	English
Shell[87,112,182,183]	General, solubility maps	English
Shen[184]	General	Chinese
Shinoda[185]	General	English, Japanese
Shvarts[186,187]	Polymer-polymer, polymer-solvent	Russian, English
Skaarup[188]	Hansen parameters, pigments	English
Skelly[189]	Textile processing	English
Slepakova et al.[190]	Polymer compatibility	Russian
Small[51]	Group contributions	English
Snyder[78–80]	General	English
Sörensen[191–193]	Pigments	English, German
Sonnich Thomsen[194,195]	General	Danish
Sonnich Thomsen[196]	General	English
Sosa[197]	General	English
Stawiszynski, Zawadski[198]	General	Polish
Tager, Kolmakova[199]	General	Russian, English
Takada[200]	General	Japanese
Taylor[201]	General	English
Teas[97,202]	Resins	English
Tlusta, Zelinger[203]	Polymer miscibility	Czechoslovaki
Union Carbide[56]	Data tables	English
van Krevelen, Hoftyzer[42]	General	English
Watanabe[204,205]	General	Japanese
Yoshida[206]	General	Japanese

opposite sign, is therefore positive and the cohesive pressure or cohesive energy density is the cohesive energy per unit volume.

At pressures below atmospheric (i.e., for temperatures below the normal boiling point of a material) the cohesive energy of a liquid can be equated with the enthalpy of vaporization less the volume work, which for a vapor obeying the ideal gas law is RT per mole. (R is the molar gas constant, T is the absolute temperature). This was the basis of the original definition by Joel H. Hildebrand and Robert L. Scott,[2-4] of what is now generally called the Hildebrand solubility parameter, δ:

$$\delta = c^{\frac{1}{2}} = [(\Delta_l^g H - RT)/V]^{\frac{1}{2}}$$

It is clear that the Hildebrand parameter[4] of a liquid may be evaluated readily if the molar volume, V, and the molar enthalpy of vaporization have been determined at the required temperature. The quantity c is the cohesive energy density or cohesive pressure.

The basis of the cohesion parameter approach to mixtures can be stated as follows. A material with a high δ value requires more energy for dispersal than is gained by mixing it with a material of low δ value, so immiscibility results. On the other hand, two materials with similar δ values gain sufficient energy on dispersal to permit mixing. This concept is attractive for practical use because it aims to predict the properties of a system knowing only the properties of its components.

Originally the Hildebrand solubility parameter was expressed in units of $cal^{\frac{1}{2}} cm^{-\frac{3}{2}}$. However, as cohesive energy density is identical with cohesive pressure, and an SI unit is required, the most appropriate unit is $MPa^{\frac{1}{2}}$, which is numerically identical with $J^{\frac{1}{2}} cm^{-\frac{3}{2}}$ and is of a convenient numerical magnitude ($1 cal^{\frac{1}{2}} cm^{-\frac{3}{2}} = 2.0455 MPa^{\frac{1}{2}}$). (Table II.) It had been suggested that the unit $cal^{\frac{1}{2}} cm^{-\frac{3}{2}}$ be named the 'Hildebrand', in honor of its originator, but a more suitable form of recognition is to call the original 'solubility parameter' the Hildebrand parameter. The Hildebrand parameter can now be seen as one member of a class of cohesion parameters, all with units of (pressure)$^{\frac{1}{2}}$, which has been developed to describe the cohesion and adhesion properties of a wide range of materials.

TABLE II
Cohesion parameter units and conversion factors

	$cal^{\frac{1}{2}} cm^{-\frac{3}{2}}$	$MPa^{\frac{1}{2}}(J^{\frac{1}{2}} cm^{-\frac{3}{2}})$	$atm^{\frac{1}{2}}$
$1 cal^{\frac{1}{2}} cm^{-\frac{3}{2}}$	—	2.0455	6.4260
$1 MPa^{\frac{1}{2}}(J^{\frac{1}{2}} cm^{-\frac{3}{2}})$	0.48888	—	3.1415
$1 atm^{\frac{1}{2}}$	0.15562	0.31832	—

HILDEBRAND PARAMETERS

The Hildebrand parameter was originally intended only for non-polar, non-ionic, non-associating liquid systems which formed 'regular' solutions. The concept has been extended so that it may be evaluated for any material (although any theoretical significance it may have is then less obvious). The Hildebrand parameter is fundamentally a liquid or amorphous state property. When gases are considered, they are treated as hypothetical 'liquid' solutes at atmospheric pressure, and substances which are crystalline solids at normal temperatures are treated as subcooled liquids. This is true also of most of the 'partial' cohesion parameters derived from the Hildebrand parameter. In practice, the distinction between amorphous and crystalline polymers is not usually important. Hildebrand parameters can be evaluated directly by thermodynamic means only for vaporisable materials, so for polymers it is always necessary to use indirect methods of determination.

REGULAR SOLUTIONS AND THE GEOMETRIC MEAN RULE

Hildebrand[5,6] defined a *regular solution* as a solution 'involving no entropy change when a small amount of one of its components is transferred to it from an ideal solution of the same composition the total volume remaining unchanged'. In other words, a solution or mixture described as 'regular' is one which, despite a non-ideal (non-zero, either positive or negative) enthalpy of formation has an ideal entropy of formation. On the molecular level, an ideal mixture is one in which the different types of molecules, i and j for example, behave exactly as if they were surrounded by molecules of their own kind, that is, all intermolecular interactions are equivalent. A regular mixture can occur only if the random distribution of molecules persists even in the presence of $i-j$ interactions which differ from the original $i-i$ and $j-j$ interactions. The concept has proved valuable in the development of an understanding of miscibility criteria and of deviations from ideality. A modification to the entropy of mixing term allows the concept to be extended to polymer solutions, where molecules of different sizes are mixed.

The requirement that a regular solution has a random molecular distribution despite the existence of interactions which lead to a non-ideal enthalpy of formation effectively restricts regular mixtures to those systems in which only dispersion forces are important, because the orientation effects of polar molecules cause non-random molecular distributions. The geometric mean rule, based partly on theoretical principles and partly on observation, states that the dispersion cohesion energy $^{ij}U_d$ of a mixture of i

and j is given by

$$^{ij}U_d = {}^iU_d^{\frac{1}{2}}\,{}^jU_d^{\frac{1}{2}}$$

This equation can be expected to hold only in situations where dispersion forces provide the only significant cohesive energy.

The regular solution equation for the energy of mixing at constant volume, based on the pioneering work of J. D. van der Waals and J. J. Van Laar, was derived on semi-theoretical grounds by G. Scatchard[7] and J. H. Hildebrand,[5,8,9] Hildebrand and Scott.[2] In deriving and discussing this equation, it is useful to define the exchange energy density or interchange cohesive pressure,

$$^{ij}A = {}^ic + {}^jc - 2{}^{ij}c$$

where c is the coehsive pressure. The ^{ij}c term is the cohesive pressure characteristic of the intermolecular forces acting between molecules of type i and type j. This definition of ^{ij}A can be appreciated in a simple fashion by considering what happens when unit volumes of components i and j are mixed: two i–j interactions are formed for each pair of i–i and j–j interactions broken.

The interchange cohesive pressure when used simply as an empirical parameter for a particular i–j pair or series of mixtures does not make full use of the opportunity to develop a method to provide information on mixtures from data on the individual components. This is where the relationship between ^{ij}A and the individual cohesion parameters is important. For those liquids where only dispersion forces are significant, the geometric mean assumption can be used:

$$^{ij}c = ({}^ic\,{}^jc)^{\frac{1}{2}},$$

and ^{ij}A can be expressed in terms of Hildebrand parameters:

$$^{ij}A = ({}^ic^{\frac{1}{2}} - {}^jc^{\frac{1}{2}})^2 = ({}^i\delta - {}^j\delta)^2$$

If account is taken of the major deviations from regular behavior due to the substantial size difference between polymer and solvent molecules, cohesion parameters may be applied to polymer solutions just as they are to mixtures of liquids.

POLYMER SOLUTIONS

One of the first and best-known attempts to formulate a model explaining the deviations arising in polymer solution properties due to the large differences in size between solute and solvent molecules was the lattice model of Flory,[10,11] Flory and Rehner[12] and Huggins[13-15] in which the entropy of mixing the long-chain molecules was calculated with the assumption that each polymer segment and each single monomeric molecule occupied a lattice site. There arise in this approach a polymer liquid interaction parameter, σ, originally introduced as an enthalpy of ilution term but considered subsequently as Gibbs free energy parameter[16] which can be divided into entropy and enthalpy components:

$$\sigma = \chi_S + \sigma_H$$

Through the chemical potential, χ can be related to several thermodynamic properties of polymer solutions, and it may be obtained experimentally by measuring the activity of the solvent with a variety of techniques such as solvent vapor pressure lowering, freezing point depression, boiling point elevation, osmotic pressure, light scattering, viscosity, equilibrium swelling, stress-strain behavior of swollen polymers, or gas-liquid chromatography.

It can be shown that

$$\chi_H = {}^{ij}A\,{}^iV/RT,$$

and in the special case where ${}^{ij}c = {}^i\delta\,{}^j\delta$,

$$\chi_H = ({}^iV/RT)({}^i\delta - {}^j\delta)^2$$

The entropy term σ_S has an almost constant value of about 0.3 for many systems.[10,11,17] Flory[10,11] showed that a polymer j and a liquid i are expected to be completely miscible through the entire composition range provided that

$$\sigma \leqslant \tfrac{1}{2}[1 + ({}^iV/{}^jV)^{\frac{1}{2}}]^2$$

There is thus a critical polymer liquid interaction parameter value

$$\chi_c = \tfrac{1}{2}[1 + ({}^iV/{}^jV)^{\frac{1}{2}}]^2,$$

and for $({}^iV/{}^jV) \to 0$,

$$\sigma_c = 0.5.$$

If χ must be less than 0.5 for full polymer-liquid miscibility, and χ_S is about 0.3, it follows that χ_H must be very small to meet the miscibility criterion, and that ${}^i\delta$ and ${}^j\delta$ must have very similar values. Specific interactions (such as hydrogen bonding between molecules of type i and j) add positive terms to ${}^{ij}c$ and so decrease ${}^{ij}A$, with the result that χ_H also decreases and the mutual solubility is enhanced.

Summarising, it can be seen that cohesion parameters reflect the enthalpy aspects of interactions, and for regular solutions the entropy aspects are ideal. At the other extreme, for polymer-solvent systems the entropy part is described by Flory–Huggins theory and may be evaluated in terms of χ_S.

MIXED SOLVENTS

Practical solvents are usually blends, and it is therefore important to be able to evaluate the effective cohesion parameters of solvent mixtures. Although other factors such as viscosity, volatility and cost must also be considered in solvent formulation, the effective cohesion parameters are particularly valuable. On the basis of the assumptions made in the derivation of cohesion parameter expressions, the effective Hildebrand parameter $\bar{\delta}$ of a binary solvent mixture is

$$\bar{\delta} = \frac{{}^i\phi\,{}^i\delta + {}^j\phi\,{}^j\delta}{{}^i\phi + {}^j\phi}$$

where ϕ is the volume fraction, predicting that the effective Hildebrand parameter of a mixture is volume-wise proportional to the Hildebrand parameters of its components. When the value of the Hildebrand parameter ${}^k\delta$ of solute k in the mixed solvent i–j lies between the values ${}^i\delta$ and ${}^j\delta$ of the component solvents, the solute should be completely miscible when the ratio ${}^i\phi\,{}^j\phi$ is adjusted so that ${}^k\delta = \bar{\delta}$, even if liquids i and j individually are nonsolvents for k.

SOLVENT SPECTRA

A list of solvents may be compiled with gradually increasing Hildebrand parameter values to form a 'solvent spectrum'.[18-25]

In its most common form (Table III) it includes subdivision into categories of hydrogen bonding ability. The Hildebrand parameter of a solute is taken as the midpoint of the range of solvent Hildebrand parameters which provides complete miscibility. 'Fine control' over solvent Hildebrand parameter values can be provided using mixtures of solvents, the effective Hildebrand parameter of a mixture being calculated on the basis of the volume fraction composition, as in the ASTM D3132[26,27] test methods for polymer solubility ranges. For polymers, either the dissolution behavior of a polymer or the swelling of a slightly cross-linked analog of the polymer of interest in a series of solvents may be studied, the polymer being assigned the δ value of the liquid providing the greatest solubility or the maximum swelling coefficient. Other physical properties which can be studied include viscosity and, in the case of greases, dropping points.

In solubility studies, the selected solvent is added to a 1 or 2 gram sample of solid polymer in a test-tube such that the final solution has about the correct solids content for the expected use (for example, 50% for alkyds, 20% for vinyl polymers). The mixture may be warmed and stirred to increase the

TABLE III
Liquids for Hildebrand parameter spectra, divided according to hydrogen bonding capability.
[Adapted from Burrell.[18-22]]

Liquids with poor hydrogen bonding capability	$\delta/MPa^{\frac{1}{2}}$
n-Pentane	14.3
n-Heptane	15.1
Methylcyclohexane	16.0
Solvesso® 150	17.4
Toluene	18.2
Tetrahydronaphthalene	19.4
o-Dichlorobenzene	20.5
1-Bromonaphthalene	21.7
Nitroethane	22.7
Acetonitrile	24.1
Nitromethane	26.0
Liquids with moderate hydrogen bonding capability	
Diethyl ether	15.1
Diisobutyl ketone	16.0
n-Butyl acetate	17.4
Methyl propionate	18.2
Dibutyl phthalate	19.0
Dioxane	20.3
Dimethyl phthalate	21.9
2,3-Butylene carbonate	24.8
Propylene carbonate	27.2
Ethylene carbonate	30.1
Liquids with strong hydrogen bonding capability	
2-Ethylhexanol	19.4
Methylisobutylcarbinol	20.5
2-Ethylbutanol	21.5
1-Pentanol	22.3
1-Butanol	23.3
1-Propanol	24.3
Ethanol	26.0
Methanol	29.7

rate of dissolution, but it is cooled and observed at room temperature, the polymer being judged insoluble if there are gel particles or cloudiness present. By successive choices, upper and lower pairs of two adjacent solvents are found, one of which dissolves the polymer and one which does not, thus defining the solubility range in each hydrogen bonding class.

Cohesion parameters may also be assessed by comparing the temperatures at which dilute polymer suspensions (often 1% or 0.1% by volume) become homogeneous, and the dissolution rates of polymers have been used for the same purpose.[28] An indirect indication of polymer-solvent interaction is the peel strength adhesion of polymer films cast from a solvent; this also exhibits a curve with a maximum when plotted against solvent δ values.[29]

Closely related to the solubility of the polymer in a solvent is the corresponding solubility of solvent in polymer, with resultant polymer swelling. A method which may be applied to a polymer which is cross-linked (or which can be cross-linked before testing, for example by irradiation) and also to a semicrystalline material such as poly(vinyl chloride) is to expose it to a series of solvents. The polymer is swollen (but does not dissolve because of the crosslinks).

Variation in polar and hydrogen bonding properties of both polymer and solvent cause considerable 'scatter' in solubility spectrum plots, but the general trend is usually clear, and an approximate single value or a range of Hildebrand parameter values can be obtained from the maximum.

Cohesion parameters can be calculated in other ways from swelling measurements. Thus Eskin and co-workers[30] have used the fact that the relative differences between the partial specific volumes of a swollen polymer in a given solvent and in one for which $^i\delta = {}^j\delta$ is a linear function of $(^i\delta - {}^j\delta)^2$. Instead of relying on a simple swelling-Hildebrand parameter curve, some investigators have employed polymer-liquid interaction parameter (χ) relationships which take account of volume differences between polymer and solvent. Using Flory's equation linking equilibrium swelling and the χ interaction parameter, Gee[31,32] showed that the swelling coefficient, Q, (the volume of liquid taken up per unit volume of polymer) when plotted against $^iV(^i\delta - {}^j\delta)^2$ is gaussian:

$$Q = Q_{max}\exp[-K^iV(^i\delta - {}^j\delta)^2]$$

and a plot of $(^iV^{-1}\ln Q_{max}/Q)^{\frac{1}{2}}$ against $^i\delta$ gives a straight line with intercept equal to $^j\delta$.

Alternatively, χ may be evaluated from swelling measurements by the equation of Flory and Rehner,[33] in the manner suggested by Bristow and Watson,[34-36] Dudek and Bueche:[37]

$$\left(\frac{^i\delta^2}{RT} - \frac{\chi}{^iV}\right) = \left(\frac{2^i\delta}{RT}\right)^j\delta - {}^j\delta^2 - \frac{\chi_s}{^iV}$$

The contribution of χ_s is neglected, and a plot of $[(^i\delta^2/RT) - (\chi/^iV)]$ against $^i\delta$ yields straight lines with slopes of $2^j\delta/RT$ and intercepts of $-{}^j\delta^2/RT$. A closely related empirical equation is

$$^i\delta = {}^j\delta \pm \left(\frac{RT}{K}\right)^{\frac{1}{2}}\left[\frac{(\chi - \chi_a)}{^iV}\right]^{\frac{1}{2}}$$

in which $^i\delta$ is plotted against $[(\chi - \chi_s)/^iV]$, $^j\delta$ being determined from the intercept.[38-40] More simply, χ shows a minimum when plotted as a function of $^i\delta$, and this minimum can be identified with the polymer Hildebrand parameter.[41]

POLYMER SOLUTION VISCOSITY

The viscosities of dilute solutions of polymers are highest in the 'best' solvent, that is, in one in which the cohesion parameters of solvent and polymer are comparable. Equilibrium configuration of molecular chains, both in swelling situations and in polymer solution viscosity studies, depend on the balance between the Gibbs free energy change due to mixing and that due to elastic deformation. The extent of deformation depends on the relative strengths of intramolecular (segment–segment) and intermolecular (segment–solvent) interactions. In a 'good' solvent the polymer is unfolded, obtaining to the maximum extent the more favorable polymer–solvent interactions (and therefore the greatest viscosity); and in a 'poor' solvent the polymer molecule remains folded because of the more favorable intra-molecular interactions.

The intrinsic viscosity,

$$[\eta] = \lim_{c \to 0} \left(\frac{\eta - {}^i\eta}{{}^i\eta c} \right)$$

where c is the polymer concentration, $^i\eta$ is the viscosity of the pure solvent i, and η is the solution viscosity, depends on the properties of isolated polymer molecules. Cohesion parameter methods have been used for correlating and predicting intrinsic viscosities.[30,42] They have been measured as functions of the Hildebrand spectrum of a series of liquids, or as functions of relative concentration ratios of a pair of mixed solvents.[43]

By analogy with the equilibrium swelling expression, the results of polymer solution viscosity measurements can be interpreted more easily when rearranged to yield a linear graph and to take the solvent molar volume into account by means of the Flory–Huggins equation:[44-46]

$$\left[{}^iV^{-1} \ln\left(\frac{[\eta]_{max}}{[\eta]} \right) \right]^{\frac{1}{2}} = K^{\frac{1}{2}}({}^i\delta - {}^j\delta)$$

Closely related expressions are

$$\left[\left(\frac{1 - [\eta]}{[\eta]_{max}} \right) {}^iV^{-1} \right] = K^{\frac{1}{2}}({}^i\delta - {}^j\delta)$$

and

$$\log[\eta] = B\log|{}^i\delta - {}^j\delta| + A.$$

Variations in terms of Hansen parameters are possible. Bagda[47] used the alternative equations

$$\eta_{sp} = B(^i\delta - {}^j\delta)^2 + A$$

and

$$\eta_{\iota\kappa} = B[(^i\delta_\beta - {}^j\delta_\beta)^2 + (^i\delta_\kappa - {}^j\delta_\kappa)^2 + (^i\delta_\rho - {}^j\delta_\rho)^2] + A$$

where B and A are dependent on the type of polymer and its concentration, but independent of the nature of the solvent.

Although the viscosities of *dilute* polymer solutions show minima in poor solvents because the polymer molecules remain 'coiled' as the polymer concentration is increased there is a 'changeover' in behavior: the viscosity exhibits a *maximum* in poor solvents at higher concentration.[48,49] There is an aggregation or clustering of polymer molecules preliminary to phase separation when solvents are used which have cohesion parameters near the limits of the miscibility range for the polymer, and it is probable that the clusters are not completely broken down to individual molecules even in 'good' solvents with cohesion parameters in the middle of the miscibility range.

This rather complex behaviour of polymer solutions can be described in terms of entropy and free volume effects which are not present in liquid–liquid systems, and it should not be expected that it can be fully explained by means of cohesion parameters, which can provide direct information only on the enthalpy aspect of the interactions.

The above discussion relates to flexible polymer molecules which exist in solution as random coils. 'Stiff' polymers such as cellulose show less change in intrinsic viscosity as the solvent is changed, but the limited solubility in a variety of solvents restricts this investigation. It appears that a poly-yne polymer containing a transition metal in the main chain retain a rod-like shape in solution, as it exhibits an intrinsic viscosity independent of solvent Hildebrand parameter in the range 13.5–20 MPa$^{\frac{1}{2}}$ and a uniformly high solubility in these solvents.[50]

CALCULATION OF POLYMER COHESION PARAMETERS

Scatchard[7] and Small[51] observed that there are parallel linear relationships among several homologous series when the square root of the product of the molar volume and molar cohesion energy, $(-UV)^{\frac{1}{2}}$ or the equivalent quantity (δV) has additive properties. It has been described as the molar attraction, F, and can be considered to be made up of the sum of the group molar attraction constants ^zF of all the molecular groups z:

$$-U = \frac{(\Sigma_z {}^zF)^2}{V}$$

so

$$c = \left(\frac{\Sigma_z {}^zF}{V}\right)^2$$

and

$$\delta = \frac{\Sigma_z {}^zF}{V} = \frac{\Sigma_z {}^zF}{\Sigma_z {}^zV}.$$

Molar cohesive energies and Hildebrand parameters can thus be estimated for any molecular compound once the individual group molar attraction constants and group molar volumes are known. Although the additivity of these group constants on an incremental basis is questionable on theoretical grounds, the method is adequate for many applications.[52,53]

Hoy,[54–56] van Krevelen,[57] Askadskii et al.,[58,59] and others have compiled lists of group contributions; these have been summarized by van Krevelen and Hoftyzer.[42] The group contribution values reported by different authors show considerable variation[60–64] and values suitable for polymer molecules are not necessarily identical to those for smaller molecules. Despite this, the *cohesion parameters* obtained by adding group contributions do not differ so widely, and each system usually provides values acceptable for most practical purposes. It is essential, therefore, to use a *self-consistent* set of group contributions.

The Hildebrand parameter of a polymer is obtained by considering the properties of the repeating unit (superscript u), not those of the monomer. The molar enthalpy of vaporization of an oligomer or polymer made up of n repeating units may be subdivided according to

$$\Delta_l^g H = 2{}^e\Delta H + (n-2){}^u\Delta H$$

where $^e\Delta H$ and $^u\Delta H$ are the contributions of the end groups and repeating units, respectively. The corresponding molar mass is

$$M = 2{}^eM + (n-1){}^uM$$

Therefore

$$\frac{\Delta_l^g H}{M} = \frac{2{}^e\Delta H}{M} + \frac{(n-2){}^u\Delta H}{2{}^eM + (n-2){}^uM}$$

In polymers where M is large, $2{}^e\Delta H/M$ and $2{}^eM$ are negligible, so

$$\frac{\Delta_l^g H}{M} = \frac{{}^u\Delta H}{{}^uM}$$

If the density, $\rho = M/V$, is inserted,

$$\frac{\Delta_l^g H}{V} = \frac{\rho\,^u\Delta H}{^u M}$$

and from the definition of the Hildebrand parameter,

$$\delta = \left[\frac{(\Delta_l^g H - RT)}{V}\right]^{\frac{1}{2}}$$

with RT/V negligible because V is large,

$$\delta = \left(\frac{\rho\,^u\Delta H}{^u M}\right)^{\frac{1}{2}}$$

If this concept is extended[65,66] to group molar attraction constants,

$$\delta = \frac{\Sigma_z\,^z F}{V} = \frac{\left[\dfrac{^u F(V - {}^e V)}{^u V + {}^e F}\right]}{V}$$

where the relative number of repeat units is given by $(V - {}^e V)/^u V$. Using $^u F/^u V = {}^u\delta$ and $^e F/^e V = {}^e\delta$,

$$\delta = {}^u\delta\left(\frac{1 - {}^e V}{V}\right) + \frac{{}^e\delta\,{}^e V}{V}$$

and for $^e V \ll V$ in high polymers,

$$= {}^u\delta = \frac{^u F}{^u V}.$$

From this it can be seen that $^u\delta$ can be evaluated from directly observable δ and V values of liquid and volatile monomers and oligomers, using the molar volume of the smallest member of the homologous series as $^e V$ and plotting δ against $^e V/V$ to provide intercept $^u\delta$.

HYDROGEN BONDING AND PRACTICAL LIQUID SOLUBILITY SCALES

One of the first attempts to deal with the hydrogen bonding factor in the application of Hildebrand parameters to practical systems was made by H. Burrell.[23] He did it in a very simple way, dividing solvents into three classes according to their hydrogen bonding capacities, on the assumption that complete miscibility can occur only if the degree of hydrogen bonding is comparable in the components, as indicated in Table III:

(i) weak or poor hydrogen bonding capacity liquids, including hydrocarbons, chlorinated hydrocarbons and nitrohydrocarbons;

(ii) moderate hydrogen bonding capacity liquids, including ketones, esters, ethers and glycol monoethers; and

(iii) strong hydrogen bonding capacity liquids, such as alcohols, amines, acids, amides and aldehydes.

Typical results for polymers appear in Table IV. An alternative classification in the CRC 'Handbook of Chemistry and Physics'[69] distinguishes between nonpolar solvents, moderately polar solvents, and hydrogen-bonded solvents.

Parameters describing and correlating the solvent action of liquids have been based on a great variety of chemical and physical properties. Some are measures of solvent basicity, while others are direct determinations of the solubility of a representative solute in a range of liquids. Once a solubility scale has been established, it is necessary to determine the position on it of any required solvent. If the solubility scale has a theoretical basis, it may be possible to calculate values from information on other properties, but if it is empirical or semi-empirical, direct testing is required.

Cohesion parameters (particularly Hildebrand parameters) have been used in many such solubility scales, either on their own or combined with other properties. Because cohesion parameter values are obtained from various sources, both theoretical and empirical, it should not be expected that they will necessarily be in good agreement. The aim should be to select a solubility scale which is appropriate for a particular task, and to determine the parameters for all the necessary materials on that scale. Comparison with values obtained by other methods on other scales may be useful, but are less reliable. The greater the extent of polar and specific interactions that occur, the greater the divergence between different solubility scales is likely to be.

E. P. Lieberman[70] and his du Pont colleagues[71-72] developed the classification of Burrell by assigning quantitative values to the hydrogen bonding capabilities, and plotting graphs of Hildebrand parameters against hydrogen bonding capacities. Subsequently, spectroscopic hydrogen bonding parameters were used (see below), as well as sonic velocity in solvent-paper systems.

EXPANDED COHESION PARAMETER FORMULATION

To be generally useful, theories or models attempting to systematize the behavior of matter must deal with molecular interactions by providing

TABLE IV

Approximate Hildebrand parameter ranges for some common polymeric materials, classified by hydrogen bonding capability and in order of increasing δ values.
[Adapted from Seymour.[67,68]]

Polymer	Hildebrand parameter ranges $\delta/MPa^{\frac{1}{2}}$ in solvents with hydrogen bonding capability which is		
	poor	moderate	strong
Polytetrafluorocarbon	12–13	—	—
Ester gum	14–22	15–22	19–22
Alkyd 45% soy soil	14–22	15–22	19–24
Silicone (DC-1107)	14–19	19–22	19–24
Poly(vinyl ethyl ether)	14–23	15–22	19–29
Poly(butyl acrylate)	14–26	15–24	—
Poly(butyl methacrylate)	15–23	15–20	19–23
Silicone (DC-23)	15–17	15–16	19–21
Polyisobutylene	15–16	—	—
Polyethylene	16–17	—	—
Gilsonite®	16–19	16–17	—
Poly(vinyl butyl ether)	16–22	15–21	19–23
Natural rubber	17	—	—
Chlorosulfonated polyethylene (Hypalon® 20)	17–20	17–18	—
Ethyl cellulose (N-22)	16–23	15–22	19–30
Chlorinated rubber	17–22	16–22	—
Dammar gum	17–22	16–21	19–22
Polyamide (Versamid® 100)	17–22	17–18	19–23
Polystyrene ·	17–22	19	—
Poly(vinyl acetate)	17–19	—	—
Poly(vinyl chloride)	17–23	16–22	—
Phenolic resins	17–24	16–27	19–28
Butadiene-acrylonitrile copolymer (Buna N)	18–19	—	—
Poly(methyl methacrylate)	18–26	17–27	—
Poly(ethylene oxide) (Carbowax® 4000)	18–26	17–30	19–30
Poly(ethylene sulfide) (Thiokol®)	18–21	—	—
Polycarbonate	19–22	19–21	—
Pliolite® P-1230	19–22	—	—
Poly(ethylene terephthalate) (Mylar ᴿ)	19–22	19–20	—
Vinyl chloride-acetate copolymer	19–23	16–27	—
Polyurethane	20–21	—	—
Styrene-acrylonitrile copolymer	22–23	19–20	—
Vinsol® [rosin derivative]	22–24	16–27	19–26
Epoxy (Epon® 1001)	22–24	17–27	—
Shellac	—	21–23	19–29
Polymethacrylonitrile	—	22–23	—
Cellulose acetate	23–26	21–30	—
Nitrocellulose	23–26	16–30	26–30
Polyacrylonitrile	—	25–29	—
Poly(vinyl alcohol)	—	—	25–27
Poly(hexamethylene adipamide) (Nylon 66)	—	—	28–31
Cellulose	—	—	30–33

information about their natures or origins as well as about their magnitudes. The cohesive pressures ^ic, ^jc, ^{ij}c and the corresponding Hildebrand parameters represent the resultant effect of all types of force acting between molecules of types i and j, but this kind of parameter is useful when only dispersion forces are present. A more general cohesion parameter system identifies the various cohesive contributions.[73-77]

The dispersion cohesive pressure of a pure material i is denoted ^ic_d, and the corresponding cohesion parameter, δ_d, is defined by

$$-\frac{^iU_d}{^iV} = {}^ic_d = {}^i\delta_d^2$$

The dispersive interactions between unlike molecules of type i and type j provide a contribution to the cohesive pressure which is based on geometric mean behavior because the interaction is of a 'symmetrical' nature: each number of a pair of molecules interacts by virtue of the same property, the polarizability:

$$^{ij}c_d = ({}^ic_d{}^jc_d)^{\frac{1}{2}} = {}^i\delta_d{}^j\delta_d$$

It follows that the interchange cohesive pressure due to dispersion is

$$^{ij}A_d = {}^i\delta_d^2 + {}^j\delta_d^2 - 2{}^i\delta_d{}^j\delta_d = ({}^i\delta_d - {}^j\delta_d)^2$$

For nonpolar molecules, dispersion forces should make the only contributions to the cohesive pressure, so

$$^i\delta_d = \left(\frac{\Delta_l^g U}{V}\right)^{\frac{1}{2}}$$

The orientation cohesive pressure of a pure material i is denoted $^ic_0$:

$$-\frac{^iU_0}{^iV} = {}^ic_0 = {}^i\delta_0^2$$

Like dispersion forces, these are 'symmetrical' interactions, depending on the same property of each molecule, which in this case is the dipole moment. It follows that the geometric mean rule is obeyed well for orientation interactions between unlike molecules,

$$^{ij}c_0 = ({}^ic_0{}^jc_0)^{\frac{1}{2}} = {}^i\delta_0{}^j\delta_0$$

The interchange cohesive pressure due to the orientation is

$$^{ij}A_0 = ({}^i\delta_0 - {}^j\delta_0)^2$$

In contrast to dispersion and orientation interactions, dipole induction interactions are 'unsymmetrical', involving the dipole moment of one molecule and the polarizability of the other. Consequently, the cohesive pressure term for induction in a pure material i involves the product ${}^i\delta_i {}^i\delta_d$, where ${}^i\delta_i$ and ${}^i\delta_d$ are the induction cohesion parameter and dispersion cohesion parameter, respectively. Similarly, in a mixture of i and j,

$$ {}^{ij}c = {}^i\delta_i {}^j\delta_d + {}^j\delta_i {}^i\delta_d $$

It can be shown, therefore, that

$$ {}^{ij}A_i = 2\,{}^i\delta_i {}^i\delta_d + 2\,{}^j\delta_i {}^j\delta_d - 2\,{}^i\delta_i {}^j\delta_d - 2\,{}^j\delta_i {}^i\delta_d $$
$$ = 2({}^i\delta_d - {}^j\delta_d)({}^i\delta_i - {}^j\delta_i) $$

Lewis acid-base or electron donor-acceptor interactions occur when there are energetic advantages in electrons from one molecule being donated to another, and differ from a 'normal' chemical bond in that only one molecule supplies the pair of electrons which enable the co-ordination of the molecules to take place. These interactions are 'unsymmetrical', involving a donor and an acceptor with different roles, so two separate cohesion parameters are required: a Lewis acid cohesion parameter δ_a and a Lewis base cohesion parameter δ_b, in a manner analogous to that for induction interactions:

$$ {}^{ij}A_{ab} = 2({}^i\delta_a - {}^j\delta_b)({}^i\delta_b - {}^j\delta_a) $$

It can be seen that the maximum interaction occurs when ${}^{ij}A_{ab}$ is large and negative. This makes *exothermic* mixing possible, in contrast to mixing being restricted to athermic or endothermic processes which is the case when only dispersion and polar forces exist. Hydrogen bonding interactions are another kind of donor-acceptor interaction or association, a special type of Lewis acid-base reaction with the electron acceptor being a Bronsted acid, and the same equation for ${}^{ij}A_{ab}$ applies.

One of the assumptions central to the cohesion parameter approach is that the various contributions to the cohesive energy (and therefore to the cohesive pressure) of a pure or mixed material are additive, so the interchange cohesive pressure or exchange energy density for a mixing process is

$$ {}^{ij}A = {}^{ij}A_d + {}^{ij}A_0 + {}^{ij}A_i + {}^{ij}A_{ab} $$
$$ = {}^i\delta_t^2 + {}^j\delta_t^2 - 2\,{}^i\delta_d {}^j\delta_d - 2\,{}^i\delta_i {}^j\delta_d - 2\,{}^j\delta_i {}^i\delta_d - 2\,{}^i\delta_a {}^j\delta_b - 2\,{}^j\delta_a {}^i\delta_b $$

where ${}^i\delta_t$ is Hildebrand or total cohesion parameter for component i, as evaluated from vaporization energy, and related to the individual interaction parameters by

$$ {}^i\delta_t^2 = {}^i\delta_d^2 + {}^i\delta_0^2 + 2\,{}^i\delta_i {}^i\delta_d + 2\,{}^i\delta_a {}^i\delta_b $$

Numerical values of these interaction cohesion parameters for a few common liquids are presented in Table V.

In systems containing ions, ion-dipole and ion-induced dipole interactions (which are long range) dominate all other effects. It would be possible to develop ionic or electrostatic cohesion parameters, and compared with the contributions the other interaction cohesion parameters would be negligible. This has not been done, although non-ionic cohesion parameters remain useful criteria in certain ionic systems, such as in solvents of low relative permitivity where ions exist largely as ion-pairs. Extensions of cohesion parameters to fully ionic systems may prove as instructive as the addition of orientation, induction and Lewis acid-based terms to the original Hildebrand parameters.

THREE-COMPONENT HANSEN PARAMETERS

C. M. Hansen[81–85] and Hansen and Beerbower[86] proposed an extension of the Hildebrand parameter which uses only three parameters and which therefore could be represented by a solid model. It was assumed that dispersion (δ_d), polar (δ_p) and hydrogen bonding (δ_h) parameters were valid simultaneously, related by the equation

$$\delta_t^2 = \delta_d^2 + \delta_p^2 + \delta_h^2$$

each component parameter being determined empirically on the basis of many experimental observations. Hansen's total cohesion parameter, δ_t, corresponds to the Hildebrand parameter, although the two quantities may not be identical because they are determined by different methods.

For liquids (Table VI), the dispersion component was obtained by comparison with the properties of similar liquids with polar or hydrogen bonding effects, and the remainder was split into a hydrogen bonding and polar contributions in such a way as to optimize the description of the solubility behavior of all the systems investigated, comprising several polymers and many liquids.

Once the three component parameters for each solvent was evaluated, the set of parameters for each polymer could be obtained (Table VII). Solubility data were found by visual inspection of polymer mixtures at concentrations of 10% w/v. Viscous solvents in which the solute did not appear to dissolve at room temperature were heated, the degree of solubility being assessed after the mixture was cooled to room temperature. When plotted in three dimensions, the Hansen parameters provide an approximately spherical volume of solubility for each polymer. The scale on the dispersion axis is usually doubled to improve the spherical nature of this volume. The distance

TABLE V

Interaction cohesion parameters of liquids in order of increasing Hildebrand parameter. [Adapted from Karger et al.[75-77] and Snyder.[78-80]]

Liquid	$\delta/MPa^{\frac{1}{2}}$						$V/cm^3 \, mol^{-1}$
	δ_t	δ_d	δ_0	δ_i	δ_a	δ_b	
Perfluoroalkanes	ca. 12	ca. 12	—	—	2.1	—	—
n-Pentane	14.5	14.5	—	—	—	—	115
Diisopropyl ether	14.5	14.1	2.1	0.2	—	6.1	102
n-Hexane	14.9	14.9	—	—	—	—	131
Diethyl ether	15.3	13.7	4.9	1.0	—	6.1	105
Triethylamine	15.3	15.3	—	—	—	9.2	140
Cyclohexane	16.8	16.8	—	—	—	—	108
Propyl chloride	17.2	14.9	5.9	1.2	—	1.4	88
Carbon tetrachloride	17.6	17.6	—	—	—	1.0	97
Diethyl sulfide	17.6	16.8	3.5	0.5	—	5.3	108
Ethyl acetate	18.2	14.3	8.2	2.1	—	5.5	98
Propylamine	18.2	14.9	3.5	0.4	3.7	11.3	82
Ethyl bromide	18.2	16.0	6.3	1.2	—	1.6	77
Toluene	18.2	18.2	—	—	—	1.2	107
Tetrahydrofuran	18.6	15.5	7.2	1.6	—	7.6	82
Benzene	18.8	18.8	—	—	—	1.2	89
Chloroform	19.0	16.6	6.1	1.0	13.3	1.0	81
Ethyl methyl ketone	19.4	14.5	9.6	2.5	—	6.5	90
Acetone	19.6	13.9	10.4	3.1	—	6.1	74
1,2-Dichloroethane	19.8	16.8	8.6	1.0	—	1.4	79
Anisole	19.8	18.6	4.3	0.8	—	3.5	109
Chlorobenzene	19.8	18.8	3.9	0.6	—	2.1	102
Bromobenzene	20.2	19.6	3.1	0.4	—	2.1	105
Methyl iodide	20.2	19.0	5.1	0.6	—	1.4	62
Dioxane	20.7	16.0	10.6	2.1	—	9.4	86
Hexamethylphosphoramide	21.5	17.2	7.0	3.5	—	8.2	176
Pyridine	21.7	18.4	7.8	2.1	—	10.0	81
Acetophenone	21.7	19.6	5.5	1.4	—	6.8	117
Benzonitrile	21.9	18.8	7.0	2.1	—	4.7	103
Propionitrile	22.1	14.1	13.5	3.7	—	4.3	71
Quinoline	22.1	21.1	3.7	0.6	—	8.6	118
N,N-Dimethylacetamide	22.1	16.8	9.6	3.3	—	9.2	92
Nitroethane	22.5	14.9	12.3	4.5	—	2.1	71
Nitrobenzene	22.7	19.4	7.4	2.3	—	2.1	103
Tricresylphosphate	23.1	19.6	5.1	3.1	—	(?)	316
Dimethylformamide	24.1	16.2	12.7	4.9	—	9.4	77
Propanol	24.5	14.7	5.3	0.8	12.9	12.9	75
Dimethylsulfoxide	24.5	17.2	12.5	4.3	—	10.6	71
Acetonitrile	24.7	13.3	16.8	5.7	—	7.8	53
Phenol	24.7	19.4	4.7	0.8	19.0	4.7	92
Ethanol	26.0	13.9	7.0	1.0	14.1	14.1	59
Nitromethane	26.4	14.9	17.0	6.1	—	2.5	54
γ-Butyrolactone	26.4	16.4	14.7	6.5	—	(?)	77
Propylene carbonate	27.2	20.0	12.1	4.9	—	(?)	85
Diethylene glycol	29.2	16.8	8.2	1.2	10.8	10.8	96
Methanol	29.7	12.7	10.0	1.6	17.0	17.0	41
Ethylene glycol	34.8	16.4	13.9	2.3	12.5	12.5	56
Formamide	39.3	17.0	(?)	(?)	(large)	(large)	40
Water	47.9	12.9	(?)	(?)	(large)	(large)	18

TABLE VI
Hansen parameters for solvents at 25°C. [Selected from Union Carbide.[56]]

Liquid	$M/\mathrm{g\,mol^{-1}}$	$\rho/\mathrm{g\,cm^{-3}}$	$\delta_d/\mathrm{MPa^{\frac{1}{2}}}$	$\delta_p/\mathrm{MPa^{\frac{1}{2}}}$	$\delta_h/\mathrm{MPa^{\frac{1}{2}}}$	$\delta_t/\mathrm{MPa^{\frac{1}{2}}}$
Acetic acid	60.1	1.044	13.9	12.2	18.9	26.5
Acetone	58.1	0.785	13.0	9.8	11.0	19.7
Benzene	78.1	0.874	16.1	8.6	4.1	18.7
Bromobenzene	157.0	1.486	18.4	8.2	0.0	20.1
Butane	58.1	0.572	13.5	0.0	0.0	13.5
Cellosolve®	90.1	0.925	13.0	9.1	15.2	21.9
Chloroform	119.4	1.477	11.0	13.7	6.3	18.7
Cyclohexane	84.2	0.774	16.5	3.1	0.0	16.8
1,4-Dioxane	88.1	1.028	16.3	10.1	7.9	20.7
Ethyl acetate	88.1	0.894	13.4	8.6	8.9	18.2
Ethylene glycol	62.1	1.110	10.1	15.1	29.8	34.9
Ethyl ether	74.1	0.707	13.4	5.3	5.6	15.4
Glycerol	92.1	1.258	9.3	15.4	31.4	36.2
Morpholine	87.1	0.995	16.0	11.4	10.1	22.3
Nitrobenzene	123.1	1.190	17.6	14.0	0.0	22.5
Pyridine	79.1	0.978	17.6	10.1	7.7	21.7
Styrene	104.2	0.901	16.8	9.1	0.0	19.1
Toluene	92.1	0.862	16.4	8.0	1.6	18.3
Water	18.02	0.997	12.2	22.8	40.4	48.0
m-Xylene	106.2	0.860	16.5	7.2	2.4	18.2

of the co-ordinates of any solvent ($^i\delta_d$, $^i\delta_p$, $^i\delta_h$) from the center point ($^j\delta_d$, $^j\delta_p$, $^j\delta_h$) of the solute sphere of solubility is

$$^{ij}R = [4(^i\delta_d - {}^j\delta_d)^2 + (^i\delta_p - {}^j\delta_p)^2 + (^i\delta_h - {}^j\delta_h)^2]^{\frac{1}{2}}$$

This distance can be compared with the radius ^jR of the solute sphere of solubility, and if

$$^{ij}R < {}^jR$$

the likelihood of the solvent i dissolving the solute j is high. This works well, despite the limited theoretical justification for the method. The 'sphere' can be projected on to the three planes passing through two axes and the origin, to provide circles in two-dimensional graphs.

There is considerable variation in the Hansen parameters reported for water. A study of the solubility of a range of liquids in water provides δ_h and δ_t values considerably lower than and inconsistent with those previously reported, which appear to be correct for solutions of water in organic liquids. This is a fundamental problem associated with the use of δ_h rather than the more appropriate acid-base (δ_a and δ_b) parameters, and attempts to reconcile the divergent δ_h values are futile.

An example will clarify the procedure for using Hansen parameters to predict polymer solubility.[87] In order to determine if polystyrene is expected to dissolve in a solvent mixture of 60/40 v/v methyl ethyl ketone/n-hexane, the Hansen parameters are obtained by reference to a compilation of data:

MEK: $^i\delta_d = 16.0\,\mathrm{MPa}^{\frac{1}{2}}$, $^i\delta_p = 9.0\,\mathrm{MPa}^{\frac{1}{2}}$, $^i\delta_h = 5.1\,\mathrm{MPa}^{\frac{1}{2}}$

n-hexane: $^j\delta_d = 14.9\,\mathrm{MPa}^{\frac{1}{2}}$, $^j\delta_p = 0.0\,\mathrm{MPa}^{\frac{1}{2}}$, $^j\delta_h = 0.0\,\mathrm{MPa}^{\frac{1}{2}}$

The Hansen parameters are combined on a 60/40 volume fraction basis:

$$^{ij}\delta_d = 0.6 \times 16.0 + 0.4 \times 14.9 = 15.6\,\mathrm{MPa}^{\frac{1}{2}}$$

$$^{ij}\delta_p = 0.6 \times 9.0 + 0.4 \times 0.0 = 5.4\,\mathrm{MPa}^{\frac{1}{2}}$$

$$^{ij}\delta_h = 0.6 \times 5.1 + 0.4 \times 0.0 = 3.1\,\mathrm{MPa}^{\frac{1}{2}}$$

The center-point co-ordinates and radius of the solubility sphere for polystyrene are (Table VII):

$$^k\delta_d = 21.3\,\mathrm{MPa}^{\frac{1}{2}}; \quad ^k\delta_p = 5.8\,\mathrm{MPa}^{\frac{1}{2}}; \quad ^k\delta_h = 4.3\,\mathrm{MPa}^{\frac{1}{2}}; \quad ^kR = 12.7\,\mathrm{MPa}^{\frac{1}{2}}$$

The distance of the point representing the solvent mixture from the centre of the polymer solubility sphere is

$$[4(21.3 - 15.6)^2 + (5.8 - 5.4)^2 + (4.3 - 3.1)^2]^{\frac{1}{2}} = 11.5\,\mathrm{MPa}^{\frac{1}{2}}$$

As this value is less than the radius of the polymer solubility sphere ($12.7\,\mathrm{MPa}^{\frac{1}{2}}$) the polymer is expected to be soluble. This is found to be the case in a 10% w/w solution.

TABLE VII

Hansen parameters and interaction radius of some polymers and resins
[Selected from Hansen and Beerbower.[86]]

| Polymer | $\delta/\mathrm{MPa}^{\frac{1}{2}}$ | | | |
	$^j\delta_d$	$^j\delta_p$	$^j\delta_h$	$^jR/\mathrm{MPa}^{\frac{1}{2}}$
Acrylonitrile butadiene	18.6	8.8	4.2	9.6
Alkyd, long oil	20.4	3.4	4.6	13.7
Alkyd, short oil	18.5	9.2	4.9	10.6
Cellulose acetate	18.6	12.7	11.0	7.6
Chlorine polypropylene	20.3	6.3	5.4	10.6
Coumarone-indene resin	19.4	5.5	5.8	9.6
Epoxy	20.4	12.0	11.5	12.7
Furfuryl alcohol resin	21.2	13.6	12.8	13.7
Isoprene elastomer	16.6	1.4	−0.8	9.6
Nitrocellulose	15.4	14.7	8.8	11.5
Petroleum hydrocarbon resin	19.7	11.6	14.6	12.7
Phenolic resin	23.3	6.6	8.3	19.8
Polyamide, thermoplastic	17.4	−1.9	14.9	9.6
Poly(ethyl methacrylate)	17.6	9.7	4.0	10.6
Poly(methyl methacrylate)	18.6	10.5	7.5	8.6
Polystyrene	21.3	5.8	4.3	12.7
Poly(vinyl acetate)	20.9	11.3	9.6	13.7
Poly(vinyl chloride)	18.2	7.5	8.3	3.5
Styrene-butadiene	17.6	3.4	2.7	6.6
Urea-formaldehyde	20.8	8.3	12.7	12.7

Although the Hansen parameters provide useful information, the boundary of the solubility sphere is not a clear-cut division between solvents and non-solvent. Particular caution should be exercised when there are significant donor-acceptor interactions. Hansen parameters may also be used in preliminary screening for compatible pairs of polymers. Often with polymers, ranges of cohesion parameters rather than single values are quoted, and one of the most convenient ways of doing this is to describe the Hansen volume of interaction, i.e., the co-ordinates of the center and the radius of the interaction sphere or ellipsoid.

The incorporation of the numerical factor 4 does not appear to be necessary to provide a spherical interaction volume and an equation of the form

$$ {}^{ij}r = {}^{ij}A^{\frac{1}{2}} = \left[({}^{i}\delta_{\mathrm{d}} - {}^{j}\delta_{\mathrm{d}})^2 + ({}^{i}\delta_{\mathrm{p}} - {}^{j}\delta_{\mathrm{p}})^2 + ({}^{i}\delta_{\mathrm{h}} - {}^{j}\delta_{\mathrm{h}})^2 \right]^{\frac{1}{2}} $$

is just as satisfactory.[88,89] In terms of the polymer interaction parameter, it follows that

$$ \sigma_{\mathrm{H}} = \left(\frac{{}^{i}V}{RT} \right) \left[({}^{i}\delta_{\mathrm{d}} - {}^{j}\delta_{\mathrm{d}})^2 + ({}^{i}\delta_{\mathrm{p}} - {}^{j}\delta_{\mathrm{p}})^2 + ({}^{i}\delta_{\mathrm{h}} - {}^{j}\delta_{\mathrm{h}})^2 \right] $$

Instead of assuming spherical Hansen volumes for polymer solubilities, it is possible to use 'boxes'.[90] Thus all liquids represented by Hansen co-ordinates inside a 25% swell contour box swell the polymer by more than 25%; those outside the box do not. The information may be presented diagrammatically by projecting the boxes on the co-ordinate planes, or numerically by listing the co-ordinates of the corners and the centroid. Information may also be presented in the form of diagrams showing irregular 'volumes' rather than spheres or cuboids, as illustrated in the publications of Hansen,[83] Meyer zu Bexten,[91] and Walz and Emrich.[92]

There is considerable theoretical and experimental justification for considering surface free energies and related interfacial properties in terms of additive components analogous to partial cohesion parameters. Hansen,[84,93] Hansen and Pierce,[94] and Zorll[95] have considered the characterization of surfaces in terms of the liquids which spread spontaneously on them, in contrast to those liquids which yield contact angles, by means of Hansen parameters. The results were expressed in the same way as those of solubility studies, with values of ${}^{s}\delta_{\mathrm{d}}, {}^{s}\delta_{\mathrm{p}}, {}^{s}\delta_{\mathrm{h}}$ and ${}^{s}R$ reported for each surface. These can be used in conjunction with liquid data to evaluate ${}^{sl}R$:

$$ {}^{sl}R^2 = 4({}^{s}\delta_{\mathrm{d}} - {}^{l}\delta_{\mathrm{d}})^2 + ({}^{s}\delta_{\mathrm{p}} - {}^{l}\delta_{\mathrm{p}})^2 + ({}^{s}\delta_{\mathrm{h}} - {}^{l}\delta_{\mathrm{h}})^2 $$

Liquids for which ${}^{sl}R < {}^{s}R$ are expected to wet the surface, and liquids with co-ordinates on the spherical boundary surface, ${}^{sl}R = {}^{s}R$, correspond to critical surface free energies. As in the 'solubility sphere' studies, the spherical boundaries correspond to situations where there is a predicted zero free energy change for the wetting process.

Alternative forms of Hansen parameters are fractional cohesion parameters, defined by[96-98]

$$f_d = \frac{\delta_d}{\delta_d + \delta_p + \delta_h}; \quad f_p = \frac{\delta_p}{\delta_d + \delta_p + \delta_h}; \quad f_h = \frac{\delta_h}{\delta_d + \delta_p + \delta_h}$$

The triangular representation makes the simplifying assumption that the total cohesion parameter, δ_t, is constant for all materials, and that it is the relative magnitude of the three contributions which determines the extent of miscibility. Inspection of tables of Hansen parameter values shows that although there is much greater variation in δ_p and δ_h than in δ_t, δ_t is not even approximately constant.

FRACTIONAL POLARITY

The nature of molecular interactions has been discussed in terms of the fraction of total interaction due to dipole–dipole or orientation attractions (p), induction (i), and dispersion (d) effects, such that

$$p + i + d = 1$$

In particular, the fractional polarity, p, may be calculated from the dipole moment, polarizability and ionization potential[99,100] and $p - \delta$ maps constructed. The criterion for solubility in a pure or mixed solvent is that the solvent (p, δ) point lies within the solubility region, volume fraction averages being used for the calculation of p and δ in mixed systems.

ELECTRON DONOR-ACCEPTOR PARAMETERS

A hydrogen bonding parameter with a spectroscopic origin has been widely used.[101-106] This was based on the effect that a solvent has on a small amount of alcohol introduced into the solvent: the greater the extent of hydrogen bonding between the solvent and the alcohol protons, the weaker the O—H bond and the lower the frequency of the infrared radiation absorbed. Deuterated alcohol was used because the O—D stretch band is in a region with little interference, permitting detection at low alcohol concentrations. The extent of the shift to lower frequencies of the O—D stretching infrared absorption of deuterated methanol in the liquid under study thus provides a measure of its hydrogen bond acceptor capability. The spectrum of a solution of deuterated methanol in the test liquid is then compared with that of a solution in benzene, and the hydrogen bonding parameter is defined

$$\gamma_C = \frac{\Delta v}{10}$$

where the OD absorption shift is expressed in wave numbers. Thus, replacing benzene with nitrobenzene as solvent shifts the peak from $2681\,cm^{-1}$ to $2653\,cm^{-1}$, indicating a γ_C value of 2.8 for nitrobenzene.

More precise measurements on the deuterated methanol system were made by Nelson et al.,[107,108] and it appears that although there is hydrogen bonding with either benzene or tetrachloromethane, cyclohexane is a suitable reference system. Another definition of the hydrogen bonding parameter,[109,110] yielding what is described here as γ_D, is

$$\gamma_D = 0.0359\Delta v + 2.2$$

Both sets of data are included in the ASTM D3132[26,27] standard test method for the 'Solubility Range of Resins and Polymers', but in fact they may be interconverted by means of the equations

$$\gamma_D = 0.359\gamma_C + 2.2$$
$$\gamma_C = 2.79\gamma_D - 6.13$$

The hydrogen bonding parameter has been used together with the Hildebrand and dipole moment to construct three dimensional models representing solubility behavior. It is also possible to construct two dimensional contour maps of this information; in many cases the dipole moment is not required, and a Hildebrand parameter–hydrogen bonding parameter map results. In the case of three solvent constituents, the triangle formed by the lines joining the points representing the three pure liquids on a δ–γ diagram provides the entire array of Hildebrand parameter–hydrogen bonding index combinations that can be obtained by blending. When combined with a polymer solubility map, this information provides a good starting point for solvent formulation. Contours may be included in Hildebrand parameter–hydrogen bonding index to denote the maximum permissible fractional polarity for a solvent. Examples of such two-dimensional maps and three-dimensional models are provided in the publications of Barton Crowley,[104,105] Eastman Kodak,[111] Shell,[112] Cosaert,[109] du Pont,[110] Zunker,[113] Ellis et al.,[114] Ramsbotham,[115–118] Izumi and Miyake,[88] and Beerbower et al.[106]

CORRELATION WITH OTHER PROPERTIES

Cohesive pressure, surface free energy and ideal tensile strength are all manifestations of the energy constant of materials. It is therefore reasonable to expect correlations to exist between the observed mechanical properties of materials such as polymers and their cohesion parameters. In particular, stress-strain properties, environmental stress cracking, glass transition temperatures, solvent-induced crystallization, gas and liquid permeation rates, and polymer-plasticizer interactions have been discussed in terms of cohesion parameters. Natural polymers have also been investigated: cellulose, cellulose derivatives (particularly membranes) wool keratin, lignin, coal and petroleum products such as bitumen. Even dynamic processes such as reaction kinetics, viscosity and diffusion depend on cohesion properties. The range of characteristic properties studied in this way is extremely broad, but the full potential of the method has not been realised in most areas.

Details of these applications of cohesion parameters cannot be provided here, but may be found in a recent publication.[1]

CONCLUSIONS

(i) Hildebrand parameters (δ) provide a broad, qualitative indication of behavior for most polymer-solvent systems.

(ii) Hansen parameters (δ_d, δ_p, δ_h) and similar sets of partial parameters give an approximate quantitative measure of the extent of interaction for all polymer systems.

(iii) A set of interaction cohesion parameters (δ_d, δ_0, δ_i, δ_a, δ_b) including separate Lewis acid and Lewis base terms is necessary for a reliable evaluation of the behavior of most systems.

(iv) In a few areas (e.g., chromatography) the subject has developed to stage (iii); in others (e.g., polymer-solvent interactions in fields such as paint formulation and reverse osmosis membrane performance) it has reached stage (ii), but in many others (notably biocompatible materials) progress has not been made past stage (i).

References

1. A. F. M. Barton, *Handbook of Solubility Parameters and Other Cohesion Parameters*. CRC Press, Boca Raton, Florida (1983).
2. J. H. Hildebrand and R. L. Scott, Solubility of Non-Electrolyles, 3rd ed, Reinhold, New York (1950).
3. J. H. Hildebrand and R. L. Scott, *Regular Solutions*, Prentice-Hall, Englewood Cliffs, N.J. (1962).
4. J. H. Hildebrand, J. M. Prausnitz and R. L. Scott, Regular and Related Solutions, Van Nostrand Reinhold, Neew York (1970).
5. J. H. Hildebrand, Solubility. XII. Regular solutions, *J. Am. Chem. Soc.*, **51**, 66 (1929).
6. J. H. Hildebrand, Order from chaos, *Science*, **150**, 441 (1965).
7. G. Scatchard, Equilibria in non-electrolyte solutions in relation to the vapor pressures and densities of the components, *Chem. Rev.*, **8**, 321 (1931).

8.   J. H. Hildebrand, Solubility, *J. Am. Chem. Soc.*, **38**, 1452 (1916).
9.   J. H. Hildebrand, Solubility. III. Relative values of internal pressures and their practical application, *J. Am. Chem. Soc.*, **41**, 1067 (1919).
10. P. J. Flory, Thermodynamics of high polymer solutions, *J. Chem. Phys.*, **10**, 51 (1942).
11. P. J. Flory, *Principles of Polymer Chemistry*, Cornell University Press, Ithaca, N.Y. (1953).
12. P. J. Flory and J. Rehner, Jr., Statistical mechanics of cross-linked polymer networks. I. Rubberlike elasticity, *J. Chem. Phys.*, **11**, 512 (1943).
13. M. L. Huggins, Solutions of long chain compounds, *J. Chem. Phys.*, **9**, 440 (1941).
14. M. L. Huggins, Some properties of long-chain compounds, *J. Phys. Chem.*, **46**, 151, *Ann. N.Y. Acad. Sci.*, **43**, Art. 1, 1 (1942).
15. M. L. Huggins, Theory of solution of high polymers, *J. Am. Chem. Soc.*, **64**, 1712 (1942).
16. R. A. Orwoll, The polymer-solvent interaction parameter χ, *Rubber Chem. Technol.*, **50**, 451 (1977).
17. R. F. Blanks and J. M. Prausnitz, Thermodynamics of polymer solubility in polar and nonpolar systems, *Ind. Eng. Chem. Fundam.*, **3**, 1 (1964).
18. H. Burrell, Solubility parameters, *Interchem. Rev.*, **14**, 3–16 and 31–46 (1955a).
19. H. Burrell, Solubility parameters for film formers, *Off. Dig., Fed. Paint Varn. Prod. Clubs*, **27**, 726 (1955b).
20. H. Burrell, Solubility parameter values, in *Polymer Handbook*, BJ. Brandrup and E. H. Immergut, Eds., Wiley Interscience, New York, IV–341 (1966).
21. H. Burrell, The challenge of the solubility parameter concept, *J. Paint Technol.*, **40**, 197 (1968).
22. H. Burrell, Solubility of polymers, in *Kirk-Othmer Encyclopedia of Polymer Science and Technology*, **12**, Wiley, New York, 618 (1970).
23. H. Burrell, Solubility parameter values in *Polymer Handbook* 2nd ed., J. Brandrup and E. H. Immergrut, Eds., Wiley interscience, New York, IV—337 (1975).
24. V. A. Piskarev and R. L. Radushkevich, Method for determining compatibility of bitumens with elastomers, *Polim. Stroit. Mater.*, **42**, 163 (1972).
25. E. A. Lazurin, V. T. Samorodov and L. V. Kosmodemyanskii, Determination of solubility parameters of mineral oils, *Prom-st. Sint. Kauch*, **8**, 10 (1980).
26. ASTM Standard Test Method for Solubility Range of Repairs and Polymers, ANSI/ASTM D3132–72, Ameridan Society for Testing and Materials.
27. Philadelphia, American National Standards Ihnstitute, New York (1972, reapproved 1976).
28. E. Bagda, Determination of solution rate and solubility parameters of polymers, *Colloid. Polym. Sci.*, **255**, 384 (1977).
29. J. H. Engel, Jr. and R. N. Fitzwater, Adhesion of surface coatings as determined by the peel method, in *Adhesion and Cohesion*, Weiss, P.N.Y., Ed., Elsevier, 89 (1962).
30. V. E. Eskin, U. Zhurayev and T. N. Nekrasova, On some features of intermolecular interaction exhibited in polymer solutions, *Vysokomol. Soedin, (A)* 18, 2529, 1976 (Russian); *Polym. Sci. USSR*, **18**, 2890 (1976).
31. G. Gee, The interaction between rubber and liquids, III. The swelling of vulcanised rubber in various liquids, *Trnas. Faraday Soc.*, **38**, 418 (1942).
32. G. Gee, Interaction between rubber and liquids, IV. Factors governing the absorption of oil by rubber, *Trans. Inst. Rubber Ind* **18**, 266 (1943).
33. P. J. Flory and J. Rehner, Jr., Statistical mechanics of cross-linked polymer networks, II, Selling, *J. Chem. Phys.*, **11**, 521 (1943).
34. G. M. Bristow and W. F. Watson, Viscosity-equilibrium swelling correlations for natural rubber, *Trans. Faraday Soc.*, **54**, 1567 (1958).
35. G. M. Bristow and W. F. Watson, Cohesive energy densities of polymers. Part 1. Cohesive energy densities of rubbers by swelling measurements, *Trans. Faraday Soc.*, **54**, 1731 (1958).
36. G. M. Bristow and W. F. Watson, Cohesive energy densities of polymers. Part 2. Cohesive energy densities from viscosity measurements, *Trans. Faraday. Soc.*, **54**, 1742 (1958).

176 POLYMER YEARBOOK

37. T. J. Dudek and F. Bueche, Polymer-solvent interaction parameters and creep behaviour of ethylene-propylene rubbers, *J. Polym. Sci. A*, **2**, 811 (1964).
38. A. G. Shvarts, Comparative determination of [specific] cohesion energy of natural and polyisoprene rubber, *Zhur Fiz. Khim.*, **32**, 718 (1958).
39. A. G. Shvarts, E. K. Chefranova and L. A. Iotkovskaya, Solubility parameters of resins based on dimethylvinylethynyl-*p*- =hydroxyphenylmethane, *Kolloid. Zhur*, **32**, 603,, *Colloid J. USSR*, **32**, 606 (1970).
40. L. V. Makarova, A. G. Shvarts, N. D. Zakharov and A. M. Priborets, Determination of the cohesion energy density of some synthetic rubbers with functional groups, *Vysokomol. Soedin.*, **7**, 1056, (Russian); *Polym. Sci. USSR*, **7**, 1168 (1965).
41. W. R. Moore and G. F. Boden, Heptane-soluble material from atactic polypropylene. II. Interaction with liquids, *J. App. Polym. Sci.*, **10**, 1121 (1966).
42. D. W. van Krevelen and P. J. Hoftyzer, *Properties of Polymers: Their Estimation and Correlation with Chemical Structure*, 2nd ed., Elsevier, Amsterdam (1976).
43. G. M. Bristow and W. F. Watson, Swelling and viscosity of rubbers in mixed solvents. *Trans. Inst. Rubber Ind.*, **35**, 73 (1959).
44. M. B. Huglin and D. J. Pass, Cohesive energy density of polytetrahydrofuran, *J. App. Polym. Sci.*, **12**, 473 (1968).
45. J. M. Sosa, A. Rubio and B. H. Pérez, Viscosities of polystyrene solutions in *Structure-Solubility Relationships in Polymers* (Proc. Symp., 172nd Am. Chem. Soc. Meeting, San Francisco, 1976), F. W. Harris and R. B. Seymour, Eds., Academic Press, New York, 89 (1977).
46. J. Velickovic and J. Filipovic, Cohesive energy densities of poly(di-n-alkyl itaconates), *Angew Makromol. Chem.*, **57**, 139 (1977).
47. E. Bagda, The relation between surface tension and solubility parameter in liquids, *Farbe Lack*, **84**, 212 (1978).
48. H. Burrell, Trends in solvent science and technology, in *Solvents Theory and Practice*, *Adv. Chem. Ser.*, **124** (1973); Proc. Symp. Div. Org. Coat. Plast. Chem., 162nd Am. Chem. Soc. Meeting Washington D.C., Sept. 15–16 (1971); R. W. Tess, Am. Chem. Soc., Washington, D.C., Chap. 1.
49. D. Hoenschemeyer, The influence of solvent type on the viscosity of concentrated polymer solutions, *J. App. Polym. Sci.*, **18**, 61 (1974).
50. S. Takahashi, M. Kariya, T. Yatake, K. Sonogashira and N. Hagihara, Studies of polyyne polymers containing transition metals in the main chain. 2. Synthesis of poly[*trans*-bis (tri-*n*-butyl(phosphine) platinum 1,4-butadiynediyl] and evidence of a rodlike structure, *Macromol.*, **11**, 1063 (1978).
51. P. A. Small, Some factors affecting the solubility of polymers, *J. Appl. Chem.*, **3**, 61 (1953).
52. J. A. Ostrenga, On the additivity of molar attraction constants, *J. Pharmaceut. Sci.*, **58**, 1281 (1969).
53. A. Cammarata and S. J. Yau, Molecular estimates for molar attraction constants, *J. Pharm. Sci.*, **61**, 723 (1972).
54. K. L. Hoy, *Tables of Solubility Parameters*, Union Carbide Corporation, Chemicals and Plastics Research and Development Dept., South Charleston, W. Virginia (1969), 3rd ed. 1975).
55. K. L. Hoy, New values of the solubility parameters from vapor pressure data, *J. Paint Technol.*, **42**, 76 (1970).
56. Union Carbide Corporation, *Tables of Solubility Parameters*, 3rd ed., Chemicals and Plastics Research and Development Dept., Tarrytown, N.Y. (1975).
57. D. W. van Krevelen, Chemical structure and properties of coal. XXVIII – Coal constitution and solvent extraction, *Fuel*, **44**, 229 (1965).
58. A. A. Askadskii, L. K. Kolmakova, A. A. Tager, G. L. Slonimskii and V. V. Korshak, Evaluation of solubility parameters of low-molecular-weight liquids and polymers, *Dokl. Akad. Nauk. SSR*, **226**, 857 (Russian); *Doklady Phys. Chem.*, **226**, 99 (1976).
59. A. A. Askadskii, L. K. Kolmakova, A. A. Tager, G. L. Slonimskii and V. V. Korshak, Estimation of cohesion energy of low-molecular weight liquids and polymers,

Vysokomol. Soedin.(**A**), 19, 1004 (1977), (Russian; English translation in *Polym. Sci. USSR*).

60. J. F. Voeks, Cohesive energy density and internal pressure of high polymers, *J. Polym. Sci. A*, **2**, 5319 (1964).

61. H. Ahmad and M. Yaseen, Determination of solubility parameters of polymers by a group contribution technique, *J. Oil Col. Chem. Assoc.*, **60**, 488 (1977).

62. H. Ahmad and M. Yaseen, Evaluation of solubility parameters of hologenated polymers by a group contribution technique, *Paint Nanuf.*, **48**, 28 (1978). H. Ahmad and N. Yaseen, Application of intrinsic viscosity data for determination of solubility parameters and molecular weights of alkyds, *J. Coat. Tech.*, **50**(640), 86 (1978).

63. H. Ahmad and M. Yaseen, Application of a chemical group contribution technique for calculating solubility parameters of polymers, *Polym. Eng. Sci.*, **19**, 858 (1979). H. Ahmad and M. Yaseen, Solubility parameter determination from chemical group constituents. Solubility parameter values for poly(ethylenesulfonates) and poly(ethylenesulfonyl halides) determined by chemical groups contribution technique, *Farbe Lack*, **85**, 356 (1979).

64. H. Ahmad ad M. Yaseen, Principle of additivity of chemical group, atom, and bond used for determination of solubility parameters, *Paintindia*, **30**(4), 3 (1980).

65. B. A. Wolf, An extrapolation method for the determination of solubility parameters of polymers demonstrated for polyethylene, *Makromol. Chem.*, **178**, 1869 (1977).

66. B. A. Wolf and W. Schuch, Oligomer-oligomer incompatibility, 3. End-group effects, *Makromol. Chem.*, to be published.

67. R. B. Seymour, Plastics vs. solvents, *Mod. Plast.*, **48**, 150 (1971).

68. R. B. Seymour, *Introduction to Polymer Chemistry*, McGraw-Hill, New York, Section 2.2 (1971).

69. R. C. Weast, Ed., Solubility Parameters of organic compounds in *CRC Handbook of Chemistry and Physics*, 60th ed., C-732 (1979).

70. E. P. Lieberman, Quantification of the hydrogen bonding parameter, *Off. Dig., Fed. Soc. Paint Technol.*, **34**(444), 30 (1962).

71. E. I. Du Pont de Nemours, and Co., *A New Dimension in Solvent Formulation – Elvacite Acrylic Resins*, A44369, du Pont Electrochemicals Dept. (1965).

72. J. F. Hagman, *Solvent Systems for Neoprene-Predicting Solvent Strength*, A35117, du Pont Elastomer Chemicals Dept. (1964).

73. R. A. Keller, B. L. Karger and L. R. Snyder, Use of the solubility parameter in predicting chromatographic retention and eluotropic strength, *Gas Chromatog.*, (Proc. Int. Symp., Eur, 1970), **8**, 125 (1971).

74. R. A. Keller and L. R. Snyder, Relation between the solubility parameter and the liquid-solid solvent strength parameter, *J. Chromat. Soc.*, **9**, 346 (1971).

75. B. L. Karger, L. R. Snyder and C. Eon, An expanded solubility parameter treatment for classification and use of chromatographic solvents and adsorbents. Parameters for dispersion, dipole and hydrogen bonding interactions, *J. Chromat.*, **125**, 71 (1976).

76. B. L. Karger, L. R. Snyder and C. Eon, Expanded solubility parameter treatment for classification and use of chromatographic solvents and adsorbents, *Anal. Chem.*, **50**, 2126 (1978).

77. B. L. Karger, L. R. Snyder and C. Horvath, *An Introduction to Separation Science*, Wiley, New York (1973).
A. Hartkopf, A reconsideration of the Rohrschneider approach for characterizing gas chromatographic liquid phases. I. Review and description of basic approach. *J. Chromat. Sci.*, **12**, 113 (1974).

78. L. R. Snyder, Solvent selection for separation processes, in *Separation and Purification*, 3rd ed. (Techniques of Chemistry, **12**), E. S. Perry and A. Weissberger, Wiley-Interscience, New York, 25 (1978).

79. L. R. Snyder, Classification of the solvent properties of common liquids, *J. Chromat. Sci.*, **16**, 223 (1978).

80. L. R. Snyder, Solutions to solution problems, *Chemtech.*, **9**, 750 (1979); **10**, 188 (1980).

81. C. M. Hansen, The three dimensional solubility parameter—key to paint component

affinities: I. Solvents, plasticizers, polymers and resins, *J. Paint Technol.*, **39**, 104 (1967).

82. C. M. Hansen, The three dimensional solubility parameter—Key to paint component affinities: II. Dyes, emulsifiers, mutual solubility and compatibility, and pigments, *J. Paint Technol.*, **39**, 505 (1967).

83. C. M. Hansen, The universality of the solubility parameter, *Ind. Eng. Chem. Prod. Res. Dev.*, **8**, 2 (1969).

84. C. M. Hansen, Characterization of surfaces by spreading liquids, *J. Paint Technol.*, **42**, 660 (1970).

85. C. M. Hensen, Solubility in the coatings industry, *Skand. Tidskr. Farg. Lack*, **17**, 69 (1971).

86. C. M. Hansen and A. Beerbower, Kirk-Othmer Ecycl. Chem. Technol. 2nd Ed. (1971); Ed. A. Standen, Interscience, New York, Suppl. 889–910.

87. Shell Chemicals, *Solubility Parameters*, 2nd ed., Tech. Bull. ICS(X)/78/1 (1978).

88. Y. Izumi and Y. Miyake, Study of linear poly (*p*-chlorostyrene)-diluent systems. I. Solubilities, phase relationships, and thermodynamic interactions, *Polym. J.*, **3**, 647 (1972).

89. B. H. Knox, Bimodal character of polyester-solvent interactions. I. Evaluation of the solubility parameters of the aromatic and the aliphatic ester residues of poly(ethylene terephthalate), *J. App. Polym. Sci.*, **21**, 225 (1977).

90. A. Beerbower and J. R. Dickey, Advanced methods for predicting elastomer/fluids interactions, *Am. Soc. Lubric. Eng. Trans.*, **12**, 1 (1969).

91. J. H. Meyer zu Bexten, Solubility parameters and their practical application, *Farbe Lack*, **78**, 813 (1972).

92. G. Walz and G. Emrich, Solubility parameter diagrams of selected alkyd and melamine resins, *Kunstharz Nachr. (Hoechst)*, **34**(8), 19 (1975).

93. C. M. Hansen, Surface dewetting and coatings performance, *J. Paint Technol.*, **44**(570), 57 (1972).

94. C. M. Hensen and P. E. Pierce, Srface effects in coating processes, *Ind. Eng. Chem. Prod. Res. Dev.*, **13**, 218 (1974).

95. Zorll, U., Significance and problem of the critical surface tension, *Adhäsion*, **18**, 262 (1974).

96. J. P. Teas, Graphic analysis of resin solubilities, *J. Paint Technol.*, **40**(516), 19 (1968).

97. J. P. Teas, Re solubility parameters, *J. Paint Technol.*, **40**, 104 (1968).

98. J. L. Gardon and J. P. Teas, Solubility parameters, in *Treatise on Coatings*, **2**, *Characterization of Coatings: Physical Techniques*, Part II, R. R. Myers and J. S. Long, Eds., Dekker, New York, Chap. 8 (1976).

99. J. L. Gardon, Cohesive-energy density, in Encyclopedia of Polymer Science and Technology, **3**, 833, Wiley, New York (1964).

100. J. L. Gardon, The influence of polarity upon the solubility parameter concept, *J. Paint Technol.*, **38**(492) 43 (1966).

101. W. Gordy, Spectroscopic comparison of the proton-attracting properties of liquids, *J. Chem. Phys.*, **7**, 93 (1939).

102. W. Gordy and S. C. Stanford, Spectroscopic evidence for hydrogen bonds: comparison of proton-attracting properties of liquids. II, *J. Chem. Phys.*, **8**, 170 (1940).

103. W. Gordy and S. C. Stanford, Spectroscopic evidence for hydrogen bonds: comparison of proton-attracting properties of liquids. III. *J. Chem. Phys.*, **9**, 204 (1941).

104. J. D. Crowley, G. S. Teague, Jr. and J. W. Lowe, A three-dimensional approach to solubility, *J. Paint Technol.*, **38**, 269 (1966).

105. J. D. Crowley, G. S. Teague, Jr. and J. W. Lowe, A three-dimensional approach to solubility: II, *J. Paint Technol.*, **39**, 19 (1967).

106. A. Beerbower, L. A. Kaye and D. A. Pattison, Picking the right elastomer to fit your fluids, *Chem. Eng. (NY)*, **74**(26), 118 (1967).

107. R. C. Nelson, V. F. Figurelli, J. G. Walsham and G. D. Edwards, Solution theory and the computer. Effective tools for the coatings chemist, *J. Paint. Technol.*, **42**, 644 (1970).

108. R. C. Nelson, R. W. Hemwall and G. D. Edwards, Treatment of hydrogen bonding in predicting miscibility, *J. Paint Technol.*, **42**, 636 (1970).

109. E. Cosaert, Solubility parameters allowing the prediction of the solubility of a solid in a solvent system, *Chim. Peint.*, **34**, 169 (1971).
110. E. I. du Pont de Nemours and Co., *Solvent formulating maps for Elvacite acrylic resins.* Bulletin PA12-770, A. 70562, July 1970; PA-12-174 (1974).
111. Eastman Kodak, *Three Dimensional Approach to Solubility*, Rochester, New York.
112. Shell Chemicals, *Solvent System Design*, Tech. Bull. ICS(X)/78/2 (1978).
113. D. W. Zunker, Unique aqueous nitrocellulose/acrylic air-dry systems for industrial finishing, *J. Coat. Technol.*, **48**(616), 37 (1976).
114. W. H. Ellis, Z. Saary and D. G. Lesnini, Formulation of exempt replacements for aromatic solvents, *J. Paint. Technol.*, **41**, 249 (1969).
115. J. Ramsbotham, Quantitative polymer solubility maps, *FATIPEC*, **14**, 567 (1978).
116. J. Ramsbotham, Solvent system design for epoxy resin, *Paint Manu. Resin News.*, **49**, March 26; April 18 (1979).
117. J. Ramsbotham, Solvent loss from paint films, *J. Oil Col. Chem. Assoc.*, **62**, 359 (1979).
118. J. Ramsbotham, Solvent formulation for surface coatings, *Prog. Org. Coatings*, **8**, 113 (1980).
119. H. Ahmad and M. Yaseen, Determination of solubility parameters of low and high molecular weight compounds, *J. Col. Soc.*, **16**, 19 (1977).
120. G. Allen, G. Gee, D. Mangaraj, D. Sims and G. J. Wilson, Intermolecular forces and chain flexibilities in polymers: II. Internal pressures of polymers, *Polymer*, **1**, 467 (1960).
121. G. Allen, G. Gee and G. J. Wilson, Intermolecular forces and chain flexibility in polymers: I. Internal pressures and cohesive energy densities of simple liquids, *Polymer*, **1**, 456 (1960).
122. E. Bagda, The possibility of prediction of the viscosity of polymer solutions, *FATIPEC*, **14**, 137 (1978).
123. E. Bagda, Prediction of viscosity of concentrated polymer solutions with the solubility parameter concept, *DEFAZET (Deutsch Farben Z.)*, **32**, 372 (1978).
124. E. B. Bagley and J. M. Scigliano, Polymer solutions, in *Solutions and Solbuilities*, Part II, M. R. J. Dack, Ed., Wiley Interscience, New York, Chap. 16 (1976).
125. A. F. M. Barton, Solubility parameters, *Chem. Rev.*, **75**, 731 (1975).
126. J. J. Bernardo and H. Burrell, Plasticization, in *Polymer Science*, A. D. Jenkins, Ed., North-Holland, Amsterdam, Chap. 8 (1972).
127. R. F. Blanks, Engineering applications of polymer solution thermodynamics, *Chemtech.*, **6**, 396 (1976).
128. R. F. Blanks, Thermodynamics of polymer solutions, *Polym.-Plast. Technol. Eng.*, **8**, 13 (1977).
129. H. Burrell, Entropy: the hidden ingredient, *J. Paint Technol.*, **42**(540), 3 (1970).
130. P. J. Corish, Fundamental studies of rubber blends, *Rubber Chem. Tedhnol.*, **40**, 324 (1967).
131. P. J. Corish, Elastomer blends, in *Science and Technology of Rubber*, F. R. Eirich, Ed., Academic Press, New York, Chap. 12 (1978).
132. J. Dayantis, The Hildebrand parameter. Solubility of polymers, *Plast. Mod. Elastomeres*, **29**(2), 58 (1977).
133. R. Dernini and S. Vargiu, Solubility parameters of SIR [Societa Italiana Resine] resins for paints, *Pitture Vernici*, **45**, 452 (1969).
134. F. M. Fowkes, Dispersion force conbributions to surface and interfacial tensions, contact angles, and heats of immersion, in *Contact Angle, Wettability and Adhesion, Adv. Chem. Ser.*, **43**, 99 (1964).
135. F. M. Fowkes, Donor-acceptor interactions at interfaces, in *Adhesion and Adsorption of Polymers, Part A (Polym. Sci. Technol.* **12**, Part A), Int. Conf. Adhesion Adsorption. Honolulu (April 1979); L. H. Lee, Ed., Plenum, New York, 43 (1980).
136. J. L. Gardon, Relationship between cohesive energy derisities of Polymers and Zismanis critical surface tensions, *J. Phys. Chem.*, **67**, 1935 (1963).
137. J. L. Gardon, Cohesive energy density and adhesion, *Proc. Chem. Inst. Can. Symp. Polym. Interfaces: Focus Adhes.*, 33 (1976).
138. J. L. Gardon, Critical review of concepts common to cohesive energy density, surface

tension, tensile strength, heat of mixing, interfacial tension and butt joint strength, *J. Colloid Interface Sci.*, **59**, 582; *Prog. Org. Coat*, **5**, 1 (1977).

139.   I. Ceczy, The use of solubility parameters, as technological characteristics, in selecting industrial chemicals for the manifacture of plastics, *Koloriszt. Ertesito*, **4**, 99 (1962).

140.   G. M. Guzman, Solutions and solvents for polymers. I. *Rev. Plast. Mod.*, **15**, 489 (1964).

141.   H. Hadert, Polymer solvents, *Coating*, **2**(2) 39, (3) 83, (4) 108 (1969).

142.   C. M. Hansen, Solvent selection by computer, in *Solvents Theory and Practice, Adv. Chem. Ser.*, **124**, Proc. Symp. Div. Org. Coat. Plast., 162nd Am. Chem. Soc., Washington D.C. (Sept. 15 16, 1971), R. W. Tess, Ed., Am. Chem. Soc., Washington D.C., Chapt. 4.

143.   Minoru, Imoto, Introduction to theory of adhesion III. Cohesion and solubility parameter values, *Nippon Secchaka Kyokai Shi*, **5**, 156 (1969).

144.   Minoru Imoto, Senriyama evening talks. 125. Calculation of solubility parameter for random copolymers, *Secchaka*, **16**, 436 (1972).

145.   R. Johannsen, Importance and application of the solubility parameter, *Deut. Farben Z.*, **17**, 264 (1963).

146.   D. H. Kaelble, *Physical Chemistry of Adhesion*, Wiley Interscience, New York (1971).

147.   E. Klein and J. K. Smith, The use of solubility parameters for solvent selection in asymmetric membrane formation, in *Reverse Osmosis Membrane* Research, H. K. Lonsdale and H. E. Podall, Eds., Plenum, New York, 61; Asymmetric membrane formation. Solubility parameters for solvent selection, *Ind. Eng. Chem. Prod. Res. Dev.*, **11**, 207 (1972).

148.   D. M. Koenhen and C. A. Smolders, The determination of solubility parameters of solvents and polymers by means of correlation with other physical quantities, *J. App. Polym. Sci.*, **19**, 1163 (1975).

149.   R. Kumar and J. M. Prausnitz, Solvents in chemical technology, in *Solutions and Solubilities (Techniques of Chemistry*, **8**). Part 1, M. R. J. Dack, Ed., Wiley Interscience, New York, Chap. 5 (1975).

150.   M. T. Lucas, Solubility parameters and problems in their application, *Chim. Peint.*, **34**, 125 (1971).

151.   L. Mandik, Correlation between the KB (kauri-butanol) value and the solubility parameter, *Chem. Prumysl.*, **15**, 283 (1965).

152.   L. Mandik, Quantitative characterization of cohesion forces. I. Low molecular weight organic substances, *Chem. Prumysl.*, **22**, 243 (1972).

153.   L. Mandik and F. Kaspar, *The Selection of Solvents for Film-forming Materials*, Pardubice (1972).

154.   L. Mandik and J. Stanek, Selection of solvents for coating systems with the aid of the solubility parameter, *Chem. Prumysl.*, **15**, 223 (1965).

155.   B. Marti, Solubility parameter, *Schweiz. Arch. Angew. Wiss, Tech.*, **33**, 297 (1967).

156.   I. Mellan, *Compatibility and Solubility*, Noyes, Park Ridge N.J. (1968).

157.   T. Mitomo and T. Teshirogi, A simple method for the calculation of solvent solubility parameters. I, II. *Kagaku Kyoiku*, **27**, 449, 1979; **28**, 70 (1980).

158.   E. V. Moiseev and M. V. Pilyagin, Selection of solvents based on the solubility parameter for materials applied by spraying in an electrical field, *Lakokrasoch. Mater. Ikh. Primen*, **4**, 45 (1971).

159.   W. R. Moore, The swelling and solution of synthetic fibre-forming polar polymers in liquids, *J. Soc. Dyers Col.*, **73**, 500 (1957).

160.   C. J. Nunn, Solubility parameters: some practical considerations, *Chim. Peint.*, **34**, 215 (1971).

161.   O. Olabisi, L. M. Robeson and M. T. Shaw, *Polymer-Polymer Miscibility*, Academic Press, New York (1979).

162.   D. Patterson, Thermodynamics of non-dilute polymer solutions, *Rubber Chem. Technol.*, **40**, 1 (1967).

163.   D. Patterson, Role of free volume changes in polymer solution thermodynamics, *J. Polym. Sci. C*, **16**, 3379 (1968).

164.   D. Patterson, Free volume and polymer solubility. A qualitative view, *Macromol.*, **2**, 672 (1969).

165. D. R. Paul, Background and perspective, in *Polymer Blends*, Vol. 1, D. R. Paul and S. Newmans, Eds., Academic Press, New York, Chap. 1 (1968).

166. D. R. Paul and J. W. Barlow, A brief review of polymer blend technology, in *Multiphase Polymers, adv. Chem. Ser.*, **176**, Div. Polym. Chem. Symp., 175th Am. Chem. Soc. Meeting, Anaheim, March 1978), S. L. Cooper and G. M. Estes, Eds., Am. Chem. Soc., Chap. 17 (1979).

167. L.. Rebenfeld, P. J. Makarewicz, H.-D. Weigmann and G. L. Wilkes, Interactions between solvents and polymers in the solid state, *J. Macromol. Sci. Rev. Macromol. Chem.*, **C15**, 297 (1976).

168. C. Reichardt, Solvent scales and chemical reactivity, in *Organic Liquids: Structure, Dynamics, Chemical Properties*, A. D. Buckingham, E. Lippert and S. Bratos, Eds., Wiley, Chichester, Chap. 16 (1978).

169. A. E. Rhineck and K. F. Lin, Solubility parameter calculations based on group contributions, *J. Paint Technol.*, **40**, 611 (1968).

170. C. Robu, Solubility parameters, *Bul. Teh.-Inf. Lab. Cent. Cercet. Lacuri Cerneluri Bucuresti*, **2**, 57 (1969).

171. B. Sandholm, Evaluation of solubility parameters of polymers, *Suomen Kem. Tied.*, **79**, 14 (1970).

172. R. Sattelmeyer,The solubility parameters of synthetic resins and elastomers for the plastics sector, *Adhäsion*, **20**, 278 (1976).

173. P. J. Schoenmakers, H. A. H. Billiet, R. Tijssen and L. De Galan, Gradient selection in reversed-phase liquid chromatography, *J. Chromat.*, **149**, 519 (1978).

174. P. J. Schoenmakers, *A systematic Approach to Mobile Phase Effects in Reversed Phase Liquid Chromatography*, Thesis, Delft Technical College (1981).

175. J. Sevestre, Solubility of polymers, *Double Liasion*, **105**, 81 (1964).

176. J. Sevestre, On some correlations involving the solubility parameter δ, *Double Liaison*, **109**, 67 (1964).

177. R. B. Seymour, Solubility parameters for polymers and hydrocarbon solvents, *Austin Paint J. Austin Finish. Rev.*, **13**(10), 18 (1968).

178. R. B. Seymour, Structure/solubility relation in polymers, *Rev. Plast. Mod.*, **34**, 851 (1977).

179. R. B. Seymour and J. M. Sosa, Estimation of solubility parameters of nonelectrolytes, *Nature*, **248**, 759 (1974).

180. M. T. Shaw, Studies of polymer-polymer solubility using a two-dimensional solubility parameter approach, *J. App. Polym. Sci.*, **18**, 449 (1974).

181. C. J. Sheehan and A. L. Bisio, Polymer/solvent interaction parameters, *Rubber Chem. Technol.*, **39**, 149 (1966).

182. Shell Chemical Co., *Solvent Notes*, Bulletin IC: 67-64-SN (1967).

183. Shell Chemicals, *Solvent Power*, Tech. Bull. ICS(X)/79/2 (1979)

184. B.-E. Shen and S.-S. Jin, Theory of non-electrolyte solutions (II), *Hua Hsueh Tung Pao*, 119 (1980).

185. K. Shinoda, in collaboration with translator P. Becher, *Principles of Solution and Solubility*, (Undergrad. Chem. Ser., 5), Marcel Dekker, New York; traslated from *Yoeki to yokaido* (1966, revised 1974).

186. A. G. Shvarts, The compatibility of high p9olymers, *Kolloid. Zhur.*, **18**, 755 (1956); *Colloid J. USSR*, **18**, 753 (1956).

187. A. G. Shvarts, Evaluation of rubber-solvent interaction, *Kolloid. Zhur.*, **19**, 376 (1957); *Colloid J. USSR*, **19**, 375 (1957); *Rubber Chem. Technol.*, **31**, 691 (1957, 1958).

188. K. Skaarup, The three dimensional solubility parameter and its use. II. Pigmented systems, *Skand. Tidskr. Färg Lack*, **14**(2), 28; **14**(3), 45 (1968).

189. J. K. Skelly, Alternatives to water in textile processing, *J. Soc. Dye Col.*, **91**, 177 (1975).

190. S. I. Slepakova, B. E. Geller, N. G. Koralnik, A. A. Geller and V. K. Pshedetskaya, Correlation of compatibility and solubility parameters of polymers, *Deposited Doc. VINITI*, 519–578 (1978).

191. P. Sörensen, The solubility parameter concept in the formulation of liquid inks, *J. Oil Col. Chem. Assoc.*, **50**, 226 (1967).

G

192. P. Sörensen, The solubility parameter concept in pigment-resin-solvent systems, *Deut. Farben-Z.*, **24**, 473, (1970); *PKIRA Translation* No. 876 (PR), Print. Abs. 963/1971, Research Association for the Paper Board, Printing and Packaging Industries (1971).
193. P. Sörensen, New aspects on the application of the solubility parameters, *Skand. Tidskr. Färg Lack.*, **20**(2), 9 (1974).
194. E. Sonnich Thomsen, The eergy of mixing of non-polar liquids. Reflections upon the theories of Prigogine and Hildebrand, *Dansk. Kemi*, **47**, 35 (1966).
195. E. Sonnich Thomsen, Solubility parameters, *Farmaceuten*, **29**, 44 (1966).
196. E. Sonnich Thomsen, Present state of the solubility parameter theory, *Mitt. Chem. Gesell. DDR*, **16**, 150 (1969).
197. J. M. Sosa, A. Rubioo, and B. J. Pérez, Dilute solution viscosities of polystyrene solutions, *Abs. Pap. ACS*, **172**, POLY 101 (1976).
198. A. Stawiszynski and J. Zawadzki, Use of the solubility parameter δ for evaluation and selection of organic solvents, *Powloki Ochr.*, **7**, 26 (1979).
199. A. A. Tager and L. K. Kolmakova, Solubility parameter. Methods of its estimation. A relation with a polymer solubility, *Vysokomol. Soedin. Ser. A.*, **22**, 483 (1980).
200. M. Takada, Three dimensional solubility parameters and their applications, *Hyomen*, **9**, 177 (1971).
201. N. W. Taylor, in *Modern Chemistry for the Engineer and Scientist*, G. R. Robertson, Ed., McGraw-Hill, New York, 183 (1957).
202. J. P. Teas, *Predicting Resin Solubilities*, Ashland Chemical Co., Technical Bulletin 1206 (1971).
203. H. Tlusta and J. Zelinger, Miscibility of polymers, *Chem. Listy*, **65**, 1143 (1971).
204. A. Watanabe, Aspects of solubility theory, *Kagaku No Ryoiki*, **29**, 15 (1975).
205. A. Watanabe, Solubility theory, *Kagaku No Ryoiki*, **29**, 167 (1975).
206. T. Yoshida, Solubility parameters. Meaning and application *Shikizai Kyokaishi*, **44**, 186 (1971).

Recent Developments in Polycondensation

H. K. REIMSCHUESSEL

Corporate Research and Development, Allied Corporation, Morristown, N.J. 07960, USA

Reviewing recent literature on synthetic polymer chemistry shows that polycondensation reactions continue to be prominently represented in most areas of both basic and applied research. They constitute principal approaches for the synthesis of many categories of macromolecules. The diversity of the chemical reactions suitable for potential polycondensation processes accounts for the large number of corresponding new polymeric structures that are synthesized every year. Both variety and extent of published research in this area thus preclude an exhaustive discussion and a complete treatment for the present review which attempts to relate some of the recent research on polycondensation with current trends in polymer chemistry and to identify certain highlights. Selection of the topics to be emphasized is, of course, not free of subjectivity and it is quite unlikely that the judgement—if any—exercised with regard to omission or inclusion of the one or the other topic would be duplicated by another author.

For more than a decade polymer research in general has centered to a large extent on what may be considered material aspects and has thus been directed toward the development of macromolecular structures and compositions characterized by predetermined properties and performance characteristics. This has resulted in the development of macromolecular materials that include thermostable polymers, high modulus fibers, liquid crystalline polymers, and multiphase and ordered polymer systems. Many members of these groups have been synthesized by polycondensation reactions, and there has been, indeed, considerable overlap of research in some of these areas. Many polymer structures investigated for high temperature applications were found similar if not identical to those in polymer systems showing either lyotropic or thermotropic behavior. Thus, the approaches for their synthesis were usually similar and resembled in many cases also those used for obtaining multiphase and ordered polymer systems. The principal structures of thermostable polymers are well known. Stability aspects of most of them have been reviewed recently.[1] Interest in thermostable polymers appears to continue, there are reports on new structures such as poly(benzimidoazole imides),[2] poly(benzolene benzimidazoquinazolines),[3] and it is safe to predict that many new polymers will be synthesized in the future by condensation reactions entailing the formation of heterocyclic structures.

Since most of these types of polymers are not very soluble some attempts have been made to improve their solubility by use of N-methyl substitution products,[4] or introduction of side chains and pendent phenyl groups.[5] The interesting class of polyquinolines—polymers with glass transition temperatures in the range of 300°C to 400°C—has been reviewed recently.[5] Some members of this class formed liquid crystalline phases. Formation of anisotropic solutions was also observed with aromatic polymers containing units of benzobisoxazole and benzobisthiazole[6-8]—polymers that had been developed for high temperature applications. These are examples that demonstrate the overlap of the areas of research on high temperature stability and liquid crystal states of polymers.

The considerable interest in the phenomena of liquid crystallinity in polymers is a direct consequence of the commercial development of certain members of the aramide family and is reflected in an impressive amount of published research. Though not all but a large number of liquid crystal polymers have been synthesized by polycondensation processes yielding polyaramides, aromatic polyesters, and aromatic polymers containing heterocyclic structures. The latter class and the aramides including those modified by introduction of pyridine moieties[9] show lyotropic behavior whereas some aromatic polyesters revealed thermotropic behavior which, however, could only be recognized in polyester compositions that melted without decomposition. Such compositions may be obtained by introduction of mesogenic groups by cocondensation,[10] by use of asymmetrically substituted or non-linear comonomers,[14] and by interlinking mesogenic groups with either flexible[11,12] or rigid segments,[13] (spacer groups). The conversion of poly(ethylene terephthalate) into a thermotropic system by melt cocondensation entails the polycondensation of the acetate of *p*-hydroxybenzoic acid and results in what may be regarded a system in which poly(*p*-hydroxybenzoate) segments are interlinked by flexible PET units. Results of selective etching studies,[15] support such a more or less ordered block structure.

Compositions containing well defined spacer groups were obtained by polycondensation of both 4.4′-dihyroxy-α,ω-diphenoxyalkanes and 4.4′-dihydroxy-α,ω-dibenzyloxyalkanes with terephthalic acid, and by polycondensation of 4.4′-dicarboxy-α,ω-diphenoxydecane with hydroquinones.[11] Compositions containing oligomeric spacers were obtained by polycondensation of oligoethylene glycol with either 4.4′-azoxydibenzoyl chloride or 4.4'-azodibenzoyl chloride.[16] Increasing the flexible chain length resulted in narrowing of the mesomorphic temperature region.

The approaches for the controlled interlinking of the mesogenic units parallel some of those employed in processes for the synthesis of ordered polymer compositions particularly those entailing the use of telechelic structures and macromonomers. Well defined polymer structures—copolymers, graft copolymers, and block copolymers—continue to be of active interest in the synthesis of functional polymers. Depending on the properties of a telechelic structure, its use in polycondensation reactions can result in significant modification of polymer properties such as impact strength, solubility, and barrier and hydrophilic characteristics. Regularly sequenced polyamides containing definite numbers of oxyethylene units were synthesized.[17] Reduction of the amide groups in these polyamides should lead to regularly sequenced copolymers with iminoethylene and oxyethylene units. These polymers may be expected to form complexes with metals. Macromonomers containing functional groups offer interesting possibilities for the synthesis of well defined graft copolymers by poly-condensation. Graft copolyamides were synthesized by reaction of a dicarboxyl terminated macromonomer with a series of diamines.[18] The macromonomer was obtained by radical polymerization of methyl meth-acrylate in the presence of thiomalic acid as chain transfer agent. Another approach to ordered polymers was demonstrated by the synthesis of a poly(amide-ester) characterized by a regular amide-amide-ester-ester struc-ture. The telechelic amide bisphenol was obtained by a 'self-regulating' condensation process that is operating by virtue of the considerable greater reactivity of the aromatic amino groups relative to that of phenol groups toward aromatic diacyl chlorides at low temperatures.[19] Many of the polymer systems considered are readily obtained by conventional melt or solution polycondensation processes. Of particular significance for polymer synthesis is of course the interfacial polycondensation, a process used during the last three decades for the synthesis of polycarbonates, polyurethanes, polyesters, polysulfonates, polyamides, and other polymeric structures both in laboratory and industrial operations. This method is a valuable tool for polymer research, it is simple, convenient, and yields readily high molecular weight polymers. It will certainly continue to be a principal method for the synthesis of polymers. Some of the work recently reported was concerned with cationic surfactants in interfacial synthesis of linear aromatic polyesters,[20] colloid and interface effects of additives,[21] aqueous and nonaqueous systems,[22] synthesis of linear polythioesters derived from 4,4'-di(mercaptomethyl) benzophenone,[23] synthesis of nylon copolymers,[24] and specialty polyamides,[25] kinetic principles related so the synthesis of aromatic polyamides,[26] new polyamides derived from aromatic bisamines-chloral derivatives,[27] new polyamides from N,N'-dicarboethoxy pyromellitic di-imide,[28] and synthesis of N-methylated polyamides and copolyamides.[29]

Research concerned with interfacial polycondensation overlaps now considerably with investigations on phase transfer catalysis. Though by no means a new discovery—industrial practice commenced almost 30 years ago[30]—phase transfer catalyzed reactions have recently attracted increasing interest in both synthetic organic and polymer chemistry. Involving a two-phase system, they are generally nucleophilic substitution reactions, usually between an alkyl halide in an organic phase and an anion in an aqueous phase. A reaction impeded by phase separation can often effectively be catalyzed by addition of small amounts of quaternary ammonium or phosphonium salts. This effect is due to the ability of the organic soluble cations to transfer repeatedly anions into the organic phase for reaction. Phase-transfer catalyzed polycondensations yielded aromatic polysulfonates and polyphosphonates, aromatic polyethers, aliphatic and aromatic poly-sulfides using tetrabutyl ammonium chloride, benzyltriethylammonium chloride, cetyltrimethylammonium chloride, benzyltriphenylphosphonium chloride, cetyltributylphosphonium bromide, and a series of crown ethers.[31] Crown ethers were found to be effective in the synthesis of polycarbonates and polythiocarbonates from α,ω-dibromoalkanes and potassium carbonate and CO_2 and CS_2,[32,33] and also for the synthesis of fluorinated polyarylether-sulfones.[34,40] Phase transfer catalysis was also utilized for the synthesis of polyhydroxyethers by reaction of bisphenol-A and epichloro-hydrin[35] and for the synthesis of polycarbonate-polysiloxane block copolymers.[36] Kinetic studies have been carried out on the system bisphenol-A-1.4-dichloro-2-butene and tetrabutylammonium hydrogen sul-fate.[37] Polyether formation from bisphenols and 1.6-dibromohexane was studied to compare the catalytic activity of catalysts.[38] Aromatic polyethers were obtained from bisphenols and bis(halomethyl) benzene using tri-caprylylmethylammonium chloride and crown ethers.[39] Phase transfer agents were also employed in the synthesis of polyphosphonanhydrides,[41], and in the synthesis of organometallic condensation polymers.[42] An interesting development is the synthesis of carbon–carbon chain polymers by phase transfer catalyzed polycondensation.[43,44]

Versatility and convenience of the interfacial methods, however, have apparently not impaired the quest for additional polycondensation processes particularly those that proceed under mild reaction conditions and/or low temperatures. Corresponding research is characterized by two principal approaches. One has been classified as 'direct polycondensation' whereas the other has been denoted 'activated polycondensation'. At present these terms may suffice for identifying the two approaches, but it is already apparent that there is no clear line of demarcation and a consistent categorization according to these terms may be problematic. Generally, 'direct poly-condensation' refers to a system that is characterized by the initial presence of reactants containing free carboxyl groups and free amino- or hydroxyl groups in a reaction medium that contains also reagents that are capable of transforming one type of the reactive functions—usually the carboxyl groups—into a highly activated structure which then in turn readily undergoes reaction with the other reactive function yielding the desired polymeric product. 'Activated polycondensation', on the other hand, refers to polycondensation reactions—usually polyamide formation—that entail the use of active dicarboxylic acid derivatives—usually esters. Their activity result either from the presence of activating groups in the structure of the diacid moiety or from an active leaving group. Recent reports on the direct polycondensation continue to be mainly concerned with polyamide formation and the use of triphenyl phosphite. Corresponding recent reports were related to the synthesis—of aromatic polyamides in the presence of poly(4-vinyl pyridine) or poly(vinylpyrrolidone), to the effects of metal salts and solvents,[45–48,54] to the synthesis of a high molecular weight poly(amide-hydrazide),[49] to a molecular weight increasing effect of the presence of $CaCl_2$,[50] and to the effects of initiation conditions.[51] Aromatic polyesters and copolyesters derived from p-hydroxy benzoic acid were synthesized in pyridine by direct polycondensation with hexachlorocyclotriphosphatrio-zene as a condensing agent.[52] Tributylphosphine was effective in a reductive ring opening polycondensation resulting in a polymer with a thioester function in the main chain.[53] As direct polycondensation may also be considered processes in which reactive intermediates such as acid halides are generated *in situ*. Such a mechanism has been operative in the synthesis of both polyamides and polyesters. Polycondensation of 1.4-bis(trichloro-methyl)benzene with p-phenylenediamine sulfate in SO_3 yielded highly anistropic solutions of high molecular weight sulfonated poly(p-phenylene terephthalamide).[55] Applying thionyl chloride to solutions in organic solvents of hydroxy acids or systems consisting of diacids and either diamines or diols resulted in the formation of the corresponding polymers at low temperatures.[56,57]

One type of activated polycondensation for polyamide formation was indicated when diesters were employed that contained heteroatoms in the α or β position.[58,59] More recent reports emphasize that the rate of polycondensation entailing these esters was enhanced by the presence of polymers such as polysaccharides, poly(vinylpyrrolidone), poly(vinyl-alcohol), poly(vinylpyridine).[60,61] Since the rate enhancement became more pronounced with increasing polymer molecular weight a matrix effect has been postulated.[62] A matrix effect was also observed in the poly-condensation of diamines with active esters containing nucleic acid bases.[63]

Recent reports on activated polycondensation entailing active leaving groups dealt with the synthesis of polyamides by reaction of diamines with 2-benzothiazolyl diesters,[64] diacyl derivatives of 3-hydroxy-1,2-benzisoxazole,[65] and 0,0'-diacyl derivatives of N-hydroxy compounds.[66] Aliphatic polythioamides could be obtained in solution without any catalyst by polycondensation of bis(dithioesters) with diamines.[67]

Skimming the published literature concerned with polycondensation one finds that, apart from the topics referred to thus far, there are many recent reports that are concerned with polyesters particularly with aspects related to commercially important systems and operations. There are, of course, also reports on new synthetic procedures and new polyester structures. They include those dealing with the synthesis of polyesters of hindered phenols[68,69] and of a new structure, poly[p-benzoyl(1,1-dimethyl-2-ethyl)oxy] that is characterized by an unusual combination of its main transition temperatures resulting in all probability from an intramolecular interaction between the gem-dimethyl and ester carbonyl moieties.[70]

Reported theoretical treatments were concerned with application of 'correlation analysis to non-equilibrium polycondensation processes' pertaining to the synthesis of polyesters, polyethers, and polyamides by solution and interfacial processes,[71] with an application of the Macosko-Miller approach for developing an expression for the polydispersity index in linear step polymerization,[72] with a kinetic approach for the size distributions in multifunctional polycondensation systems, and for the onset of gelatin,[73] with the derivation of a general expression for the average molecular weights in polycondensation that involve elimination of a condensation by-product,[74] and with re-examinations of general polyesterification kinetics.[75,76] Kinetic and mechanistic aspects of polyesterification have been subject of a number of recent investigations. Most were related to the commercially very important terephthalic acid/ethylene glycol system. Since the overall process entailed in this system is actually a rather complex one, a number of process models have been considered for its description. Approaches based upon reaction-diffusion models[77,78] are quite sound but of somewhat limited value when over simplified rate expressions are used. Mechanistic and kinetic schemes for the direct esterification process were presented recently together with thermodynamic and kinetic parameters for the principal equilibrium reactions entailed in this process. These equilibria were identified as esterification-hydrolysis, polycondensation-glycolysis, and transesterification-acidolysis. The equilibrium constants and rate constants were derived from low molecular weight model systems.[79] This information was then assimilated in subsequent work concerned with reactor modelling.[80]

The solid state polycondensation of poly(alkylene terephthalate) is a commercially practiced operation and has thus received considerable attention. Recent reports were concerned with the influence of prepolymer properties such as molecular weight, particle size, end group concentration, ratio of OH/COOH, and comonomer concentration on rates of polycondensation and ester interchange, and final polymer composition.[81-84] The continued interest in the performance of polyesterification catalysts was reflected in reports on studies on the influence of the number of hydroxyethoxy ligands on antimony on the degree of polymerization,[85] the effect of phosphoric acid on the activity of antimony triacetate,[86] the influence of metal catalysts on formation of ether structures,[87] and on the mechanism of catalysis.[88]

References

1. C. Arnold, Jr., *J. Polym. Sci. Rev.*, **14**, 265 (1979).
2. V. V. Korshak, D. M. Mognonov, A. I. Batotsyrenova, V. V. Nikiteev, K. E. Batlaev and A. A. Izyneev, *Vysokomol. Soedin.*, *Ser. A*, **80**, 22(6), 1209.
3. K. Niume, F. Toda, K. Uno, M. Hasegawa and Y. Iwakura, *J. Polym. Sci. Chem. Ed.*, **19**, 1745 (1981).
4. T. D. Greenwood, R. A. Kahley, J. F. Wolfe, A. St. Clair and N. J. Johnson, *J. Polym. Sci. Chem. Ed.*, **18**, 1047 (1980).
5. J. K. Stille, *Macromolecules*, **14**, 870 (1981).
6. J. F. Wolfe and F. E. Arnold, *Macromolecules*, **14**, 909 (1981).
7. J. F. Wolfe, B. H. Loo and F. E. Arnold, *Macromolecules*, **14**, 915 (1981).
8. E. W. Choe and S. N. Kim, *Macromolecules*, **14**, 920 (1981).
9. N. Ogata, K. Sanui and T. Koyama, *J. Polm. Sci. Chem. Ed.*, **19**, 151 (1981).
10. W. J. Jackson, Jr., *Brit. Polym. H.*, **12**, 154 (1980).
11. J.-I. Jin, S. Antoun, C. Ober and R. W. Lenz, *Brit. Polym. J.*, **12**, 132 (1980).
12. S. Antoun, R. W. Lenz and J.-I. Jin, *J. Polym. Sci., Chem. Ed.*, **19**, 1901 (1981).
13. R. W. Lenz and J.-I. Jin, *Macromolecules*, **14**, 1405 (1981).
14. B. P. Griffin and M. K. Cox, *Brit. Polym. J.*, **12**, 147 (1980).
15. E. Joseph, G. L. Wilkes and D. G. Baird, *Polym. Prepr. ACS, Div. Polym. Chem.*, **22**, 2, 359 (1981).
16. K. Iimura, N. Koide, R. Ohta and M. Takeda, *Makromol. Chem.*, **182**, 2563 (1981).
17. H. Sato, S. Iwabuchi, V. Bohmer and W. Kern, *Makromol. Chem.*, **182**, 755 (1981).
18. Y. Yamashita, Y. Chujo, H. Kobayashi and Y. Kawakami, *Polymer Bulletin*, **5**, 361 (1981).
19. Y. Imai, S. Abe, T. Takahashi and M. Ueda, *J. Polym. Sci., Chem. Ed.*, **20**, 683 (1982).
20. E. Z. Casassa, D.-Y. Chao and M. Henson, *J. Macromol. Sci. Chem.*, **A15**(5), 799 (1981).
21. E. Z. Casassa, *J. Macromol. Sci., Chem.*, **A15**(5), 797 (1981).
22. N. Ogata, K. Sanui, T. Onozaki and S. Imanishi, *Org. Coat. Plast. Chem.*, **42**, 390 (1980); *J. Macromol. Sci. Chem.*, **A15**(5), 1059 (1981).
23. W. Podkoscielny and A. Kultys, *J. Appl. Polym. Sci.*, **26**, 1143 (1981); *J. Polym. Sci. Chem. Ed.*, **19**, 2167 (1981).
24. T. Kiyotsukuri and K. Jamshidi, *Kobunshi Ronbunshu*, **37**(7), 471 (1980).
25. W. Deitz, S. Grossman and O. Vogl, *J. Macromol. Sci. Chem.*, **A15**(5), 1027 (1981).
26. N. I. Kuz'min, B. Zhizdyuk and A. S. Chegolyz, *Khim. Volokna*, **3**, 4 (1980).
27. Z. K. Brzozowski, Z. Przybysz and A. Grabowska-Rostek, *J. Macromol. Sci. Chem.*, **A16**(7), 1207 (1981).
28. K. Taguchi, *J. Polym. Sci. Lett.*, **18**(8), 525 (1980).
29. S. J. Huang and J. Kozakiewicz, *J. Macromol. Sci. Chem.*, **A15**(5), 821 (1981).

30. P. W. Morgan, *J. Macromol. Sci. Chem.*, **A15**(5), 683 (1981).
31. Y. Imai, *J. Macromol. Sci. Chem.*, **A15**(5), 833 (1981).
32. K. Soga, S. Hosoda and S. Ikeda, *J. Polym. Sci. Chem. Ed.*, **17**, 517 (1979).
33. K. Soga, Y. Toshida, I. Hattori, K. Nagata and S. Ikeda, *Makromol. Chem.*, **181**, 979 (1980).
34. R. Kellman, D. Gerbi, R. F. Williams and J. L. Morgan, *Polymer Prepr. ACS Div. Polymer Chem.*, **21**(2), 164 (1980).
35. A. K. Banthia, D. Lunsford, D. C. Webster and J. E. McGrath; *J. Macromol. Sci. Chem.*, **A15**(5), 943 (1981).
36. J. S. Riffle, R. G. Freellin, A. K. Banthia and J. E. McGrath, *J. Macromol. Sci. Chem.*, **A15**(5), 967 (1981).
37. T. D. N'Guyen and S. Boileau, *Polym. Prepr. ACS Div. Polymer Chem.*, **23**(1), 154 (1981).
38. J.-I. Jin and Jin-H. Chang, *Polym. Prepr. ACS Div. Polymer Chem.*, **23**(1), 156 (1981).
39. G. G. Cameron and K. S. Law, *Makromol. Chem. Rapid Commun.*, **3**, 99 (1982).
40. R. Kellman, D. J. Gerbi, J. C. Williams, R. F. Williams and R. B. Bates, *Polym. Prepr. ACS Div. Polym. Chem.*, **23**(1), 174 (1982).
41. C. E. Carraher, Jr., R. J. Linville and H. S. Blaxall, *Polym. Prepr. ACS Div. Polym. Chem.*, **23**(1), 160 (1981).
42. C. E. Carraher, Jr. and M. D. Naas, *Polym. Prepr. ACS Div. Polym. Chem.*, **23**(1), 158 (1981).
43. Y. Imai, A. Kameyama, T.-Q. Nguyen and M. Ueda, *J. Polym. Sci. Polym. Chem. Ed.*, **19**, 2997 (1981).
44. U. Imai and M. Ueda, *Polym. Prepr. ACS Div. Polym Chem.*, **23**(1), 164 (1982).
45. F. Higashi, M. Goto, Y. Nakano and H. Kakinoki, *J. Polym. Sci. Chem. Ed.*, **18**, 851 (1980).
46. F. Higashi, M. Goto and H. Kakinoki, *J. Polym. Sci. Chem. Ed.*, **18**, 1711 (1980).
47. F. Higashi, K. Sano and H. Kakinoki, *J. Polym. Sci. Chem. Ed.*, **18**, 1841 (1980).
48. F. Higashi and Y. Taguchi, *J. Polym. Sci. Chem. Ed.*, **18**, 2875 (1980).
49. F. Higashi and M. Ishikawa, *J. Polym. Sci. Chem. Ed.*, **18**, 2905 (1980).
50. F. Higashi, Y. Aoki and Y. Taguchi, *Makromol. Chem., Rapid Commun.*, **2**, 329 (1981).
51. F. Higashi, Y. Taguchi, N. Kokubo and H. Ohta, *J. Polym. Sci. Chem. Ed.*, **19**, 2745 (1981).
52. F. Higashi, K. Kubota, M. Sekizuka and M. Higashi, *J. Polym. Sci. Chem. Ed.*, **19**, 2681 (1981).
53. Y. Nambu, T. Endo and M. Okawara, *Makromol. Chem., Rapid Commun.*, **1**, 603 (1980).
54. G. Wu, H. Tanaka, S. Sanui and N. Ogata, *J. Polym. Sci. Lett. Ed.*, **19**, 343 (1981).
55. F. M. Silver, *J. Polymer Sci., Chem. Ed.*, **18**, 1787 (1980).
56. Y. Amai, S. Aoyama, T.-Q. Nguyen and M. Ueda, *Makromol. Chem. Rapid Commun.*, **1**, 655 (1980).
57. H.-G. Elias and R. J. Warner, *Makromol. Chem.*, **182**, 681 (1981).
58. N. Ogata, *J. Macromol. Sci. Chem.*, **A13**, 477 (1979).
59. N. Ogata, K. Sanui, T. Ohtake and H. Nakamura, *Polym. J.*, **11**, 827 (1979).
60. N. Ogata, K. Sanui, H. Nakamura and M. Kuwahara, *J. Polym. Sci., Chem. Ed.*, **18**, 939 (1980).
61. N. Ogata, K. Sanui, H. Nakamura and H. Kishi, *J. Polym. Sci., Chem. Ed.*, **18**, 933 (1980).
62. N. Ogata, K. Sanui, H. Tanaka, H. Matsuo and F. Iwaki, *J. Polym. Sci., Chem. Ed.*, **19**, 2609 (1981).
63. M. Hattori, H. Nakagawa and M. Konoshita, *Makromol. Chem.*, **181**, 2325 (1980).
64. M. Ueda, A. Sato and Y. Imai, *J. Polym. Sci., Chem. Ed.*, **17**, 2013 (1979).
65. M. Ueda, T. Harada, S. Aoyama and Y. Imai, *J. Polym. Sci., Chem. Ed.*, **19**, 1061 (1981).
66. C. Lu and P. Liu, *J. Polym. Sci., Chem. Ed.*, **19**, 2091 (1981).
67. J. C. Gressier and G. Levesque, *Eur. Polym. J.*, **11**, 1101, (1980).
68. R. W. Stackman, *Ind. Eng. Chem. Prod. Res. Dev.*, **20**, 336 (1981).
69. W. Deits and O. Vogl, *J. Macromol. Sci. Chem.*, **A16**(6), 1145 (1981).
70. H. K. Reimschussel and B. T. DeBona, *Macromolecules*, **13**, 1582 (1980).

71. A. K. Mikitaev, *Acta Polymerica*, **32**, 453 (1981).
72. E. Ozizmir and G. Odian, *J. Polym. Sci. Chem. Ed.*, **18**, 2281 (1980).
73. J. W. Stafford, *J. Polym. Sci., Chem. Ed.*, **19**, 3219 (1981).
74. D. Durand and C.-M. Bruneau, *Brit. Polym. J.*, **13**, 33 (1981).
75. S.-A. Chen and J.-C. Hsiao, *J. Polym. Sci., Chem. Ed.*, **19**, 3123 (1981).
76. A. Fradet and E. Marechal, *J. Polym. Sci., Chem. Ed.*, **19**, 2905 (1981).
77. G. Rafler, E. Bonatz, G. Reinisch, H. Gajewski and K. Zacharias, *Acta Polymerica*, **31**, 578 (1980).
78. H. Gajewski, K. Zacharias, G. Rafler and E. Bonatz, *Acta Polymerica*, **32**, 57 (1981).
79. H. K. Reimschussel, *Ind. Eng. Chem. Prod. Res. Dev.*, **19**, 117 (1980).
80. K. Ravindranath and R. A. Maselkar, *J. Appl. Polym. Sci.*, **26**, 3179 (1981; **27**, 471 (1982).
81. L. H. Buxbaum, *J. Appl. Polym. Sci., Sym.*, **35**, 59 (1979).
82. E. Schaaf, H. Zimmermann, W. Dietzel and P. Lohmann, *Acta Polymerica*, **32**, 250 (1981).
83. M. Droscher, *Makromol Chem.*, **181**, 789 (1980); *J. Appl. Polym. Sci., Sym.*, **36**, 217 (1981); *Ind. Eng. Chem. Prod. Res. Dev.*, **21**, 126 (1982).
84. M. Droscher and F. G. Schmidt, *Polym. Bull.*, **4**, 261 (1981).
85. S. B. Maerov, *J. Polym. Sci. Chem. Ed.*, **17**, 4033 (1979).
86. H. Kamatani, S. Konagaya and Y. Nakamura, *Polymer J.*, **12**, 125 (1980).
87. V. Hornof, *J. Macromol. Sci. Chem.*, A**15**(3), 503 (1981).
88. F. Pilati, P. Manaresi, B. Fortunato, A. Munari and V. Passalacqua, *Polymer*, **22**, 799 (1981).

Copolymerization

A. RUDIN

Guelph-Waterloo Centre for Graduate Work in Chemistry, University of Waterloo, Waterloo, Ontario, Canada N2L 3G1.

This survey includes the copolymerization topics which have attracted the most attention, in terms of numbers of recent publications. As a result of this selection rule all the references quoted here are to free radical systems. The topics covered include the determination of reactivity ratios, copolymerization kinetics, complexed copolymerization and reactions of vinyl with divinyl monomers. Not all the articles on a given topic have been cited here.

The so-called simple copolymerization scheme was first enunciated in 1944[1,2] but many of the reactivity ratio data in the literature are still compromised to some extent by the use of statistically incorrect estimation methods.[3,4] Advances are still being made in methods for calculating reactivity ratios from experimental data. Linear least squares methods are generally invalid because the error variance in a linearized comonomer feed-copolymer composition equation is not distributed normally. Computer-assisted non-linear least squares fitting to experimental points provides statistically sound estimates of reactivity ratios. Proper experimental design can, however, allow the use of graphical linear least squares methods of analysis.[5] Several new linear regression methods have been published recently[6,7] along with an improved method for estimating confidence limits for a particular linear technique.[8]

Most non-linear least squares methods are restricted to the differential form of the simple copolymer equation, which limits the useable data to those obtained at less than about 20 per cent conversion. Recent work has produced algorithms for estimation of reactivity ratios from integrated forms of this kinetic model[9,10]

Computer modelling and control of copolymerization reactions are severely hindered by lack of suitable kinetic schemes for the process. To derive an expression for the instantaneous rate of copolymerization of two comonomers we assume stationary state concentrations for each of the two possible radicals and rate constants which are independent of the length or concentration of the macromolecular species. (The latter assumption is not strictly true for termination rate constants,[11] but the error is large only under auto-acceleration conditions.) An overall termination rate

k_{to}, can be defined as a function of copolymer (and hence of the corresponding feed) composition. The rate of copolymerization, R_p of two monomers M_1 and M_2 can then be expressed quite generally[12] as:

$$R_p = \frac{d([M_1] + [M_2])}{dt} = \frac{R_i^{\frac{1}{2}}(r_1[M_1]^2 + 2[M_1][M_2] + r_2[M_2]^2)}{k_{to}^{\frac{1}{2}} \dfrac{r_1[M_1]}{k_{p11}} + \dfrac{r_2[M_2]}{k_{p22}}} \quad (1)$$

where R_i is an overall initiation rate and k_{p11} and k_{p22} are the respective rate constants for homopolymerization of M_1 and M_2. Application of Eq. (1) is limited by the scarcity of k_p values and especially by lack of a reliable method for estimating k_{to} from homopolymerization rate data.

The one-parameter, 'chemically-controlled' model of Melville and co-workers[13] defines k_{to} in terms of a geometric mean of the respective k_t's for homopolymerization. The deficiencies of this scheme have been well documented.[14] Other empirical formulations for k_{to}[12,15] have had greater success in accounting for the limited available experimental data, but they lack any theoretical basis. The kinetic scheme of Russo and Munari[16,17] employs two parameters which are evaluated by least squares fits to rate data from reactions over a range of feed compositions. This model has been reported to describe copolymerizations rates for monomer pairs like styrene and methyl methacrylate[14] but not to be suitable for systems like styrene-acrylonitrile which differ greatly in polarity.[18]

To follow the course of an isothermal batch copolymerization one must integrate Eq. (1). This can be accomplished numerically if R_i does not depend on the comonomer feed composition (i.e., on reaction time) and if k_{to} could be estimated with allowance for the effects of polymer concentration. The latter consideration includes the gel effect, which is common, of course, to homopolymerizations. It can be seen that the modelling problems here are formidable.

Experimental information about copolymerization reaction behavior can be summarized usefully to some extent without a detailed reaction scheme. Sebastian and Biesenberger[18] used a dimensional analysis approach to handle the data in a differential scanning calorimeter study of the isothermal copolymerization of styrene and acrylonitrile. Auto-acceleration was found to increase in severity with increased acrylonitrile content in the feed and with decreased reaction temperature. Reaction rates and activation energies in bulk reactions were observed to vary with initiator type, suggesting that the assumption of a composition independent rate of initiation may be invalid in this system.

Tirrell and Gromley[19] considered composition control of a batch copolymerization of styrene and acrylonitrile by temperature manipulations.

The differential copolymerization rate equation was integrated using the chemical control model[13] for k_{to} which is reported not to be applicable to this system.[18] However, step-wise calculated temperature changes were found to produce the anticipated variations in copolymer composition, at least in styrene-rich, homogeneous phase copolymerizations in solution.

A set of equations have been developed to predict the time evolution of copolymer composition and comonomer sequence distribution in emulsion copolymerizations.[20] The solutions require rate coefficients for radical entry and exit from the monomer-swollen polymer particles, homo-termination and cross-termination, cross transfer and cross propagation, for each monomer. All these values are not available. They can perhaps be estimated from appropriate bulk or solution copolymerizations and emulsion homopolymerizations but the effort involved would be ardous.

Lin and co-workers[21] have reported a model which predicts conversion and product molecular weight in azeotropic emulsion copolymerizations. The system studied was again styrene-acrylonitrile and an invariant termination rate constant could be assumed because the feed and copolymer compositions were constant during the batch reaction. Auto acceleration was noticed to increase with lower initiator or emulsifier concentrations. This is a useful conclusion because polymer production rates are enhanced in the gel effect region.

Vinyl monomers with electron-rich double bonds produce one-to-one copolymers with electron-poor monomers in reaction systems which also contain certain Lewis acids. The latter are halides or alkyl halides of non-transition metals, including $AlCl_3$, $ZnCl_2$, BF_3, $Al(CH_2CH_3)Cl_2$, alkyl boron halides and other compounds. The electron acceptor monomer generally has a cyano or carbonyl group conjugated to a vinyl double bond (e.g., acrylonitrile, methacrylic esters, vinyl ketones). The wide variety of donor monomers includes various olefins, styrene, conjugated dienes, vinyl halides and allyl compounds.[22] Novel copolymers can be made in such processes and a few have been commercialized. Applications of complexed copolymerizations have been hindered, however, by the high concentrations of metal ions which are required in these reactions. Mole ratios of Lewis acid to acceptor monomers of 0.3 to 1.0 or more are the general rule. By contrast, a recent publication[23] reports that completely alternating copolymerizations of styrene and methyl methacrylate can be achieved at $-20°C$ in the presence of as little as 0.01 mole BCl_3 per mole of methacrylate. This is reminiscent of earlier work on propylene-alkyl acrylate copolymerizations in which equimolar copolymers were produced in the presence of a free radical initiator and excess propylene with a BF_3/acrylate ester mole ratio as low as 1/7.[24] In this case the copolymer yields and polymerization rates were reduced at lower BF_3 concentrations due to the slower decomposition rate of the initiator (AIBN) and the lower concentration of the BF_3 complex of the acrylate monomer. The effects of Lewis acid concentrations on yields and rates in the styrene-methyl methacrylate case were not given.[23]

The mechanism of complexed alternating copolymerizations is still unclear,[25] although the first reactions were reported almost twenty years ago.[26] Recent studies[27,28] indicate that alternation may be due to enhanced cross-propagation rates which result from the much greater reactivity of complexed acceptor monomers.

Acetylenic donor and acceptor monomers have been copolymerized using ethyl-aluminum chloride as complexing agent. These reactions, which were accelerated by transition metal catalysts like WCl_6, produced low molecular weight colored copolymers.[29]

The free radical copolymerization of vinyl and divinyl monomers is important in the manufacture of ion-exchange resins, chromatographic packings, cross-linked latex polymers and other products and there have been a relatively large number of publications on this topic recently.

The reactivity ratios for reactions of styrene with divinyl benzene and ethylene glycol dimethacrylate have been reassessed.[30] The values reported refer to the copolymerization of the first double bonds of the divinyl monomer and are valid for about the first half of a suspension copolymerization with divinyl benzene when the styrene mole fraction in the feed is high. The subsequent stage of the reaction is characterized by auto acceleration and an apparent increase in the reactivity ratio of styrene.[31] This phenomenon may result from neglect of intramolecular cyclization in handling the experimental data.[32]

The pendant double bonds in solution copolymerizations of vinyl and divinyl monomers are reported to be capable of reaction up to very late stages of conversion but the probability of network inhomogeneities increases with concentration of the divinyl monomer.[33,34] The reports cited refer to batch copolymerizations. Different behavior in the pregel region is expected between batch or continuous plug flow reactors and continuous flow stirred tank reactors.[35]

REFERENCES

1. F. R. Mayo and F. M. Lewis, *J. Am. Chem. Soc.*, **66**, 1594 (1944).
2. T. Alfrey Jr. and G. Goldfinger, *J. Chem. Phys.*, **12**, 322 (1944).
3. D. W. Behnken, *J. Polym. Sci.*, **A2**, 645 (1964).
4. P. W. Tidwell and G. A. Mortimer, *J. Macromol. Sci.-Revs.*, **C4**, 281 (1970).
5. R. C. McFarlane, P. M. Reilly and K. F. O'Driscoll, *J. Polym. Sci. Chem. Ed.*, **18**, 251 (1980).
6. D. G. Watts, H. N. Linssen and J. Schrivjer, *J. Polym. Sci. Polym. Chem. Ed.*, **18**, 1285 (1980).

7. J.-F. Kuo and C.-Y. Chen, *J. Appl. Polym. Sci.*, **26**, 1117 (1981).
8. T. Kelen, F. Tudos and B. Turcsanyi, *Polym. Bull.*, **2**, 71 (1981).
9. S. M. Shawki and A. E. Hamelec, *J. Appl. Polym. Sci.*, **23**, 3155 (1979).
10. H. Patino-Leal, P. M. Reilly and K. F. O'Driscoll, *J. Polym. Sci. Polym. Lett.*, **18**, 219 (1980).
11. H. K. Mahabadi and K. F. O'Driscoll, *Macromolecules*, **10**, 55 (1977).
12. J. N. Atherton and A. M. North, *Trans. Far. Soc.*, **58**, 2040 (1962).
13. H. W. Melville, B. Noble and W. F. Watson, *J. Polym. Sci.*, **2**, 229 (1947).
14. P. Wittmer, *Makromol. Chem. Suppl.*, **3**, 129 (1979).
15. S. S. M. Chiang and A. Rudin, *J. Macromol. Sci. Chem.*, **A9**, 237 (1975).
16. S. Russo and S. Munari, *J. Macromol. Sci. Chem.*, **A2**, 1321 (1968).
17. G. Bonta, B. Gallo and S. Russo, *J. Chem. Soc. Farad. Trans.*, **1**, 69, 328 (1973).
18. D. H. Sebastian and J. A. Biesenberger, *J. Macromol. Sci. Chem.*, **A15**, 553 (1981).
19. M. Tirrell and K. Gromley, *Chem. Eng. Sci.*, **36**, 367 (1981).
20. M. J. Ballard, D. H. Napper and R. G. Gilbert, *J. Polym. Sci. Polym. Chem. Ed.*, **19**, 939 (1981).
21. C. C. Lin, H. C. Ku and W. Y. Chiu, *J. Appl. Polym. Sci.*, **26**, 1327 (1981).
22. M. Hirooka, H. Yobiuchi, S. Kawasumi and K. Nakaguchi, *J. Polym. Sci. A-1*, **11**, 128 (1973).
23. H. Hirai and K. Takeuchi, *Makromol. Chem. Rapid Commun.*, **1**, 541 (1980).
24. A. L. Logothetis and J. M. McKenna, *J. Polym. Sci. Polym. Chem. Ed.*, **16**, 2797 (1978).
25. T. Saegusa, *Makromol. Chem. Suppl.*, **3**, 157 (1979).
26. U.S. Patent 3,326,870, June 14, 1963, Sumitomo Chemical Co.
27. H. Bamford and P. J. Malley, *J. Polym. Sci. Polym. Lett.*, **19**, 239 (1981).
28. M. Hirooka, *Pure Appl. Chem.*, **53**, 681 (1981).
29. J. Furakawa, E. Kobayashi and T. Wakui, *Polymer J.*, **12**, 17 (1980).
30. C. D. Frick, A. Rudin and R. H. Wiley, *J. Macromol. Sci.-Chem.*, **A16**, 1275 (1981).
31. P. W. Kwant, *J. Polym. Sci. Polym. Chem. Ed.*, **17**, 1331 (1979).
32. L. Mrkvickova and P. Kratochvil, *J. Polym. Sci. Polym. Phys. Ed.*, **19**, 1675 (1981).
34. G. D. Patterson, J. R. Stevens, J. P. Jarry and C. P. Lindsey, *Macromolecules*, **14**, 86 (1981).
35. C. Ver Strate, C. Cozewith and W. W. Graessley, *J. Appl. Polym. Sci.*, **25**, 59 (1980).

Ring Opening Polymerization

S. PENCZEK

Polish Academy of Sciences, Centre of Molecular and Macromolecular Studies, 90-362 Łódz Boczna 5, Poland

Ring-opening polymerization is one of the important methods of synthesis of macromolecules and has been used for the production of polymers on a commercial scale. However in this application it has not reached the significance for synthesis of macromolecules that have free radical and ionic polymerization of unsaturated monomers.

On the other hand, the ring-opening polymerization is unsurpassed in its versatility in preparing various and unusual polymeric structures. Indeed, almost any mer can be bent into a ring, which, upon polymerization, would give the parent polymer

Usually the ring-opening polymerization means that heterocyclic monomers are involved; more recently, however, it has been shown that the unsaturated hydrocarbon rings do also polymerize. Thus, we may visualize the ring-opening polymerizations by the following scheme:

where X is a double bond or heteroatom, and the structures on the right-hand side can belong to the linear and/or cyclic macromolecules. The arrows indicate that the involved monomers polymerize reversibly. This is because only the three- and four-membered rings are highly strained (ca. $20 \, \text{kcal mole}^{-1}$) and polymerize to completion whereas larger rings polymerize with pronounced reversibility.

SOME HISTORICAL INFORMATION

One of the first important achievements of the ring-opening polymerization came at the turn of the century (1906), when Leuchs prepared poly-α-aminoacids by the ring-opening polymerization of N-carboxyanhydrides of α-aminoacids (NCA), called also, after him, the Leuchs anhydrides:

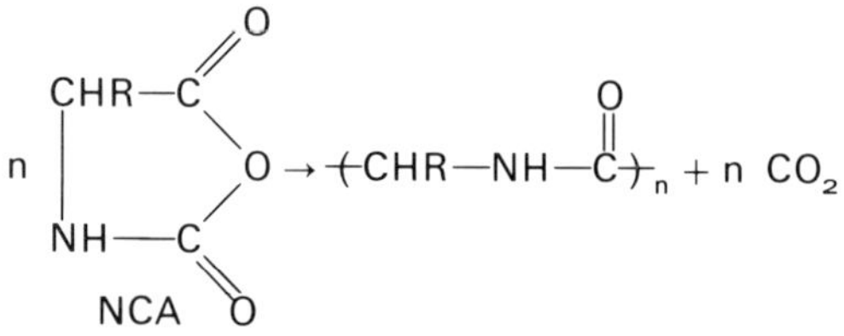

$$n \; \underset{\text{NCA}}{\begin{array}{c} CHR-C \\ | \\ NH-C \end{array}} \Big\rangle O \rightarrow (CHR-NH-\overset{O}{\overset{\|}{C}})_n + n \; CO_2$$

Thus, linear, high-molecular weight polypeptides were prepared before the macromolecular state of matter became accepted.

The NCA polymerization is even today a method of choice in preparing homopolymers of α-aminoacids.

The 'inorganic rubber'—polydichlorophosphazene:

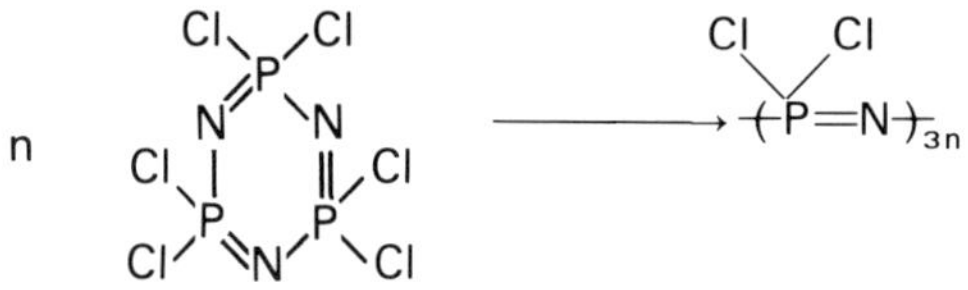

$$n \; \underset{\text{}}{\begin{array}{c} Cl \diagup P \diagdown Cl \\ N \quad N \\ Cl-P \quad P-Cl \\ | \quad \| \\ Cl \diagup N \diagdown Cl \end{array}} \longrightarrow (P=N)_{3n}^{Cl \quad Cl}$$

was first prepared even earlier, at the end of the nineteenth century and its rubbery features were then recognized. It is only however in 1960, that the proper modification converted this inherently unstable polymer into the modern, high performance elastomer (H. R. Allock, Pennsylvania).

The real milestone in science has also been erected with the aid of the ring-opening polymerization: Hermann Staudinger has established the macro-molecular features of the products of polymerization by using poly(ethylene oxide) as one of his major models:

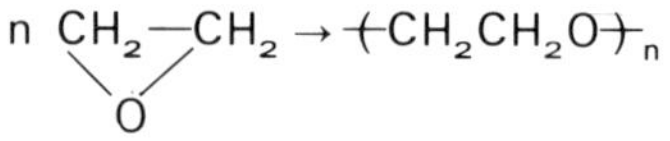

$$n \; CH_2 \diagdown_O \diagup CH_2 \rightarrow (CH_2CH_2O)_n$$

SOME TECHNICAL POLYMERS BASED ON THE RING-OPENING POLYMERIZATION

Today there is a number of polymers that are obtained on the technical scale by the ring-opening polymerization. Below the most important examples are given:

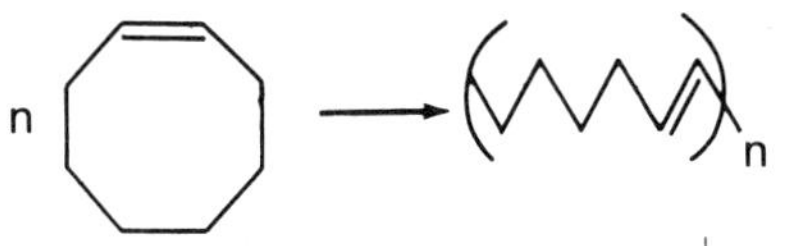

polypentenamer, rubber, several companies (also polyoctenamer, Hüls, FRG; polynorbornene)

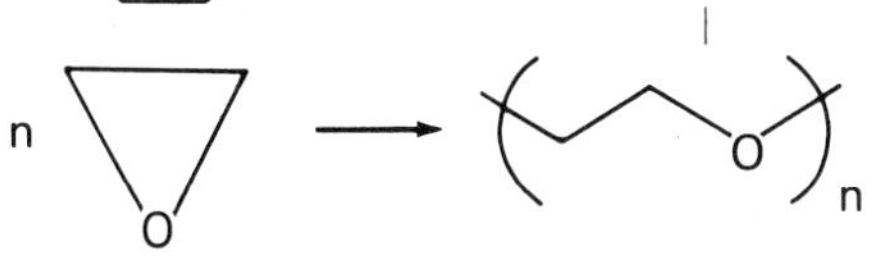

polyethyleneoxide, high mol.-wt., water soluble polymer, Union Carbide, USA also low mol.-wt. homo- and co-polymers e.g. with propylene oxide: surface active compounds, soft components of PUR, many producers

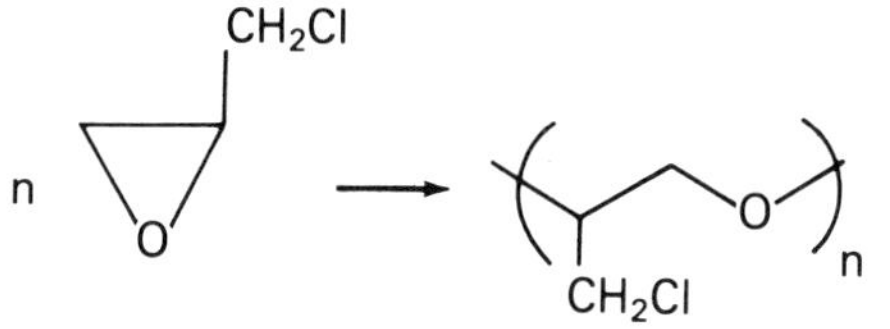

poly-α-epichlorohydrin elastomer, Hydrin®, Hercules Powder, Goodrich, USA

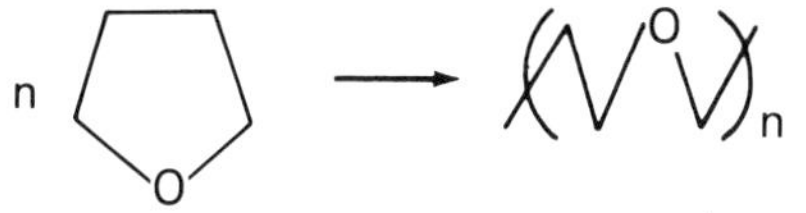

polytetrahydrofuran —OH ended oligomer; used for PUR and other elastoplastics, several producers (DuPont, USA; BASF, FRG)

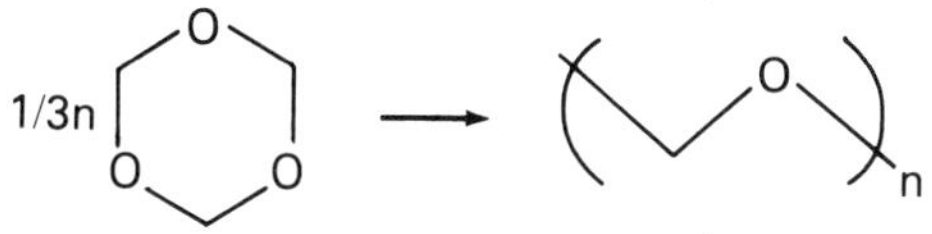

polyoxymethylene, crystalline thermoplastic (copolymer with ethylene oxide or 1,3-dioxolane) Hostaform®, Hoechst, FRG; Celcon®, Celanease, USA; Tarnoform®, ZPA Azoty, Poland

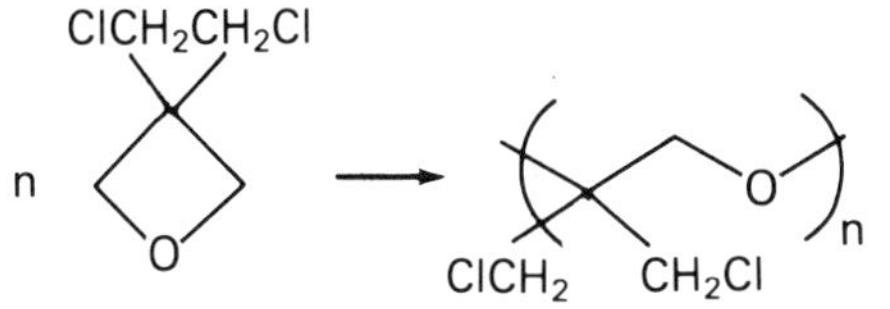

chlorinated polyether, crystalline thermoplastic, corrosion resistant, Penton®, Hercules Powder, USA; Pentaplast®, USSR

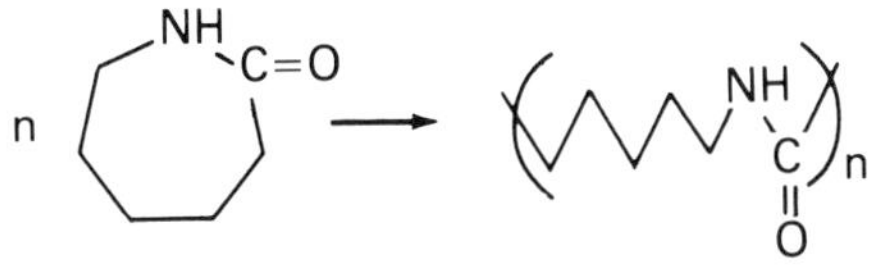

Polyamid 6, Nylon 6, fiber thermoplastic material; many producers all over the world (also polyamid 12)

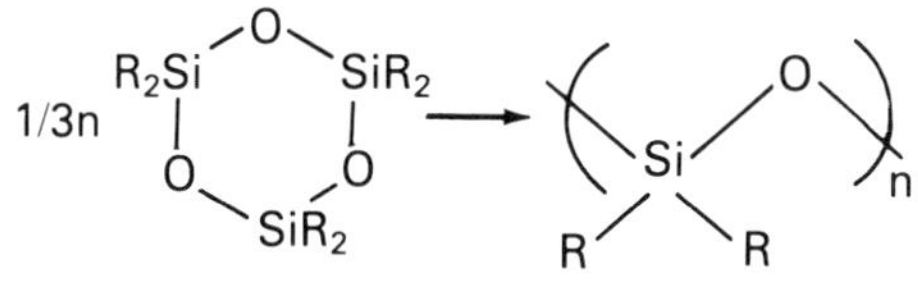

polydialkyl(aryl)siloxanes heat resistant fluids, rubbers (alkyl) or plastics (aryl); many producers

Studies of the ring-opening polymerization are still of full activity in both mechanistic and synthetic aspects. The new developments are being systematically summarized at the specialized international symposia. The first one was held in 1975 in Poland, organized by the present writer's group together with P. Rempp and E. Frantar, CRM, Strasburg, second, organized by T. Saegusa and E. J. Goethals within a framework of the ACS Meeting (New Orleans, USA 1977), and the third one as the IUPAC Micro-symposium, organized by J. Sebenda in Karlovy Vary (Czechoslovakia).

Proceedings of these three meetings encompass all of the major achievements of the ring-opening polymerization of the last years. Although the choice of examples and their order as given below is highly personal but the author believes that it fairly well describes both the present state of research in ring-opening polymerization and the major achievements of the last few years.

PRESENT UNDERSTANDING AND RECENT PROGRESS IN THE RING-OPENING POLYMERIZATION: THERMODYNAMICS, KINETICS AND MECHANISMS

Thermodynamics

The major driving force of the ring-opening polymerization is the ring strain. The corresponding values, expressing ring strain (in kcal/mol), and obtained by direct measurement of the heat of polymerization or by other methods, are given below together with the formulae of the monomer.

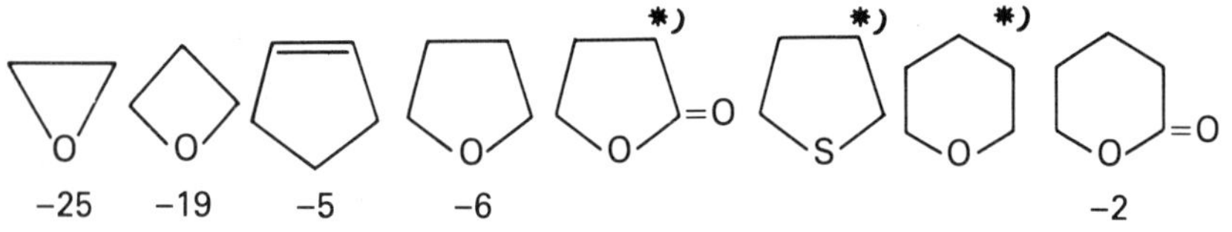

*These monomers could not be polymerized.

Thus, small structural differences decide in the group of the 5- and 6-membered monomers about their polymerizability. Larger rings, mostly because of the transannular strain, due to the repulsion of eclipsed hydrogen atoms, are becoming strained again.

Some monomers with larger rings, when transannular strain does not operate, are strainless and the positive change of entropy becomes then responsible for their polymerizability. The positive change of entropy is due to the increased conformational freedom in the chain, when compared with a cycle. Sulfur, some cyclic siloxanes and, as it was found recently, cyclic phosphates belong to this group. The corresponding values are shown below:

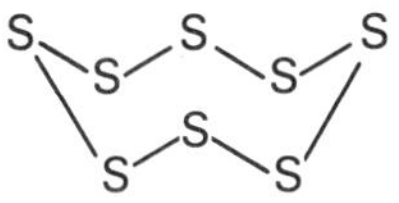

$\Delta H_p = +3\,\text{kcal/mol}$
$\Delta S_p = +4.7\,\text{e.u.}$

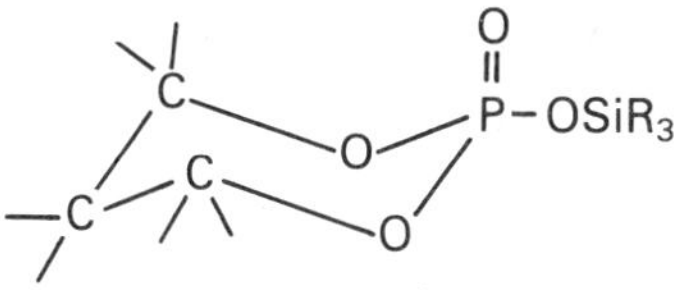

$\Delta H_p = +1\,\text{kcal/mol}$
$\Delta S_p = +7\,\text{e.u.}$

Substitution usually decreases polymerizability, this has been determined for the ring-substituted lactams and 1,3-dioxolans, and more recently for the cyclic esters of phosphoric acid. In these groups of monomers substitution decreases the exothermicity as well as entropy of polymerization.

These and related thermodynamic problems have been discussed in the past by F. S. Dainton and K. J. Ivin, and more recently by K. J. Ivin (Belfast) again.

Kinetics

The major understanding of the polymerization mechanisms comes from systems behaving as so called living systems, i.e. when the propagation step can be studied independently of other complicating side reactions. Thus, the studies of the living anionic polymerization of vinyl monomers has had its impacts on both anionic and cationic ring-opening polymerization.

There are three monomers in the anionic ring-opening polymerization studied in detail by using the living polymerization approach. These are: ethylene oxide (P. Sigwalt and S. Boileau in Paris, K. S. Kazanski in Moscow), propylene sulfide (P. Sigwalt and S. Boileau) and β-propiolactone (S. Penczek and S. Slomkowski in Lodz).

Structure of the growing species have been determined indirectly, mostly on the basis of the analysis of the end groups. The corresponding polymerizations can be written as follows:

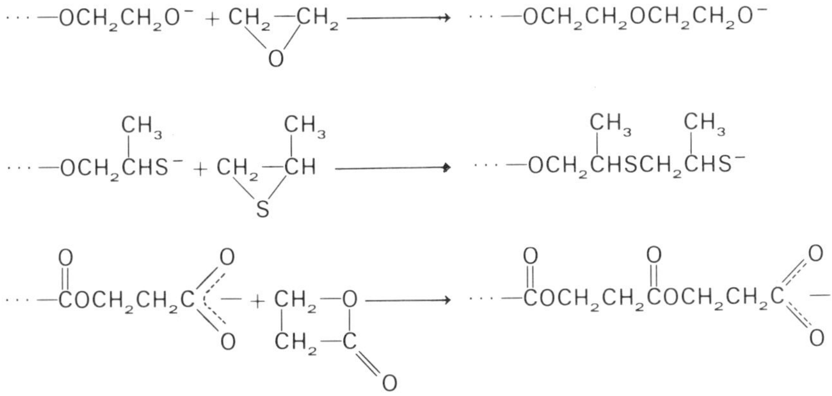

In some monomers the ambient reactivity is observed; thus, although β-propiolactone polymerizes exclusively with carboxylate anions (as shown above), ε-caprolactone polymerizes only with alcoholate anions as the growing species, but some other lactones may open at both sites.

In the cationic polymerization the living systems have also been observed. It has been established that some cyclic amines, and sulfides, particularly bearing large substituents (e.g. N-tert-butylaziridine, E. J. Goethals, Ghent), tetrahydrofuran (M. Levy and D. Vofsi, Rehovot; S. S. Medvedev and B. A. Rosenberg, Moscow), and oxazolines (T. Kagiya, T. Saegusa, Kyoto) give living systems. Cyclic acetals, when polymerized with special initiators give living polymers with two active ends (P. Kubisa and S. Penczek, Lodz). The active species in the cationic polymerization could directly be observed by NMR. Several research groups established in this way structures of the growing species to be onium ions as shown below:

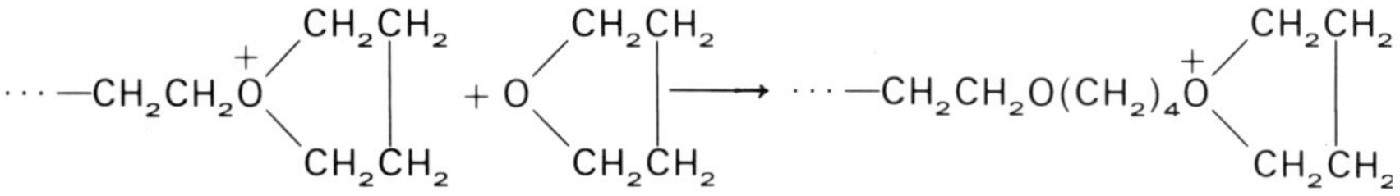

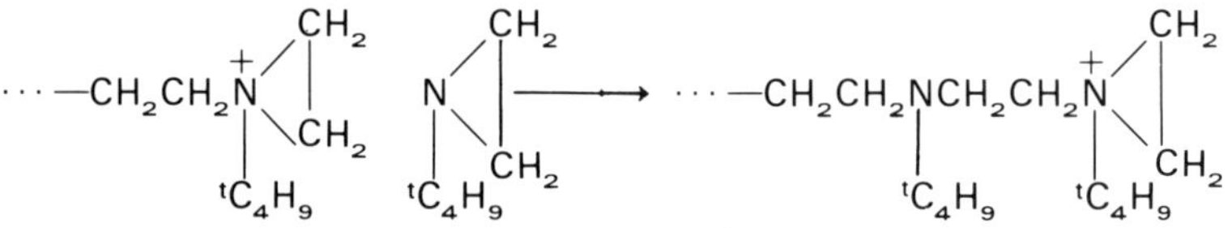

In all of the given examples and in the large majority of the other ring-opening polymerizations the ionic mechanisms operate. Thus, the simple propagation reactions shown above are in reality much more complicated. Ions can exist in various forms and for a given chemical structure one can write a number of physical forms. Thus, it has been shown that the elementary reactions proceed on macroions, macroion-pairs, higher aggregates of macroion-pairs and, at least in some systems, reactive covalent end groups in macromolecules coexist with their ionic counterparts. This can be illustrated by the scheme of chain propagation on various species in the cationic polymerization of tetrahydrofuran when anion (counterion) is able to form covalent species:

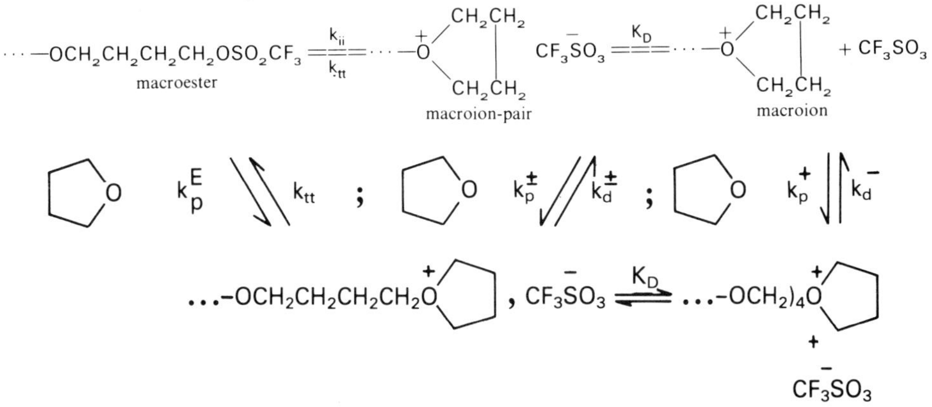

Recently this complex set of parallel and consecutive reactions has been solved and all of the included rate constants and/or equilibrium constants determined, giving the first clear picture in the cationic polymerization (understanding of any of the vinyl cationic systems is not comparable with the level of apprehension of the THF polymerization).

Chain propagation in the anionic polymerization can be in principle presented in the same way. There are, however some peculiarities, due to the formation of various ionic aggregates and strong solvation of cations by the chain (K. S. Kazanski, Moscow; M. Mazurek and J. Chojnowski, Lodz).

The following generalities can be formulated for the anionic and cationic chain growth in the ring-opening polymerization of heterocyclics:

● active species are highly solvated by the constituents of the systems; in contrast to the vinyl polymerizations, monomers and polymers are the strongest solvating agents. In the cationic polymerization the growing macrocations are solvated whereas in the anionic polymerization solvated are counterions (cations)

● due to the high solvation the macroions and macroion-pairs in the cationic polymerization are of similar reactivity

● in the anionic polymerization, proceeding with alcoholate anions, the large portion of the active species is highly aggregated and nonreactive. Macroions are more reactive than the macroion-pairs, with a few exceptions, when the push–pull mechanism operates. Solvation of the macroanions is less pronounced than the solvation of macrocations in the cationic polymerization

● macromolecules tend to solvate cations (counterions) in the anionic polymerization of heterocyclics; this effect is particularly apparent in the polymerization of ethylene oxide, where the chain can form solvating shells bearing resemblance of crown ethers. Similar phenomenon, but less pronounced, is observed in the polymerization of cyclic siloxanes.

Another specific feature of the ring-opening polymerization is the formation of macrocyclic oligomers. The thermodynamics of this process, has been elaborated some time ago (Jacobson and Stockmayer, Dartmouth; J. A. Semlyen, York) whereas the kinetic aspects of macrocyclization have only recently been discussed (R. C. Schulz, Mainz; Y. Yamashita, Nagoya; S. Slomkowski, S. Penczek, Lodz).

Thus, in every ring-opening polymerization there is a certain distribution of linear macromolecules and quite independent of it a distribution of macrorings. The proportions of these two populations depend for a given monomer mainly on its starting concentration.

NEW SYNTHESES

In one of the previous sections the major synthetic achievements of the ring-opening polymerization are presented. These are the systems that led to the industrial applications.

On the other hand there is a number of other polymers prepared by ring-opening polymerization, that, either recently developed or known already for some time, are important for other areas of science and/or are promising for the future industrial developments. To this group we would include the following systems:

A. Some new organic polymers

Ring-opening polymerization is a useful method of preparing optically-active polymers. Understanding of the general laws of stereoselection and stereoelection is particularly important, because all of the biological macromolecules are optically active and ring-opening polymerization is being used in the synthesis of models of biopolymers. Basic principles of preparing optically active polymers from chiral NCA, oxiranes, thiiranes a.o. were discussed thoroughly at the international meetings (T. Tsuruta, Tokyo; P. Sigwalt and N. Spassky, Paris) and many of the new developments are still relying on the older discoveries of the tacticity in poly(propylene oxide) (C. C. Price, Philadelphia, E. Vandenberg, Wilmington).

Another important direction is the polymerization of some spiro and bicyclic monomers, that expand during polymerization (Bailey, Maryland, H. K. Hall, Jr., Tucson). This expansion is due to the replacement of the shorter distances by the longer ones upon polymerization. It is hoped that this expansion shall be used in several modern technologies.

Understanding of the polymerization of oxazolines and related monomers has led to the synthesis of the N-acylaziridines, convertable to the first linear, crystalline polyethyleneimines ($+CH_2CH_2NH+_n$) (T. Saegusa, Kyoto).

These polymers were further used in the synthesis of various derivatives, including nucleic acids models by attaching nucleic acids bases (C. G. Overberger, Ann Arbor).

There is further progress in synthesis of ionomers based on the ring opening polymerization, e.g. synthesis of ionomeric polyacetals or polyethers (O. Vogl, Amherst–Brooklyn, N.Y.)

$$\big[\!\!-(CH_2O)_{\overline{x}}-(CH_2CH-O)_{\overline{y}}\,\big]_{\overline{n}}\!\!-$$
$$(CH_2)_n$$
$$COO^-Mt^+$$

$$CH_3$$
$$\big[\!\!-(CH_2CH-O)_{\overline{x}}-(CH_2CHO)_{\overline{y}}\,\big]_{\overline{n}}\!\!-$$
$$(CH_2)_n$$
$$COO^-Mt^+$$

Another new development is the synthesis of crystalline poly-lactones from unsymmetrically α,α'-disubstituted β-lactones (R. W. Lenz, Amherts).

Polymerization of cyclic acetals is further explored. Several new macrocyclic acetals have been polymerized and the dependence of the polymer properties on the polymer unit structure is systematically being studied (R. C. Schulz, Mainz).

New possibilities are being open with a first successful preparation of the ABA block copolymers based on the dicationically living polyacetal, constituting the internal block (P. Kubisa, Lodz):

e.g.

$$(CF_3SO_2)_2O + O \quad O \longrightarrow \;^+\!\!>OCH_2O \;\text{\textbackslash\textbackslash\textbackslash\textbackslash}\; CH_2 \quad ^+OCH_2O<$$
$$\diagdown CH_2 \qquad\quad CF_3SO_3^- \qquad\qquad\qquad CF_3SO_3^-$$

Another important contribution of the recent years is related to the new initiators, based on Al- porfirine complexes, giving living polymers and therefore di- and multiblock copolymers in the polymerization of some cyclic ethers and cyclic esters (S. Inoue, Tokyo).

On the bases of the cationic (THF) and anionic (ethylene oxide) polymerization of heterocyclic monomers the first macromonomers (macromers) have recently been prepared and characterized. These medium size macromolecules with polymerizable end-groups are perspective elements of the well organized graft copolymers leading e.g. to the new elastoplastics (E. Franta, P. Rempp, Strasburg).

B. Inorganic ring-opening polymerizations

Elemental sulfur has been known for a long time to polymerize on heating, presumably by the free radical mechanism. The homopolymer is thermodynamically unstable at room temperature and tends to depropagate by unzipping mechanism. In order to set up a barrier to this disastrous depropagation, the copolymerization of elemental sulfur with cyclic sulfides was recently elaborated, based on the recent discovery of the anionic ability of elemental sulfur to copolymerize with a variety of monomers (A. Duda and S. Penczek, Lodz):

e.g.:

$$n \ S_8 + n \ \underset{S}{\overset{CH_2-CH}{\diagup}} \overset{CH_3}{\diagdown} \longrightarrow +CH_2-CH-S \overset{}{\underset{x}{\diagup}}_n$$

When larger excess of S_8 is used some S_8 is left unreacted, thus $x < 9$. Further developments of this original approach may lead to the new polymers based on the cheap and still available raw material.

The polymers of $+SN+_x$ are also getting a new life. The S_2N_2 (prepared from S_4N_4), known already for some time to polymerize in the gaseous state to the solid polymer with metallic properties, has recently achieved much attention. The revival of the work on polymerization and properties (M. M. Labes, McDiarmid, Philadelphia) is based on the renewed hopes that electrically conducting or even superconducting polymers can be prepared in this way.

Another group of polymers of increasing importance are the poly-phosphazenes (H. R. Allcock, Pennsylvania). These polymers are mostly based on the poli(dichlorophosphazene), prepared by thermal and/or catalytic polymerization of the parent monomer, prepared from PCl_5 and NH_4Cl:

$$\underset{\underset{Cl}{\overset{N}{|}}\underset{Cl}{\overset{P}{\diagdown}}\underset{N}{\overset{P}{\diagdown}}\underset{Cl}{\overset{Cl}{}}}{\overset{Cl}{\overset{P}{\diagup}}\overset{Cl}{\overset{N}{\diagdown}}} \xrightarrow[\text{cat.}]{\Delta} +P=N+_{3n}$$

Further substitution at the P atoms (mostly perfluoroalkoxy derivatives) permitted to obtain a large family of polymers that have already found applications as heat resistant rubbers. Water soluble polymers, heat resistant plastic, grafted materials, etc. can be expected for the near applications.

C. Models of biopolymers

Models of biopolymers, namely polypeptides, polysaccharides, and poly-phosphates (nucleic acids, teichoic acids) can be prepared by the ring-opening polymerization. In the synthesis of polypeptides the stereoelection is permitting to prepare optically active polymers from racemic monomer (T. Tsuruta, Tokyo). Several new polysaccharides and polysaccharide models were prepared using the method elaborated in the sixties (C. Schuerch,

Syracuse) and using new, more recently developed methods (Uryu, Tokyo; M. Okada, H. Sumitomo, Nagoya).

Ring-opening polymerization of cyclic phosphorous containing monomers has led to the preparation of two important models of biopolymers, namely teichoic acid, and polyphosphate bearing deoxyribose moiety, i.e. a polynucleotide chain devoid of base (P. Klosinski, G. Lapienis, and S. Penczek, Lodz):

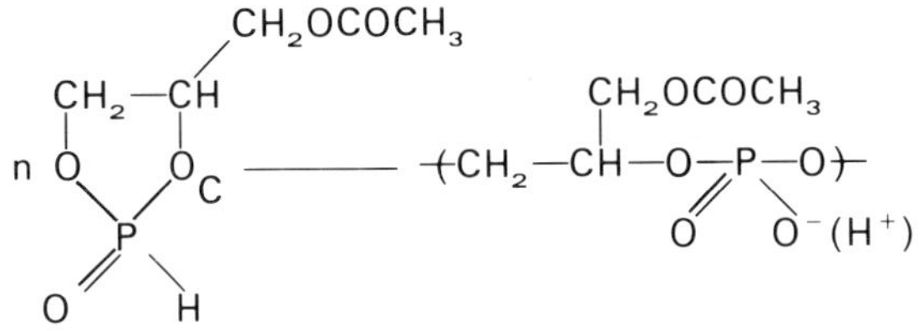

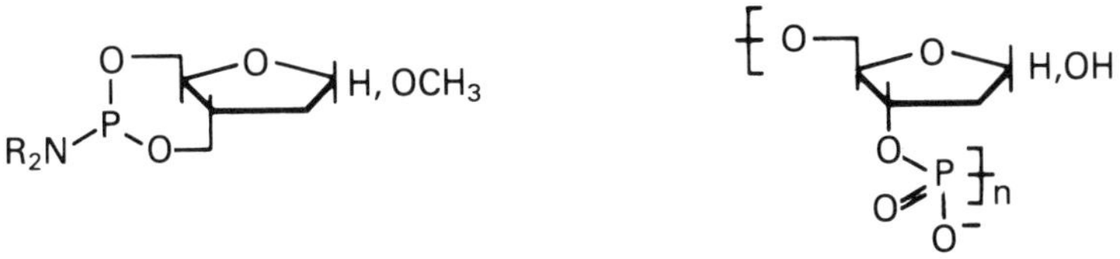

These and related novel polymer may open new directions in research of biomedical polymers and biologically active polymers, using the unusual combination of properties of polyphosphates shown above: their water solubility, biodegradability, expected non-toxicity of components and their metabolites, as well as the anionic character of polymers.

The large majority of problems and examples given in this short review can be found, as indicated above, in the proceedings of three international meetings devoted to the ring-opening polymerization over the past few years. Thus, only the references to these proceedings are given here.

References

1. S. Penczek, Ed., Polymerization of heterocycles (ring opening), IUPAC Symposium, Warsaw Jablonna, Poland, Pergamon (1975).
2. T. Saegusa and E. Goethals, Eds., Ring-opening polymerization, ACS Symposium Ser. 59, ACS, Washington DC (1977).
3. J. Sebenda, Ring-opening polymerization of heterocycles, IUPAC-PMM Microsymposium, Inst. of Macromol. Chem., Prague (1980).

Quo Vadis Cationic Polymerization?

J. P. KENNEDY

Institute of Polymer Science, The University of Akron, Akron, Ohio 44325, USA.

Where are you going Cationic Polymerization?

You just emerged from a dark morass where signposts were far and few, and where only the fearless dared to penetrate, and many of the venturesome who searched your labyrinths were invariably mauled by the minotaurs of irreproducibility or impaled by the draculas of impurity effects. But the few who preserved and survived the long, almost two decades journey can now start to relax; they have reached a firm beachhead and can look ahead toward more meaningful excursions.

This very brief and wholly subjective essay on carbocationic polymerizations focuses on important research developments that occurred during the last few years. This writer has very recently completed a book on this subject[1] which forced him to assemble and digest a lot of information. During this process, he gained insight into this field and came to some surprising/disturbing conclusions that he wishes very briefly to record:

Carbocationic polymerization is undergoing a silent revolution; the times of haphazard and trivial kinetic studies are coming to an end and the era of Macromolecular Engineering by cationic techniques is upon us. We no longer have to accept what nature wants to give us, we can control nature by disturbing our systems and direct our results with a high degree of sophistication toward useful end-products. We can confidently tailor polymer segments, design head groups and/or end groups, adjust molecular weight or molecular weight distributions, derivatize preformed polymers, assemble complicated polymer topologies, etc. Recently we went beyond product tailoring and started efforts to reduce the cost of synthesis and processing by working with liquid or oligomeric prepolymers in low shear equipment.

No doubt, the key to these developments was our increased understanding of the mechanisms of the underlying polymerization events: initiation, propagation, chain transfer and termination.

The era of macromolecular engineering started by gaining insight into the mechanism of *initiation*. This research led to a clearer formulation of this process, in particular to the view that initiation embraces two chemically totally dissimilar events:

(1) ionization and (2) cationation,

$$RX \quad + \quad LA \quad \rightleftharpoons \quad R^{\oplus} \quad + \quad LAX^{\ominus} \qquad (1)$$

cationogen   Lewis acid   carbocation   counteranion

$$R^{\oplus} \quad + \quad C{=}C \quad \rightarrow \quad R{-}C{-}C^{\oplus} \qquad (2)$$

olefin   initiated chain

and that these processes can be carried out in many 'open' systems, i.e., in the presence of moisture under conventional laboratory conditions. The reduction to practice of the above *controlled initiation* sequence led to polymers with tailor-designed head groups. Obviously, the nature of the head unit was controlled by the initiator ≡ cationogen. Many initiating systems giving rise to rapid and efficient controlled initiation have been found and were used to introduce useful head groups (e.g., allyl, benzyl, chlorine, bromine, silyl) into polymers, e.g., allyl chloride/Et_2AlCl, benzyl chloride/Me_3Al.[2]

Older conventional initiating systems BF_3, $TiCl_3$, $AlCl_3$ do not require the purposeful addition of cationogens and kick off polymerizations in the presence of traces of moisture. These systems produce uninteresting CH_3-head groups

$$H_2O/BF_3 + CH_2 = \overset{|}{\underset{|}{C}} \rightarrow CH_3 - \overset{|}{\underset{|}{C}}{}^{\oplus}BF_3OH^{\ominus}$$

at haphazard rates that depend on the impurity (moisture) level in the charge.

The principle of controlled initiation led to the discovery of a new family of graft and block copolymers.[3] It was demonstrated that polymeric cationogens PX, e.g., poly(vinyl chloride), polychloroprene, are just as effective initiators as their small molecule counterparts RX and that these initiators in conjunction with certain Lewis acids give rise to graft or block copolymers:

$$PX + LA \rightarrow [P^{\oplus}LAX^{\ominus}] \xrightarrow{+C=C} P{-}C{-}C^{\oplus}$$

and depending on the respective nature of the prepolymer PX and the grafted or blocked polymer sequence (rubbery, glassy, crystalline), the widest variety of unique materials can be assembled. Table I lists some of these composites.

Propagation is a repetitive cationation of monomer (olefin) by the growing polymer chain: $\sim C_n^{\oplus} + C{=}C \rightarrow \sim C_{n+1}^{\oplus}$. The aim of the macromolecular

TABLE I
Graft copolymers prepared by carbocationic techniques

I. *Elastomeric Backbones*

    a)   *Elastomeric Branches*
        poly(butadiene-*g*-isobutylene)[a]
        poly(chloroprene-*g*-isobutylene)
        poly[chloroprene-*g*-(isobutylene-*co*-isoprene)]
        poly[chloroprene-*g*-(isobutylene-*b*-α-methylstyrene)]
        poly(chloroprene-*g*-isobutyl vinyl ether
        poly[(styrene-*co*-butadiene)-*g*-isobutylene][a]
        poly[(isobutylene-*co*-isoprene)-*g*-chloroprene][a]

    b)   *Glassy Branches*
        poly[(ethylene-*co*-propylene)-*g*-styrene][a]
        poly[(ethylene-*co*-propylene)-*g*-indene][a]
        poly[(isobutylene-*co*-isoprene)-*g*-styrene][ab]
        poly[(isobutylene-*co*-isoprene)-*g*-α-methylstyrene][a]
        poly[(isobutylene-*co*-isoprene)-*g*-p-chlorostyrene][a]
        poly[(isobutylene-*co*-isoprene)-*g*-indene][a]
        poly[(isobutylene-*co*-isoprene)-*g*-(indene-*co*-α-methylstyrene)][a]
        poly[(isobutylene-*co*-isoprene)-*g*-acenaphthylene][a]
        poly[(isobutylene-*co*-p-chloromethylstyrene)-*g*-indene]
        poly[isobutylene-*co*-p-chloromethylstyrene)-*g*-α-methylstyrene]
        poly(chloroprene-*g*-styrene)
        poly(butadiene-*g*-α-methylstyrene)[a]
        chlorosulfonated polyethylene-*g*-polystyrene
        poly[(p-chloromethylstyrene-*co*-butadiene)-*g*-styrene]

    c)   *Two Branches (Bigrafts)*
        1. *A Glassy and an Elastomeric Branch*
        poly[(ethylene-*co*-propylene-*co*-1,4-hexadiene)-*g*-styrene-*g*-isobutylene][ab]
        poly[(ethylene-*co*-propylene-*co*-1,4-hexadiene)-*g*-α-methylstyrene-*g*-isobutylene][ab]
        2. *Two Glassy Branches*
        poly[(ethylene-*co*-propylene-*co*-1,4-hexadiene)-*g*-styrene-*g*-α-methylstyrene]*[a]

II. *Glassy Backbones*

    a)   *Elastomeric Branches*
        poly(vinyl chloride-*g*-isobutylene)
        poly[vinyl chloride-*g*-(isobutylene-*co*-isoprene)]
        chloromethylated polystyrene-*g*-polyisobutylene
        chlorinated polyvinyl chloride-*g*-polybutadiene
        chlorinated polycyclopentadiene-*g*-polybutadiene

    b)   *Glassy Branches*
        poly(vinyl chloride-*g*-styrene)

[a]Lightly chlorinated backbone used.
[b]Lightly brominated backbone used.

engineer is to control initiation (see above) and to maintain uninterrupted propagation until he wishes to terminate the chain in a controlled manner (*controlled termination*). The most undesirable event is premature proton

H

elimination by chain transfer to monomer:

$$\sim C-C^{\oplus} + C{=}C \rightarrow \sim C{=}C + HC-C^{\oplus}$$
$$\underset{H}{|}$$

Macromolecular engineering by carbocationic techniques became a reality with the discovery of chain transferless systems, i.e., in polymerizations in which chain transfer to monomer is absent.

Transferless polymerizations can be achieved in systems in which the rate of termination is faster than that of chain transfer to monomer, $R_t < R_{tr,M}$. Under transferless conditions, *uninterrupted propagation* proceeds until *controlled termination* takes place. Concurrently with the clarification of these concepts, some fundamentals of the mechanism of termination (irreversible annihilation of propagation) was elucidated. It was found that in many systems termination involved a simple, predictable collapse of the propagating ion pair:

$$\sim C^{\oplus}GX^{\ominus} \rightarrow \sim CX + G$$

By judicious selection of $GX^{\ominus}$, controlled termination, i.e., end group control, became feasible. Today, the macromolecular engineer is capable of designing $GX^{\ominus}$ and thus is able to introduce several valuable functional end groups, e.g., —Cl, —Br, cyclopentadiene, vinyl, phenyl, into polymers.[4]

All in all, we have now systems in which both controlled initiation and termination operate and thus yield polymers with well characterized head and end groups, R and Y, respectively:

$$RX + LA \rightleftharpoons R^{\oplus}LAX^{\ominus}$$

$$R^{\oplus} + C{=}C \rightarrow R-C-C^{\oplus} \xrightarrow{+\, nC{=}C} R-C-C \sim \sim \sim C-C^{\oplus}$$

$$R-C-C \sim \sim \sim C-C^{\oplus}LAX \rightarrow R-C-C \sim \sim \sim C-C- Y + LA$$

For example, one can incorporate the following functional head groups:

$$\underset{|}{\overset{|}{C}}{=}\underset{|}{\overset{|}{C}}-\underset{|}{\overset{|}{C}}\sim, \quad \langle O \rangle - \underset{|}{\overset{|}{C}}\sim, \quad \text{or} \quad , \quad Cl-, \ Br-, \ HSi \sim$$

and end groups:[5]

$$-CH{=}CH_2, \quad -\langle O \rangle \quad , \quad Cl-$$

A very recent development is *quasiliving polymerizations.*[6] In these polymerizations, termination is reversible so that the system behaves as if it were *terminationless*. Chain transfer can be minimized by very low continuous monomer addition so that molecular weights M_n increase linearly with the amount of monomer added over relatively long periods. In truly living, usually anionic polymerizations, both termination and chain transfer are absent and $\bar{M}_n$'s increase linearly with time. In quasiliving systems, poly(α-methylstyrenes) with $\bar{M}_n$'s up to 10^5 could be obtained and many other olefins, even vinyl ethers (isobutyl and vinyl), could be 'grown'. The latest achievement was the synthesis of block copolymers, e.g., poly(α-methylstyrene)-poly(isobutyl vinyl ether).[7]

All the systems outlined until now were 'undisturbed', i.e., all the action involved only three actors: the monomer (olefin), the initiator (cationogen), and coinitiator (Lewis acid); and optionally a solvent. If the director was good and knew how to cast his characters and set the stage, he could end up with quite a variety of well defined products.

Lately, advanced practitioners have started to modify the basic play by featuring additional characters, such as chain transfer agents, inifers, proton traps. Some such 'disturbed' systems have led to extremely interesting and potentially very useful materials. For example, the recognition that the mechanism of initiation and chain transfer by chlorine migration ($\sim C_n^{\oplus}$ $+ RCl \rightleftharpoons \sim C_nCl + R^{\oplus}$) are very similar, led to the discovery of the *inifer technique*[8] that was used for the synthesis of a new family of *telechelic polymers* (linear or radial polymers with well defined terminal functionalities). Among the first telechelics obtained by this method were linear and three-arm star (radial) polyisobutylenes carrying exactly two or three tertiary chlorine termini, e.g.:[8,9]

$$Cl-\underset{\underset{CH_3}{|}}{\overset{\overset{CH_3}{|}}{C}}-CH_2 \sim \sim \sim polyisobutylene \sim \sim \sim CH_2-\underset{\underset{CH_3}{|}}{\overset{\overset{CH_3}{|}}{C}}-Cl$$

$$CH_2$$
$$|$$
$$CH_3-\underset{\underset{Cl}{|}}{C}-CH_3$$

Quantitative conversion of the tertiary chlorine termini to other functional groups led to a large variety of other telechelics, i.e., to polymers with exactly two or three terminal functions, e.g.,

$$-C{=}CH_2,^{10} \quad \underset{\underset{CH_2SO_3H}{|}}{-C{=}CH_2},^{11} \quad -OCO-NCO,^{12} \quad -CH_2OH,^{13}$$

In turn, these telechelics gave rise to many interesting new sequential copolymers,[8] networks, polyurethanes,[14] and ionomers.[11]

Other 'disturbed' systems include olefin polymerizations carried out in the presence of proton traps, i.e., materials that exhibit extraordinary specificity toward free protons. A most efficient proton trap is 2,6-di-tert-butyl pyridine (DtBP):

Very recently it has been discovered that various olefins, e.g., isobutylene, styrene, α-methylstyrene, can be readily polymerized with BCl_3, $SnCl_4$, or $TiCl_4$ in the presence of DtBP.[15,16] Since free protons cannot survive in the presence of DtBP, some time-honored and accepted views relative to proton initiated olefin polymerizations ('$H^{\oplus}BF_3OH^{\ominus}$') have to be abandoned and/or modified. Interestingly, polymers obtained in the presence of proton traps are of higher molecular weight and narrower molecular weight distribution ($M_w/M_n > 2.0$) than in the absence of these materials. It is of more practical significance that in the presence of proton traps, grafting and/or blocking of olefins (α-methylstyrene, isobutylene) from chlorinated prepolymers (polychloroprene, chlorinated butyl rubber) with conventional Lewis acids (BCl_3, $SnCl_4$) give $\sim 100\%$ grafting and blocking efficiencies.[17] Prior to this discovery, very high grafting or blocking efficiencies could be achieved only with certain alkylaluminium coinitiators (see above).

References

1.   J. P. Kennedy and E. Maréchal, *Carbocationic Polymerization*, J. Wiley-Interscience Pub.: New York (1982).

2.   The following references provide details and illustrate the use of controlled initiation:

2a. J. P. Kennedy and V. S. C. Chang, *Adv. Polym. Sci.*, **43**, (1981).

2b. V. S. C. Chang and J. P. Kennedy, *Polym. Bull.*, **5**, 379 (1981).

2c. J. P. Kennedy, S. Y. Huang and R. A. Smith, *J. Macromol. Sci. Chem.*, **A14**, 1085 (1950).

2d. L. Reibel, J. P. Kennedy and D. Y. Chung, *J. Polym. Sci., Polym. Chem. Ed.*, **17**, 2757 (1979).

2e. J. P. Kennedy and F. J.-Y. Chen, *Polym. Prepr.*, **20**, 310 (1979).

2f. J. P. Kennedy and E. Maréchal, *Marcromol. Rev.*, **16**, 123 (1981).

3.   *Cationic Graft Copolymerization*, J. P. Kennedy, Ed., *J. Appl. Polym. Sci., Symp.*, **30** (1977).

4.   The following references are illustrative of transferless systems and controlled termination:

4a. J. P. Kennedy and D. Y.-L. Chung, *Polymer Prepr.*, **21**, 150 (1980).

4b. J. P. Kennedy and K. F. Castner, *J. Polym. Sci., Polym. Chem. Ed.*, **17**, 2055 (1979).

4c. B. M. Mandal and J. P. Kennedy, *J. Polym. Chem., Polym. Chem. Ed.*, **16**, 833 (1978).

4d. J. P. Kennedy, S. Y. Huang and S. C. Feinberg, *J. Polym. Sci., Polym. Chem. Ed.*, **15**, 2801 (1977).

5.   J. P. Kennedy, *Makromol. Chem., Suppl.*, **3**, 1 (1979).

6. R. Faust, A. Fehérvári and J. P. Kennedy, *J. Macromol. Sci.-Chem.*, **A18** (9) 1209 (1983).
7. *M. Sawamoto and J. P. Kennedy, Polymer Prepr.*, **22**(2), 140 (1981).
8. J. P. Kennedy, R. A. Smith and L. Ross, U.S. Patent 4,276,394, June 30, 1981.
9. J. P. Kennedy, L. R. Ross, J. E. Lackey and O. Nuyken, *Polym. Bull.*, **4**, 67 (1981).
10. J. P. Kennedy, V. S. C. Chang, R. A. Smith and B. Iván, *Polym. Bull.*, **1**, 575 (1979).
11. J. P. Kennedy, R. F. Storey, Y. Mohajer and G. L. Wilkes, Abstracts, *IUPAC MACRO'82*, p. 905 (1982).
12. R. Wondraczek and J. P. Kennedy, *Polym. Bull.*, **4**, 445 (1981).
13. B. Iván, J. P. Kennedy and V. S. C. Chang, *J. Polym. Sci., Polym. Chem. Ed.*, **18**, 3177 (1980).
14. J. P. Kennedy, B. Iván and V. S. C. Chang, *Adv. Urethane Sci. and Tech.*, **8**, 245 (1981).
15. J. P. Kennedy and R. T. Chou, *Polymer Prep.*, **20**, 306 (1979).
16. J. P. Kennedy and R. T. Chou, *Polymer Prep.*, **21**(2), 148 (1980).
17. J. P. Kennedy and S. Guhaniyogi, *J. Macromol. Sci., Chem.*, **A18** (1), 103 (1982).

Renewable Raw Materials for the Polymer Industry

E. H. PRYDE and F. H. OTEY

Northern Regional Research Center, Agricultural Research Service, US Department of Agriculture, Peoria, Illinois 61604, USA.

INTRODUCTION

Chemicals from biomass, agricultural residues and byproducts, and agricultural crops themselves, must eventually serve as feedstock for the chemical industry. Many of the renewable resources have served and continue to serve industrial product needs, but growth has been curtailed because of petrochemical incursions into the various fields of application. Starch and fats and oils have already made notable contributions to socio-economic needs and probably will contribute greater shares to such needs in the future. Starch and fats and oils are limited resources, and great care must be taken to ensure optimum benefits to our technological society. Nevertheless, their contribution will be essential, significant, and increasing.

The following discussion briefly reviews some of the research areas in starch and fats and oils and their industrial implications as investigated mainly at the Northern Regional Research Center.

RENEWABLE RAW MATERIALS FROM STARCH

Starch is a prime candidate for replacing certain petrochemicals because it is available at a low cost and can be converted readily into a variety of useful monomeric and polymeric products. In 1979, the six major U.S. cereal grain crops contained 470 billion pounds of starch. Traditionally, only 2 to 3% of this starch is isolated and less than 1% goes for industrial applications. The need for more environmentally acceptable products coupled with increasing

---

†The mention of firm names or trade products does not imply that they are endorsed or recommended by the U.S. Department of Agriculture over other firms or similar products not mentioned.

prices and decreasing supplies of petroleum have intensified the socio-economic desirability of starch-based industrial products.

Polyols

Polyols are basic materials used in making rigid polyurethane foam, alkyd resins, and surfactants. Since 1974, the price of conventional polyols (pentaerythritol, glycerol, sorbitol) has increased from the low 20's ¢/lb range to the low 60's ¢/lb (170% increase). In contrast, starch has maintained a very stable price, increasing from 8 ¢/lb to 10.5 ¢/lb (30% increase). A process was developed and evaluated extensively for producing glycol glucosides from starch.[1] Cost estimates suggest that these starch-based polyols could be produced for 25–30 ¢/lb. Techniques were developed, in pilot-plant studies, for using these starch-derived polyols in preparing urethane foams,[2] alkyd resins,[3] and surfactants.[4] Flame resistance can be incorporated into a polyurethane foam raw material by the stepwise reaction of starch-derived glucose with allyl alcohol, propylene oxide, and various halogenating compounds.[5] A. E. Staley recently came on stream with a fermentation pilot-plant process for making methyl glucoside from corn starch.[6] The company's aim is ultimately to produce a line of corn-derived organic chemicals for a 5-billion lb/yr market for foams, adhesives, paints, and detergents.

Plastics

Techniques have been developed for substituting starch for significant amounts of petroleum-derived chemicals used to manufacture plastics. Not only is it the renewable aspect of starch that has piqued the interest of industry, but also the potential of starch to impart varying degrees of water solubility and biodegradability. One commercial process now uses starch as a partial replacement for polyvinyl alcohol to produce water-soluble laundry bags used by hospitals to keep the laundry staff from coming in contact with soiled or contaminated clothing collected in them.[7]

Combinations of starch, poly(ethylene-co-acrylic acid) and polyethylene yield biodegradable plastics that show promise for a variety of agricultural uses, especially as mulch film.[8] Farmers now use an estimated 60 million pounds of petroleum-derived polyethylene mulch film, which improves crop yields by 50 to 350% by retaining soil moisture, preventing loss of added fertilizer, and keeping down weed growth. Use of biodegradable mulch film would eliminate the estimated $100 per acre cost to remove and dispose of the nondegradable film. Farmers have also expressed a need for some 50 other new plastic materials to further improve agricultural technology.

Many of these relate to biodegradable plastics that could be used in the planting, growing, and harvesting of crops.

Rubber

Starch has application in processing crude latex to finished goods. It can be incorporated in the processed rubber to fill the same function as carbon black.[9] Also, powdered rubber can be made by starch encasement of the rubber droplets in latex.[10] The use of powdered rubber in place of conventional slab rubber would result in large savings of time and energy in the manufacture of rubber articles, because injection molding machines will permit direct handling of powders without the need for the high-energy mixing for processing slab rubber.

Purification of waste water

A sulfur-containing starch derivative called ISX (insoluble starch xanthate) is effective in removing traces of heavy metal contaminants from industrial process water.[11] About 30 companies have licensed the U.S. patent to use or produce ISX, and at least 25 metal-plating firms have installed the ISX treatment to meet EPA requirements. This technology provides the dual potential of increasing markets for agricultural products and at the same time solving a pollution problem by permitting reuse of processing waters.

Encapsulated pesticides

Technology has been discovered whereby a range of pesticides can be encapsulated within a matrix formed by a starch derivative.[12] Pesticide is blended into an aqueous alkali dispersion of either starch or starch xanthate. Upon precipitation of the starch as a metal complex by adding calcium chloride or boric acid or the starch xanthate as an oxidatively crosslinked xanthide, the pesticide is entrapped in small cells within the starch matrix. The matrix is biodegradable, and the rate of release of the pesticide can be controlled by varying the derivatizing process or by changing the encapsulation step. Encapsulation keeps the pesticide where targeted and reduces environmental impact; also, the safety for farmers handling toxic pesticides is greatly increased. Field tests with encapsulated herbicides by cooperating scientists at several universities show that efficacy of weed control is greatly improved.

Starch graft copolymers

Basic research on the grafting of starch with various synthetic monomers has yielded new products having a wide variety of potential applications.

Grafting is achieved by irradiating or chemically initiating free radicals on the starch backbone and then allowing the radicals to react with polymerizable vinyl or acrylic monomers. To yield plastic or elastomeric copolymer compositions that can be extruded or milled, monomers such as styrene, isoprene, acrylonitrile, and various alkyl acrylates were employed.[13] Super Slurper, the first major product of this study to receive worldwide attention, is made by graft polymerizing 50 parts of acrylonitrile onto 50 parts of starch followed by alkaline saponification.[14] The copolymer absorbs many hundred times its weight of water, an especially important characteristic for industrial applications. Super Slurper is now used in commercial products for the absorption of body fluids such as urine, blood, and perspiration; in controlling forest fires; in several agricultural applications for establishing seedlings and transplants; and in industrial thickening applications. It may have application in removing water from fuel alcohol mixtures and in many other agricultural, industrial, and medical areas.

The concept of chemically combining synthetic materials with starch can lead to a wide variety of other products that meet specific requirements for diverse agricultural and consumer applications.

These and many other developments for using starch to meet a wide range of socio-economic goals are now available. Some have achieved limited industrial success; others simply await appropriate technology transfer and implementation.

RENEWABLE RAW MATERIALS FROM FATS AND OILS

Of the 20-billion-lb fats and oils industry in the United States, one-half is consumed in edible oil products, one-quarter goes into nonfood products, and one-quarter is exported.[15] About 2 billion lb of vegetable oils (coconut, palm, castor, tung, and oiticica oils) are imported annually, with about one-third of this amount going into industrial products. The fats and oils contribution to the total synthetic organic chemical industry is about 2%.[16] In certain traditional markets, fats and oils (about 3 billion lb) contribute a much greater share—40% of the coatings, 45% of the surfactant, and 15% of the plastics additives markets, for example. If these shares are continued and the markets grow at the expected rates, an additional 3 billion lb of fats and oils will be required in 1990. If, as anticipated, these shares increase as the consequence of increased costs and scarcity of petroleum products, an additional 7.5 billion lb might be required by 1990.[15] Further, it is to be expected that entirely new markets can and should be developed for fats and oils to help compensate for increasing materials shortages, even though they will have lesser impact than our coal, forest, and biomass resources.

Animal fats contribute the greatest share (57%) of all fats and oils used for nonfood purposes, followed by the industrial oils (including tall oil fatty acids) (28%) and the edible oils (13%). Soybean and coconut oils are the predominant vegetable oils in the fats and oils industry. Large amounts of soybean oil are available as a consequence of its co-production along with high-protein meal for animal feeds. The demand for the latter will continue to increase at home and abroad. Of the 9 billion lb of soybean oil consumed annually in the United States, 94% goes into edible oil products and about 6% into nonfood applications. The latter applications are important to the farmer to maintain diversity in the market and to serve as a cushion for what could otherwise develop into wide price fluctuations.

Potential intermediates from commercial fats and oils

Cyclic fatty acids. Alkali treatment of linseed oil causes cyclization of linolenic acid. The cyclic acids are potentially useful as lubricants[17,18] or in alkyd resins.[19] Other types of cyclic acids can be made from linseed fatty acids[20] or soybean soapstock.[21]

Fatty aldehydes have been made by reductive ozonolytic cleavage of unsaturated fatty compounds[22,23] or by the addition of carbon monoxide and hydrogen to unsaturated fatty acids by the hydroformylation (oxo) reaction.[24] The aldehyde group is a versatile one that may be easily converted to acetals, alcohols, carboxylic acids, and amines, which variously have potential value in polymers, plastics, and plastics additives.

Acetals of aldehyde esters have potential value as plasticizers, whether from 9-carbon[25] or 19-carbon[26] aldehyde esters.

Amino acids have potential value for nylon-9 engineering thermoplastics[27] as well as polyamide resin adhesives and printing inks.[28]

Polyols formed by reaction of formaldehyde with 9-carbon aldehydes[29] or 19-carbon aldehydic acids[30] may have use in coatings, lubricants, and plasticizers with improved flexibility and water-resistance.

Polycarboxylic acids have possibilities in coatings, lubricants, plasticizers, and polyamide resins.[24,31]

Organosulfur compound research will lead to a better understanding of the sulfurization reaction and to better extreme-pressure lubricant additives[32] to take the place of products derived from sperm whale oil.

Epoxidized soybean oil is made in large quantities (87 million pounds in 1980) for use as a plasticizer/stabilizer for vinyl plastics. Such availability suggests its use as starting material for other applications such as acrylated epoxidized oils. Copolymerization of the latter with various other vinyl monomers results in copolymers having flexibility, toughness, abrasion resistance, and good tensile and flexural strength that are unusual in crosslinked copolymers.[33]

Potential applications

Coatings   High solids resins, suitable for baked coatings on metal substrates, were made from acetals of aldehydic esters, or better, the diethanolamides and commercial amino resins.[34] For making water-dispersible resins, soybean oil was caused to react with diethanolamine in a partial aminolysis reaction, and the product mix of glycerol, mono- and diglycerides, and fatty acid amides was further cooked with dibasic anhydrides and diisocyanates.[35] Such resins contain 70–80% of the vegetable oil-derived material and 20–30% of petrochemical-derived materials. Films from these resins dried rapidly at room temperature in the absence of the usual metallic driers to form flexible, adherent coatings.

Engineering thermoplastics   Nylon 9 can be made from polyunsaturated oils such as soybean oil, although the cleavage by ozonolysis would be more efficient with a high-oleic vegetable oil.[27] Nylon 9 absorbs less moisture than nylon 6 and, consequently, has better dimensional stability and dielectric properties in moist environments. Even more hydrophobic is nylon 13 or 1313, which can be made from erucic acid available from oilseed crops such as crambe and rapeseed.[36] In many ways, nylon-1313 is comparable to nylons 11 and 12, but it is lower melting, slightly less dense, and more hydrophobic than either of those commercial nylons.

Plasticizers   Dioctyl phthalate, one of the most important commercial plasticizers for vinyl plastics, has become an ubiquitous and undesirable environmental contaminant.[37] There are several vegetable oil-based plasticizers that could serve in place of dioctyl phthalate. The aldehyde groups in hydroformylated vegetable oils are easily converted to acyloxy,[38] carboalkoxy,[39] and acetal[26] compounds that have useful properties as plasticizers. Bis-acetoxymethyl compounds from hydroformylated oleic acid have interesting plasticizing value.[40] Such compounds should have better properties with regard to biodegradability and potential toxicity than dioctyl phthalate. However, such properties have not been assessed for the vegetable oil materials. Excellent low-temperature plasticizers with exceptional light stability have been made from esters of brassylic acid, the dibasic acid that is made by oxidative ozonolysis of erucic acid from crambe oil.[41]

Poly(ester-acetals) and poly(amide-acetals)   These spirocetal, heterochain polymers are made from the pentaerythritol acetals of either C-9 aldehyde esters[42,43] or C-19 aldehyde esters.[44] Linear polymers are readily formed from the spiroacetal diesters; the polymers have latent functionality and can be crosslinked under appropriate temperatures and with certain metal oxide or acidic catalysts. The polymers from the C-19 aldehyde esters have less crystallinity, better solubility, and crosslink at lower temperatures than those from the C-9 aldehyde esters.

Simultaneous interpenetrating networks A 100% natural polymer, synthesized from castor oil and sebacic acid, has been used as the basis for synthesis of simultaneous interpenetrating networks (SINs) with crosslinked polystyrene.[45] Such SINs can be used to make tough, impact-resistant plastics and reinforced elastomers that compare favorably with commercial products.

Potential oilseed crops

Many oilseed plants have been screened for their economic potential.[46,47] Some of the more promising ones include *Crambe* and *Limnanthes* species for their long-chain fatty acid content, *Cuphea* species for their short-chain fatty acid content, *Lesquerella* species for their hydroxy fatty acid content, and *Vernonia* species for their epoxy fatty acids. *Vernonia galamensis* seed oil has been evaluated for its film-forming properties; excellent flexibility, adhesion to substrate, and chemical and solvent resistant were found.[48]

References

1. F. H. Otey, B. L. Zagoren and C. L. Mehltretter, *Ind. Eng. Chem., Prod. Res. Dev.* **2**, 256–259 (1963).
2. R. H. Leitheiser, C. N. Impola, R. J. Reid and F. H. Otey, *Ind. Eng. Chem., Prod. Res. Dev.*, **5**, 276–282 (1966).
3. W. J. McKillip, J. N. Kellen, C. N. Impola, R. W. Buckney and F. H. Otey, *J. Paint Technol.*, **42**, 312–319 (1970).
4. P. E. Throckmorton, R. R. Egan, D. Aelony, G. K. Mulberry and F. H. Otey, *J. Am. Oil Chem. Soc.*, **51**, 486–494 (1974).
5. F. H. Otey, R. P. Westhoff and C. L. Mehltretter, *J. Cell. Plast.*, **8**, 156–159 (1972).
6. Anon., *Chem. Week*, **130**, 30 (1982).
7. F. H. Otey, A. M. Mark, C. L. Mehltretter and C. R. Russell, *Ind. Eng. Chem., Prod. Res. Dev.*, **13**, 90–92 (1974).
8. F. H. Otey, R. P. Westhoff and W. M. Doane, *Ind. Eng. Chem., Prod. Res. Dev.*, **19**(4), 592–595 (1980).
9. R. A. Buchanan, W. F. Kwolek, H. C. Katz and C. R. Russell, *Staerke*, **23**(10), 350–359 (1971).
10. T. P. Abbott, W. M. Doane and C. R. Russell, *Rubber Age*, **105**(8), 43–49 (1973).
11. R. E. Wing, L. L. Navickis, B. K. Jasberg and W. E. Rayford, *EPA Interagency Agreement Final Report (Technical Research Report)*, EPA-600/2-78-085, 106pp (1978).
12. B. S. Shasha, D. Trimnell and F. H. Otey, *J. Polym. Sci., Poly Chem. Ed.*, **19**, 1891–1899 (1981).
13. E. B. Bagley, G. F. Fanta, R. C. Burr, W. M. Doane and C. R. Russell, *Polym. Eng. Sci.*, **17**(5), 311–316 (1977).
14. M. O. Weaver, E. B. Bagley, G. F. Fanta and W. M. Doane, *Appl. Polym. Symp.*, **25**, 97–102 (1974).

15. E. H. Pryde, *J. Am. Oil Chem. Soc.*, **56**, 849–854 (1979).
16. E. H. Pryde, *J. Am. Oil Chem. Soc.*, **56**, 719A–725A (1979).
17. J. P. Friedrich, E. W. Bell and L. E. Gast. *J. Am. Oil Chem. Soc.*, **42**, 643–645 (1965).
18. J. P. Friedrich, and R. E. Beal, *J. Am. Oil Chem. Soc.*, **39**(12), 528–533 (1962).
19. W. R. Miller, H. M. Teeter, A. W. Schwab and J. C. Cowan, *J. Am. Oil Chem. Soc.*, **39**, 173–176 (1962).
20. E. W. Bell and L. E. Gast, *J. Coat. Technol.*, **50**(636), 81–87 (1978).
21. R. E. Beal, L. L. Lauderback and J. R. Ford, *J. Am. Oil Chem. Soc.*, **52**, 400–403 (1975).
22. P. E. Throckmorton and E. H. Pryde, *J. Am. Oil Chem. Soc.*, **49**, 643–648 (1972).
23. E. H. Pryde and J. C. Cowan, *Topics in Lipid Chemistry*, F. D. Gunstone, Ed., **2**, Logos Press Limited (now available from John Wiley and Sons, Inc., New York), Chapter 1 (1971).
24. E. N. Frankel and E. H. Pryde, *J. Am. Oil Chem. Soc.*, **54**, 873A–881A (1977).
25. E. H. Pryde, D. J. Moore, J. C. Cowan, W. E. Palm and L. P. Witnauer, *Polym. Eng. Sci.*, **6**, 60–65 (1966).
26. R. A. Awl, E. N. Frankel, E. H. Pryde and J. C. Cowan, *J. Am. Oil Chem. Soc.*, **49**, 222–228 (1972).
27. R. B. Perkins, Jr., J. J. Roden III and E. H. Pryde, *J. Am. Oil Chem. Soc.*, **52**, 473–477 (1975).
28. W. R. Miller, W. E. Neff, E. N. Frankel and E. H. Pryde, *J. Am. Oil Chem. Soc.*, **51**, 427–432 (1974).
29. D. J. Moore and E. H. Pryde, *J. Am. Oil Chem. Soc.*, **45**, 517–519 (1968).
30. W. R. Miller and E. H. Pryde, *J. Am. Oil Chem. Soc.*, **55**, 469–470 (1978).
31. W. L. Kohlhase, E. N. Frankel and E. H. Pryde, *J. Am. Oil Chem. Soc.*, **54**, 506–510 (1977).
32. A. W. Schwab, L. E. Gast and H. E. Kenney, U.S. Patent 4,218,332 (August 19, 1980).
33. C. S. Nevin and B. F. Moser, *J. Appl. Polym. Sci.*, **7**, 1853–1866 (1963).
34. F. L. Thomas and L. E. Gast, *J. Coat. Technol.*, **51**(657), 51–59 (1979).
35. W. J. Schneider and L. E. Gast, *J. Coat. Technol.*, **51**(654), 53–57 (1979).
36. H. J. Nieschlag, J. A. Rothfus, V. E. Sohns and R. B. Perkins, Jr., *Ind. Eng. Chem., Prod. Res. Dev.*, **16**, 101–107 (1977).
37. Anonymous, *Chem. Week*, 14 (October 22, 1980).
38. E. N. Frankel, W. E. Neff, F. L. Thomas, T. H. Khoe, E. H. Pryde and G. R. Riser, *J. Am. Oil Chem. Soc.*, **52**, 498–504 (1975).
39. E. J. Dufek, F. L. Thomas, E. N. Frankel and G. R. Riser, *J. Am. Oil Chem. Soc.*, **53**, 198–203 (1976).
40. W. R. Miller, E. H. Pryde and G. R. Riser, *J. Am. Oil Chem. Soc.*, **55**, 469–470 (1978).
41. H. J. Nieschlag, J. W. Hagemann, I. A. Wolff, W. E. Palm and L. P. Witnauer, *Ind. Eng. Chem. Prod. Res. Dev.*, **3**, 146–149 (1964).
42. E. H. Pryde, D. J. Moore, H. M. Teeter and J. C. Cowan, *J. Polym. Sci.*, **58**, 611–620 (1962).
43. E. H. Pryde, R. A. Awl, H. M. Teeter and J. C. Cowan, *J. Polym. Sci.*, **59**, 1–11 (1962).
44. R. A. Awl, W. E. Neff, D. Weisleder and E. H. Pryde, *J. Am. Oil Chem. Soc.*, **53**, 20–26 (1976).
45. N. Devia, J. A. Manson, L. H. Sperling and A. Conde, *J. Appl. Polym. Sci.*, **24**, 569–582 (1979).
46. L. H. Princen, *J. Am. Oil Chem. Soc.*, **56**, 845 (1979).
47. L. H. Princen, *Econ. Bot.*, in press.
48. K. D. Carlson, W. J. Schneider, S. P. Chang and L. H. Princen, *New Sources of Fats and Oils*, E. H. Pryde, L. H. Princen and K. D. Mukherjee, Eds., American Oil Chemists' Society, Champaign, Illinois (1981).

Polymer Degradation

H. H. G. JELLINEK

Department of Chemistry, Clarkson College of Technology, Potsdam, NY 13676, USA.

As soon as polymers became commercial products, the problem of their stability as affected by the environment (i.e. the sun's radiation in conjunction with the oxygen of the air) arose. Research in this area started about eighty years ago and soon became quantitative. Work was undertaken with the view of developing the interpretation of degradation processes systematically. This was possible only for the relatively simple polymers such as polystyrene, polymethylmethacrylate and other similar ones. However, polymers having more complicated structures such as the high temperature polymers are not amenable to such a systematic treatment. However, many generalizations pertaining to the stability of such polymers have emerged.

Probably, one of the first scientists engaged in quantitative polymer degradation problems was H. F. Mark[1] (1922) one of the founders of modern polymer science. The polymer was a natural one, i.e. starch; random chain scission due to hydrolysis was elaborated. Kuhn and Freudenberg[2] studied the random degradation of cellulose. One of the first degradation studies on a synthetic polymer (i.e. polystyrene) was carried out by H. Staudinger and A. Steinhofer (1935). H. Staudinger fought successfully for the concept of macromolecules. Two workers, who contributed much to the field of polymer degradation, should not be forgotten. The late A. V. Tobolsky[4] (chemrheology) and the late L. A. Wall[5] (degradation kinetics, mechanisms and thermodynamics).

(1) THERMAL DEGRADATION KINETICS

Soon degradation mechanisms were developed for polymers of relatively simple structures. Here, a whole 'degradation spectrum'[6] applies. At the one extreme end, the random chain scission process prevails (valid for many thermal, oxidative, photo-, photo-oxidative, high energy and hydrolysis degradation reactions). A monodisperse chain polymer is considered subject to main chain link scission purely according to the rules of probability. All links are assumed to be of equal strength and accessibility irrespective of their position in the polymer chain. The rate of rupturing links is proportional to the number of main chain links in the sample at time t.

An important concept here is the degree of degradation α which is the probability of rupturing one main chain link at a certain stage of the degradation process. Another one is the average number of scissioned main chain links $\bar{s}$ in one original chain at the same stage of degradation,

$$\alpha = \frac{\bar{s}}{DP_0 - 1} \cong \frac{\bar{s}}{DP_0} \tag{1}$$

where $DP_0 - 1$ is the number of main chain links in an original polymer chain DP_0. Further,

$$\alpha = \frac{1}{DP_{n,t}} - \frac{1}{DP_{n,o}} = k_{ir}t \tag{2}$$

where $\overline{DP}_{n,t}$ and DP_0 are the number average chain lengths at time t and $t = 0$, respectively; k_{ir} is the random chain scission rate constant. The number size distribution at t or the so-called 'random chain scission distribution' n_{DP} which is almost identical with the most probable distribution, is ($\bar{s} \geqslant 6$),

$$n_{DP} = \alpha^2(1 - \alpha)^{DP - 1} \tag{3}$$

The random chain scission process is characterized by a very steep decrease in number average chain-length with time while scarcely any monomer is generated.

At the other extreme end of the 'degradation spectrum' is the so-called 'reverse polymerization' or 'depolymerization'. It consists in the simplest case of initiation at chain ends, followed by a zip or rapid depropagation reaction consisting of monomer coming off the chain end. This is the exact opposite process of propagation and is referred to as depropagation. The polymer chain can be zipped completely or only partly. In the latter instance, a second order termination process comes into play which is exactly identical with that operating in polymerization. This reaction is characterized by leaving the chain-length completely untouched in the residue or by a moderate decrease in chain-length accompanied in either case by generation of plenty of monomer.

If $\bar{\varepsilon}$ is the average kinetic chain length (i.e. monomer zipped off on the average from one initiated chain end) then only monomer, m, is formed, which is given by ($\bar{\varepsilon} \gg DP$),

$$\ln \frac{m_0}{m_0 - m_{t,total}} = 2k_{ie}t \tag{4}$$

m_0 and m_t is the monomer at $t = 0$ and t respectively and k_{ie} is the depolymerization rate constant.

For $\bar{\varepsilon} < DP_0$ and as long as the number of chains remain constant, the rate of monomer formation is for this case (second order termination),

$$\frac{dm_{1,\text{total}}}{dt} = 2k_d \left(\frac{k_{ie}}{k_t}\right)^{\frac{1}{2}} \left(\frac{m_0 V_m}{DP_0}\right)^{\frac{1}{2}} (m_0 - m_{1,\text{total}})^{\frac{1}{2}} \tag{5}$$

Here k_d and k_t are the rate constants for depropagation and second order termination, respectively, V_m and DP_0 are the unit molar volume and the initial chain length of the monodisperse polymer, respectively. Equation (5) is only valid for the initial stages of the depolymerization process.

Table I gives the % monomer yields and overall energies of activation for various polymer.

TABLE I

% Monomer Yields and Energies of Activation

	% Monomer Yield	Energy of Activation (kcal/mol)
(1) Polyoxymethylene	100	—
(2) Poly-(α-methylstyrene)	100	—
(3) Polystyrene	41	55
(4) Poly-(m-methylstyrene)	45	52
(5) Polymethylmethacrylate	95	52
(6) Polymethylacrylate		34
(7) Polytetrafluoroethylene	96	81
(8) Polychlorotrifluoroethylene	26	57
(9) Poly(methyacrylonitrile)	85	—
(10) Polyacrylonitrile	0	58
(11) Polyisobutene	20	49
(12) Polybutadiene	20	62
(13) Poly(ethylene oxide)	4	46
(14) Poly(propylene oxide)	3	20
(15) Polypropylene	0	63
(16) Polyethylene	1	—
(17) Poly(butylmethacrylate)	0	high yield of isobutene
(18) Polyvinylchloride	0	32
(19) Polyvinylacetate	0	17

(2) CEILING TEMPERATURES

Under certain given conditions and temperatures, an equilibrium may be established between propagation and depropagation.[7]

$$RH_{n-1} + m_1 \underset{\text{depropagation}}{\overset{\text{propagation}}{\rightleftharpoons}} RH_n$$

$m \sim$ monomer

If m, is in a standard state (e.g. 1 M or 1 atm), the temperature where the equilibrium is established is called the ceiling temperature, T_C, given by,

$$T_C = \frac{\Delta H_P^\circ}{S^\circ + R \ln [m_1]} = \frac{\Delta H_P^\circ}{R \ln (A_P[m_1]/d)} \tag{6}$$

If a gaseous (g) monomer polymerizes to a condensed polymer (c) and the moment is in a standard state (1 atm), then,

$$T_{g,c} = \frac{\Delta H_{P,gc}^\circ}{\Delta S_{gc}^\circ} \tag{7}$$

Here, ΔH_P° is the heat of polymerization, ΔS° the entropy of polymerization, R the gas constant, A_P and A_d the pre-exponential factors in the Arrhenius equation for propagation and depropagation, respectively. $[m_1]$ is 1 M monomer solution.

A few examples for ceiling and decomposition temperatures T_c and T_d, respectively, are given below. Polytetrafluoroethylene does not fit into the series. C—C bonds in radicals are usually weaker than those in the polymers, for $PFTE$ the reverse seems to be the case. Usually $T_d > T_c$.

Polymer:	PFTE	PE	PPR	PS	PMMA	PMAN	PIB	P-α-S
$T_d(^\circ C)$	510	400	380	360	330	7220	340	250
$T_c(^\circ C)$	680	400	300	230	320	177	50	7

(3) SCISSION ON A MOLECULAR SCALE[8]

If a main chain link ruptures, the scission is not at once a permanent one. The polymer radical-ends have to diffuse out of a medium cage either by segmental or translational diffusion before the rupture becomes permanent. Fortunately, most often the break is mended before this happens due to the repeated collisions of the radicals in the cage. Thus, escape from the cage is a relatively rare event.

(4) PHOTOLYSIS

As mentioned above, scission by u.v. radiation results in many cases in random chain scission, cross-linking can occur simultaneously. The random scission rate constant is given by,[9]

$$k_{ir} = \frac{\phi_s^- I_{abs.}}{[n_0]} \tag{8}$$

Here, ϕ_s^- is the quantum yield for chain scission (i.e. scission per light quantum absorbed); for solutions, $[n_0]$ is the concentration of initial main chain links and I_{abs} is the absorbed light intensity. If the optical cell has plane windows, Beer's Law is obeyed and the monochromatic light absorption is small, Eq. (8) becomes,

$$k_{ir} = \frac{\phi_s^- 2.303 E I_0}{[n_0]l} = \phi_s^- k_n I_0 \tag{9}$$

Here E is the absorbance, I_0 the incident light intensity, l the length of the optical cell and k_2 a constant. Solvents can play an important role in photodegradation forming radicals on irradiation and acting as transfer agents.

It is a noteworthy fact that many polymers which should not absorb near U.V. radiation according to their chemical structure nevertheless do so. This is due to impurities, defects in the polymer chains, oxidation during manufacture and later handling. All these abnormalities are of a chromophoric nature with respect to near U.V. radiation or the sun's radiation (>300 nm). Protection against photodegradation is afforded by screening and quenching compounds.

(5) RANDOM CROSS-LINKING

Some brief remarks are in order about random crosslinking. Here crosslinks are formed at random between radical-sites along two or more chains. A degree of crosslinking is defined, q, which is the ratio of the average number of crosslinked structural units in an original chain; q is analogous to $\alpha = \bar{s}/(DP_0 - 1)$ for chain scission.[10]

For an initially monodisperse sample, $\overline{DP'_n}$ tends to $2DP_{n,o}$ where DP'_n is the number average chain length of the crosslinked polymer. If the chains have initially a random distribution, then $\overline{DP'_n}$ tends towards $1.33DP_{n,o}$. Eventually, a gel point is reached where each chain has, on the average, one crosslinked unit. The sol fraction reaches zero if scission and crosslinking occur simultaneously.

(6) HIGH ENERGY RADIATION

High energy radiation (x, α, β, γ-rays) affects polymer molecules indiscriminately while U.V. radiation needs chromophores on the chains which absorb the respective radiation.[11] During high energy degradation electrons are eliminated at random from the polymer molecule leading to chain-scission and cross-linking or both. Protection from the primary process cannot readily be provided. Some additives, however, can be used to absorb part of the radiation. Aromatic molecules serve this purpose well. Interference with the subsequent chain reactions can be achieved. Radical chains are interrupted by additives. High energy radiation can lead to a random chain scission process and follows, therefore, Eq. (2) adapted to this process,

$$\frac{1}{\bar{M}_{n,D}} - \frac{1}{M_{n,0}} = \frac{G(s)D}{100N_A} \tag{10}$$

$\bar{M}_{nD}$ and $M_{n,o}$ are the number average molecular weights which has received a radiation dose D e.v. and that at $t = 0$, respectively; $G(s)$ is the G-value with respect to chain scission (i.e. the number of main chain-link scissions for each 100 e.v. absorbed by the polymer). N_A is Avogadro's number.

(7) OXIDATIVE DEGRADATION

Oxidation of polymers is an autocatalytic process which generally proceeds via hydrogen peroxide formation and decay leading to chain scission, crosslinking or both.[12] If tertiary H-atoms are present, the initial attack takes place at these atoms leading to hydrogen peroxides. Thus, the fewer H-atoms are present in the chain, the more stable it will be. Some of the high temperature resistant polymers have none or very few H-atoms.

The fundamental oxidation mechanisms was investigated by Farmer and his co-workers.[13] They noted that double bonds are not attacked directly by oxygen but tertiary or α-H-atoms next to the double bond react first. The mechanism of oxidation, which in its essence is still valid today was formulated by Bolland and Gee[14] and others of Sir Eric Rideal's School. The mechanism in a simplified form is given below.[15] It should be noted that oxygen can usually not diffuse into the crystalline regions of polymers except in cases where the crystalline part has an equal or even smaller density than the amorphous one. Thus, in general, the percentage crystallinity actually increases during oxidation.

General (simplified) Oxidation
Mechanism (Bolland 1946/48)[14]

Initiation	(1) $RH \rightarrow R^{\cdot} + (H^{\cdot})$ Rate $= v_i$	
Propagation	(2) $R^{\cdot} + O_2 \rightarrow RO_2^{\cdot}$ k_2	
	(3) $RO_2^{\cdot} + RH \rightarrow ROOH + R^{\cdot}$ k_3	
Termination	(7) $RO_2^{\cdot} + RO_2^{\cdot}$	inactive low molecular
	(8) $RO_2^{\cdot} + R^{\cdot}$	weight products
	(9) $R^{\cdot} + R^{\cdot}$	

For relatively small oxygen concentrations, the rate of oxygen consumption (which is practically equal to the rate of hydroperoxide formation) becomes independent of the oxygen concentration.

$$-\frac{d[O_2]}{dt} = \left(\frac{d(ROOH)}{dt}\right)_{formation} = k_3[RH]\left[\frac{v_i}{2k_7}\right]^{\frac{1}{2}} \tag{11}$$

At high oxygen concentrations,

$$-\frac{d[O_2]}{dt} = \left(\frac{dROOH}{dt}\right)_{formation} = v_i^{\frac{1}{2}}k_2k_9^{-\frac{1}{2}}[O_2] \tag{12}$$

Antioxidants can act in three ways. They can compete with H-abstraction, they can trap radicals (i.e. chain carriers) and they can decompose hydroperoxides leading to harmless products. Some antioxidants can retard the initiation reaction (preventive antioxidants).

(8) OXIDATIVE CATALYZED DEGRADATION

Catalytic oxidation of polymers by heavy metals in bulk and solution is quite wide-spread. Such catalysis is, for instance, of great importance for the electrical industry. Polyethylene and isotactic polypropylene are used for insulation of copper conductors. It was soon noticed that these polymers deteriorate much faster in presence of this metal or its oxides than in their absence. Polyethylene was studied by Allara and co-workers[16] and isotactic polypropylene in our laboratory.[17] The latter work will be briefly indicated. Physical-chemical methods from the kinetic-mechanistic standpoint were used (i.e. oxygen consumption, I.R. reflectance spectro-photometry, chromatography, diffusion of Cu-ions into the polymer). Also, a purely physical method was employed, i.e. ion-scattering, which confirmed results obtained by the other methods. Well defined Cu, $CuO_{0.67}$ and CuO/polymer interfaces were prepared. A mechanism for the Cu and Cu-oxide catalyzed oxidation was formulated on the basis of the experimental results. Initially, a polymer-copper complex is formed, which decomposes into radicals and Cu^+ ions,

$$(RHCu^{+2}O_2^-) \xrightarrow{k_i} R^{\cdot} + HO_2^{\cdot} + Cu^+$$

Cu^+ is at once reoxidized to Cu^{++} by the oxygen in the air. During oxidation of the polymer, carboxylated ions are formed which serve as anions for Cu-ions diffusing into the bulk of the polymer. These metal ions decompose ROOH-groups catalytically thus accelerating the oxidation process consisting of a redox reaction.

$$ROOH + Cu^{++} \rightarrow RO_2^{\cdot} + Cu^+ + H^+$$

$$ROOH + Cu^+ \rightarrow RO^{\cdot} + Cu^{++} + OH^-$$

The energy of activation for normal ROOH-decomposition is 33 kcal/mol whereas the catalyzed one is only 24 kcal/mol. This is the main catalytic effect of this oxidation process. We have also carried out measurements on Cu-ions diffusing from the interface into the bulk of the oxidized polypropylene. The diffusion coefficients are of an order of magnitude of 10^{-8} cm^2/s at a temperature just above 100°C.

The formation and decay of hydroperoxide groups was followed by I.R. reflectance spectrophotometry in presence and absence of reaction products. It should also be noted that the degradation mechanism is different in presence or absence, respectively, of reaction products such as water and CO_2. These were monitored chromatographically during oxidation.

(9) SENSITIZED PHOTOLYSIS AND PHOTO-OXIDATION

Photodegradation can be sensitized unintentionally due to impurities or intentionally by incorporating chromophores such as $>C=O$ groups. This is of significance for eliminating solid plastic waste.[18] A sensitizer may react as follows,

$$S + hv \rightarrow S^* \text{(excitation)}$$

$$S^* \rightarrow X^{\cdot} + Y^{\cdot}$$

These radicals may lead to H-abstraction from the polymer and eventually to chain scission or cross-linking.

Photo-oxidation is widespread due to the combined action of the near U.V. radiation from the sun and the presence of oxygen.[19,20] The latter accelerates photodegradation and prevents crosslinking as O_2 reacts with radicals. Oxygenated groups are generated and gases such as CO_2, CO and H_2O are evolved. Hydroperoxide groups are formed and decay. Keto-groups decompose by the Norrish type I reaction:

$$R_1-\underset{\underset{O}{\|}}{C}-R_2 \rightarrow R-CO^{\cdot} + R_2^{\cdot}.$$

(C = O) is the main chromophore for photo-oxidation.

Photo-oxidation of polystyrene has been studied in detail.[21] At wavelengths 280 nm the phenyl groups are the cause of degradation, but at

300 nm, degradation in preferentially due to oxygen groups in the polymer introduced during polymerization and later handling. The author discovered already in 1942 weak sites in polystyrene chains where main chain links scission faster than the normal ones.[22] This led to the elaboration of a 'weak link' chain scission process.[22] Polystyrene should not absorb any (solar) radiation in the near U.V. region according to its chemical composition. The absorption is due to defects as pointed out above.[23] Geuskehs and David[24] investigated the photo-oxidation of polystyrene sensitized by benzophenone. The latter is excited to the triplet state and then abstracts a tertiary H-atom from the polymer. Subsequently, hydroperoxide groups are formed which lead to deterioration.

It should be noted that singlet oxygen plays a role in photo-oxidation reactions.[23] Photo-degradation can be used for reproduction purposes (photo-resists) and also in the removal of plastic wastes.

(10) MORPHOLOGY AND POLYMER STABILITY

Morphology has been mentioned repeatedly as playing a role for the stability of polymers.[24] Polymer chains which show a decrease in flexibility exhibit increasing thermal and oxidative stability. This decrease in flexibility can be achieved by crystallization, cross-linking, incorporation of aromatic groups into the polymer chain, ladder structures and tacticity. The interfacial area between amorphous and crystalline regions is especially susceptible to scission. Here, the chains at the edge of the crystalline area bend around which adds strain energy to the links in the bends. This strain energy forms part of the energy of activation for bond scission. This was shown to be the case experimentally.[25] A whole range of Nylon-66 films of equal % crystallinity but different average spherulite diameters were prepared. According to the 'interfacial' degradation theory which was elaborated,[26] the chain scission rate constants should increase with the amorphous/crystalline interfacial area or with decreasing average spherulite diameter. This was found, by experiment, to be the case.

(11) POLLUTANT-POLYMER REACTIONS AND
TOXIC GAS EVOLUTION FROM POLYMERS[27]

The effect of pollutants (NO_2, SO_2, O_3, near U.V. radiation and O_2) has been studied extensively. It would lead too far to go into details here. Polymers can also function as pollutants giving off toxic gases.[28] Thus, on burning (fires) large quantities of CO and smoke are generated which are often fatal. But also heating polymers moderately produces toxic volatiles which can be quite dangerous especially in closed spaces. Thus, any polymer which contains nitrogen such as poly-amides, polyimides, polyurethanes and others evolve appreciable quantities of HCN. We have studied the mechanism of HCN-formation in detail. It was found that this HCN-evolution can be completely inhibited by adding finely dispersed copper powder or easily decomposable Cu-compounds to the polymer. Actually HCN-generation is not suppressed but HCN-gas is at once catalytically oxidized and decomposes.[29]

(12) IGNITION AND BURNING

The theory of the fundamental aspects of polymer ignition and burning are still in their early stages.[30] A great amount of work has been done on fire retardents,[34] but this field appears to be largely empirical.

Ignition of polymers is due to an explosion of the volatile gas-mixtures evolved from the polymer during heating (plus O_2). The polymer itself does not burn. The heat of the flame accelerates the evolution of volatiles from the polymer. Semenov[32] elaborated mathematically the conditions for ignition. These considerations have been modernized. Another pioneer in this field was Hinshelwood.[36] Recently, a novel apparatus has been constructed by us (see Figure 1) for the purpose of studying surface regression or surface degradation of polymers, ignition points, flames and their composition, processes in flames and influence of oxidant gas flow rates.[33] Diffusion processes are involved depending on the location of the regressing polymer surface. It is of interest to note that the energy of activation depends on the location of the regressing surface, on the oxidant flow rate and also on the chemical nature of the polymer. Polymers such as polyimides do not burn but the generated char glows. The degradation of the solid polymer is frequently not oxidative at moderate oxidant flow rates but purely thermal. Oxygen is practically used up in the 'flame region' and does not reach the polymer surface. Thus, the surface regresses only due to thermal degradation. The experiments have some bearing on ablation (see below).

(13) MECHANICAL DEGRADATION

Mechanical degradation, such as chemorheology, ultrasonic degradation, braid analysis etc. cannot be discussed in the space available.[37] Ultrasonic irradiation of polymer solutions is a mechanical degradation process. Here, the cavitation process plays a decisive role and leads to chain scission.

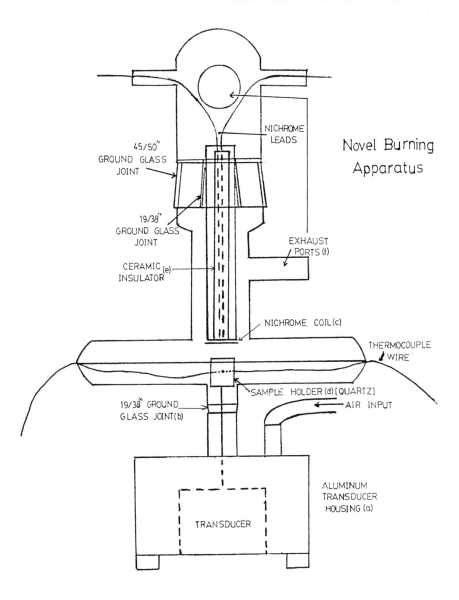

Novel Burning Apparatus

Mechanical degradation encompasses the field of chemorheology developed by Tobolsky.[4] Chain scission and crosslinking can be studied with elastomers, for instance, by stress relaxation as a function of time.

(14) ABLATION

Intensive research efforts have been made to synthesize high temperature resistant polymers for extreme environmental conditions. Ablation is an important process.[35] Space vehicles re-entering the earth's atmosphere would burn up completely without protection i.e. without ablative shields. Very high temperatures are generated by friction. Large ablation shields are fixed in front of space-vehicles which act as heat sinks. Initially polymers were used which leave no residue. Later polymers were taken which formed char-layers. Now only the latter type of polymers is employed i.e. crosslinked polymers and polymers which contain mainly aromatic structural units.

TABLE IIa[6,35]
Composition of typical ablators

Material	Resin binder	Fiber reinforcement	Low density filler	Low temperature sublimer
Silica—phenolic	Phenolic	Silica fabric		
Carbon—phenolic	Phenolic	Carbon fabric		
Nylon—phenolic; high density	Phenolic	Nylon fabric		Nylon fabric
Nylon—phenolic; low density	Phenolic		Phenolic microballoons	Nylon powder
Corkboard	Phenolic		Granular cork	
5026-39 HC/G	Phenolic— epoxy novolac	Random quartz fibers	Phenolic microballoons	
ESA 3560	Silicone	Random glass fibers	Phenolic microballoons, silica microspheres	
ESA 5500	Silicone	Random silica fibers	Phenolic microballoons	
SLA-220	Silicone	Random silica fibers	Silica microspheres	
SLA-561	Silicone	Random silica and carbon fibers	Phenolic microballoons, silica micro-spheres, granular cork	
Thermolag T-230	Phenolic— epoxy			Molybdenum hexa-carbonyl

Degradation processes (i.e. chain scission) are endothermic; in addition, products suffer phase changes melt, vaporize requiring latent heats. Also heat energy is dissipated by radiation from the hot ablators. Char-layers are good heat-insulators. Temperatures generated on re-entry reach about

3000°C. The ablative shields have to last long enough to afford full protection during re-entry. Space vehicles returning from the Moon or Mars have re-entry velocities of 11.0 km/s and 13.7 km/s, respectively. A body re-entering the earth's atmosphere at 7.3 km/s has a kinetic energy of 26.7 MJ/kg and would completely vaporize by the time the earth's surface is reached. Frictional heat starts to be generated at an altitude of 100 km where the pressure is about 10^{-6} atm.

Three polymer types are usually taken as ablatives: (1) phenolformaldehyde, (2) epoxy resins and (3) silicones. Also, fillers are added. The degradation mechanism i.e. surface degradation of ablatives is usually not the same as that of bulk degradation.

TABLE IIb[6,35]
Density and thermal conductivity of ablation materials

Density $(g\,cm^{-3})$	Material	Thermal conductivity $(W\,m^{-1}\,K^{-1})$
0.04–0.08	Isocyanurate foam	0.022
0.23	Filled silicone (SLA-561)	0.052
0.25	Filled silicone (SLA-220)	0.080
0.50	Filled silicone (ESA 3560)	0.097
0.53	Corkboard (Insulcork 2755)	0.069
0.54	Filled epoxy (5026-39 HC/G)	0.087
0.58	Nylon—phenolic (Low Density)	0.128
0.62	Foamed silicone (ESM 1004)	0.178
0.90	Filled silicone (ESA 5500)	0.225
1.15	Nylon—phenolic	0.173
1.45	Carbon—phenolic	1.040
1.60	Silica—phenolic	0.520
1.70	Graphite (ATJ)	90.3
1.96	Fused silica	0.690
2.20	Polytetrafluoroethylene	0.260

(15) HIGH TEMPERATURE RESISTANT POLYMERS

An important area is that of high temperature resistant polymers.[36] It received a great impetus from space activities especially in the 50's and 60's. One of the foremost pioneers in this field is C. S. Marvel who in his advanced 80's, continues his research with several co-workers.[37]

As pointed out above, thermal stability and also oxidative stability of polymers increase the less flexible polymer chains become and the fewer H-atoms they have. Especially, tertiary ones are vulnerable in polymer molecules. In addition, the various bond-strengths play an important role and so does polymer morphology. Resonance effects are also involved and bond angles should not be under strain. Polymer stability is often measured by thermal gravimetric analysis. However, this non-isothermal method can be misleading as in many cases isothermal degradation procedures give different stabilities. Isothermal work is better suited to a detailed kinetic and mechanistic analysis of polymers than TGA. The evaluation of TGA data can be ambiguous although it is very convenient for fast routine analysis.

The stiffness of polymer chains is very much increased by having aromatic and heterocyclic units in the chains. Stiffness is also increased in step-ladder and ladder structures. Here at least two bonds have to be ruptured in close proximity before a complete break in the chain is achieved. That these two events occur simultaneously or in rapid succession is unlikely. The first scission will have been mended before the second rupture occurs in the immediate vicinity. There are several types of high temperature polymers. Often a compromise has to be reached between stability and fabrication processes. If the chains are too stiff, fibers, for instance, cannot be produced. Types belonging to the high temperature polymers are as follows: (1) fluoropolymers, (2) chains with aromatic units, (3) chains with heterocyclic units, (4) boron-containing chains, (5) phosphorus containing chains, (6) silicon-containing polymers, (7) polysiloxanes, (8) polymetallosiloxanes, (9) co-ordination polymers, (10) miscellaneous polymers.

Poly-phenylenes are representative of the first group. p-polyphenylene is more stable than the m-compound. However, p-polyphenylene is too stiff for fiber forming. Introduction of a CH_3 group (polytolylene) gives more flexible chains but also less stable polymer. Perfluoropolyphenylenes, surprisingly, show similar stabilities as polyphenylene in inert atmospheres; in oxidizing surroundings fluorinated polyphenylenes appear to be less stable than the ordinary polyphenylenes. Groups which link units in chains can change the stability characteristics of polymers e.g. under oxidizing conditions:

$$-CO-O- > -CONH- > -S- > -CH_2- > -CO- > -CH_2CH_2-$$
$$> -O- > \{CF_2\}_3 \text{ (meta)}.$$

This sequence is valid for $\{p-C_6H_4-X\}$ polymers where X is the linking group. This was established by measuring the temperatures (°C) for 25% weight loss during exposure for two hours. DTA gave a different sequence:

$$> -O- > -SO_2- > -CH_2- > -CH_2CH_2- > -\underset{\underset{CH_3}{|}}{\overset{\overset{CH_3}{|}}{C}}-$$

For structures such as [heterocyclic ring—C_6H_4—x—C_6H_4] to which polypyromellitimides belong such as,

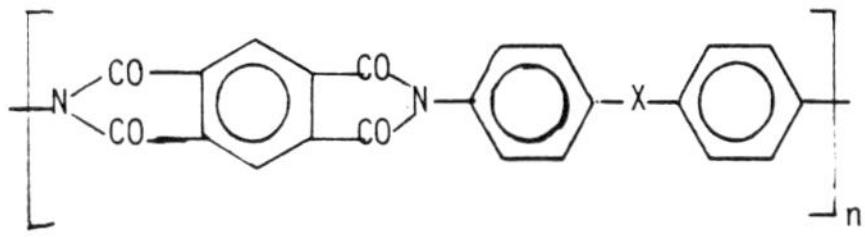

a sequence of stability for —X— is obtained during degradation in air at 100°C as follows,

Single bond > —S—

$$-SO_2 - > CH_2 - > -CO - > -SO-, \qquad -O-$$

The mellitimides can undergo profound changes in structure while only losing 1–2% in weight. This is indicated below.[36, Wright, 1981]

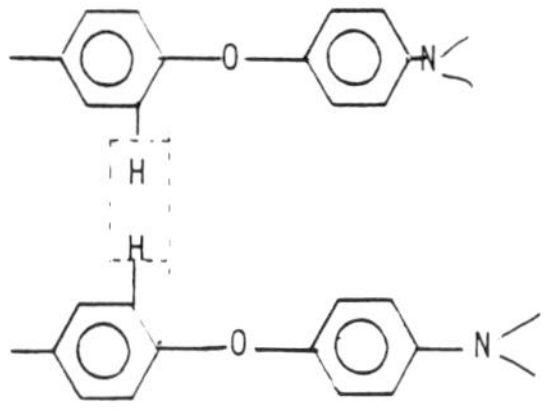

crosslinking
takes place

Such crosslinking can also happen via chain scission:

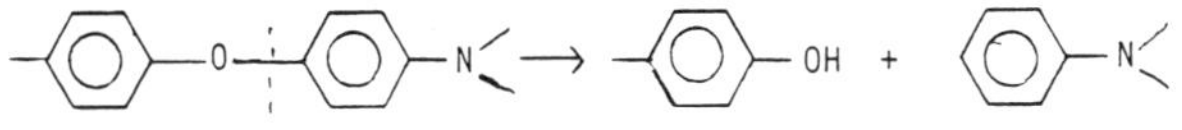

followed by

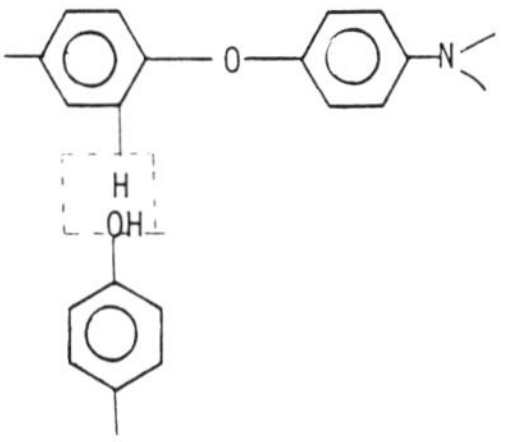

In inert atmospheres, a very large number of volatile compounds are produced on heating to 600°C. The pyro-mellitimide suffer wide spread chain scission.

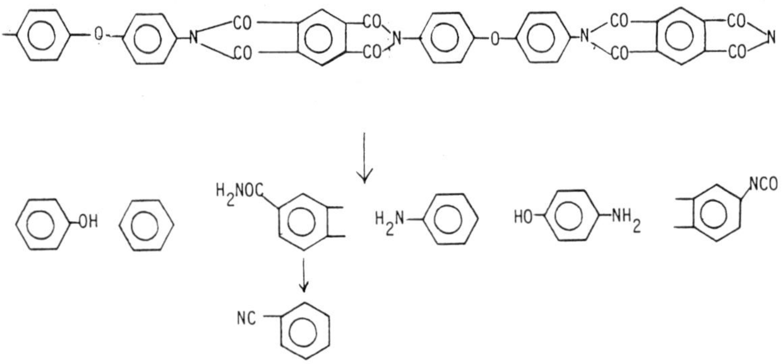

Slight variations in chemical structure can have a considerable effect on the nature of the products and also on the stability. An imide having a CH_2 group between phenyl rings is less stable than one having an oxygen instead. Table III gives a comparison of thermal stabilities of imides.

TABLE III[36, Wright, 1981]
Comparison of thermal stabilities of polyimides in inert and oxidizing atmospheres

Polyimide type	Temperature (°C) for 10% weight loss		Temperature difference (°C)
	Nitrogen	Air	
Condensation	553	518	35
Acetylenic	520	455	65
Thermoplastic	496	484	12
Norbornene	472	435	37
Bismaleimide	394	390	4

Not all types of high temperature resistant polymers can be discussed here. Reference should be made to the pertinent literature.[36] Polyimides are the most suitable stable polymers which are available. But also step-ladder and ladder polymers are very useful i.e.

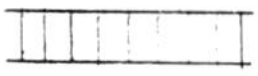

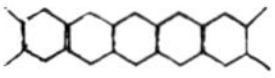

One such system is polybenzimidazopyrrolone(pyrrones),

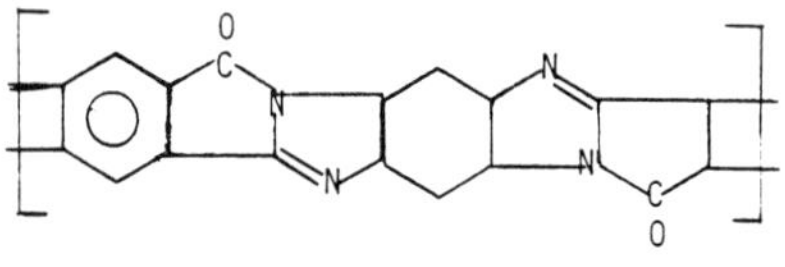

An imide has been synthesized which does not contain any H-atoms at all i.e.

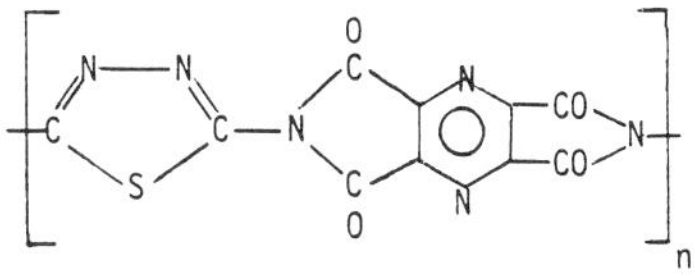

TGA does not show exceptionally high thermal stability (in air it appears slightly more stable than under N_2) but under stress this polymer is appreciably more stable than other polyimides. For recent advances of the high temperature field, Cassidy's book should be consulted.[36]

Table IV gives the thermal stabilities of some high temperature polymers under nitrogen.

TABLE IV(a and b)[36, Arnold, 1979]
Relative thermal stability of high-temperature polymers in nitrogen

(a)

Polymer Structure	Rank and Designation Number	Type	PDT[†]
	1	AC	675
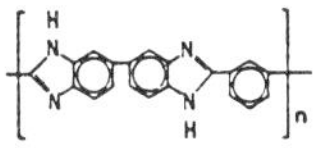	2	SL-2	650
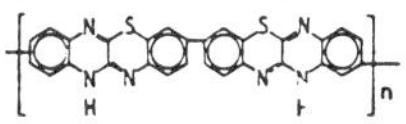	3	SL-4	650
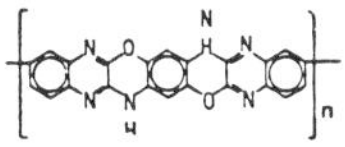	4	SL-7	~640
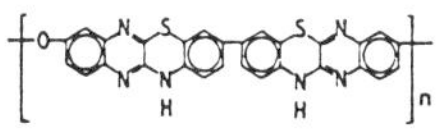	5	SL-4	600-700
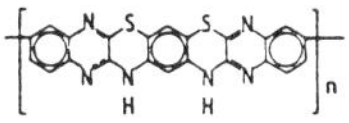	6	SL-7	500-700
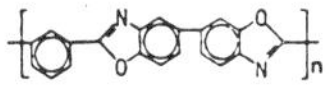	7	SL-2	590
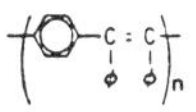	8	AC	560
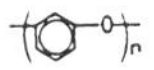	9	AF	555
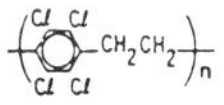	10	AA	515
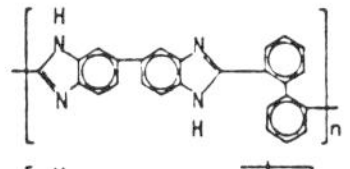	11	SL-2	510
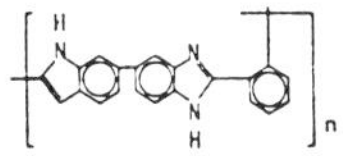	12	SL-2	510

(b)

Polymer Structure	Rank and Designation Number	Type	PDT[†]
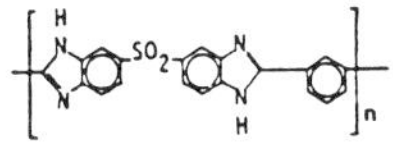	13	SL-2	490
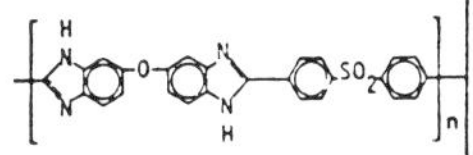	14	SL-2	480
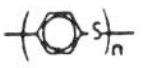	15	AF	480
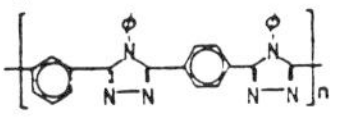	16	AC	475
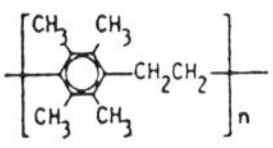	17	AA	475
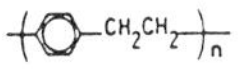	18	AA	470
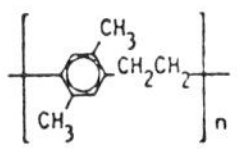	19	AA	465
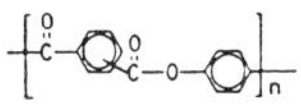	20	AF	465
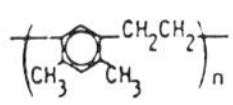	21	AA	465
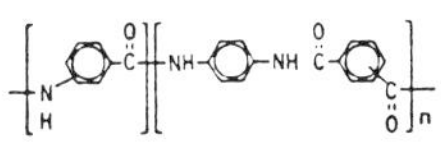	22	AF	425
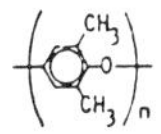	23	AF	410
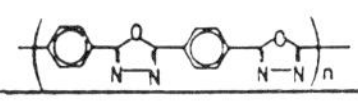	24	AC	400

†Polymer decomposition temperature.

Some thermo-oxidative stabilities are listed in Table V.

TABLE V[36, Arnold, 1979]

Relative order of thermoxidative stability in air of heat -resistant polymers

Polymer Structure	Rank and Designation Number	Type	PDT[†]
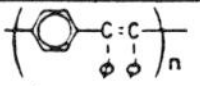 *(structure)*	1	AC	700
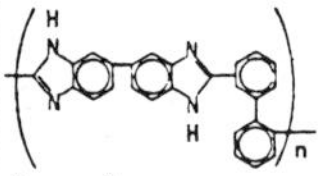 *(structure)*	2	SL-2	595
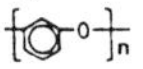 *(structure)*	3	AF	560
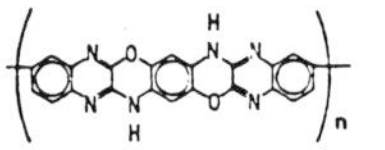 *(structure)*	4	SL-7	510
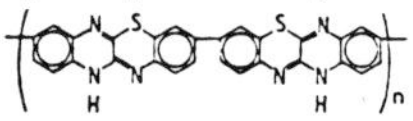 *(structure)*	5	SL-4	500
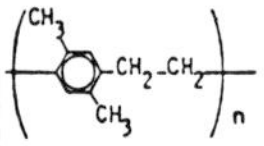 *(structure)*	6	AA	500[‡]
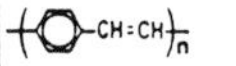 *(structure)*	7	AC	~500[‡]
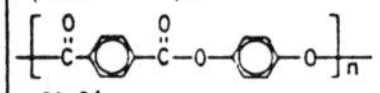 *(structure)*	8	AF	465
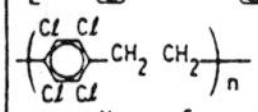 *(structure)*	9	AA	460
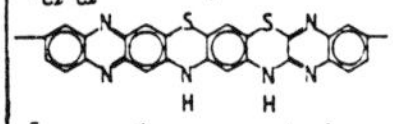 *(structure)*	10	SL-8	460
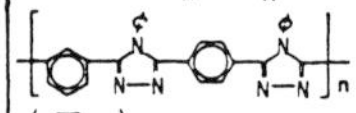 *(structure)*	11	AC	450
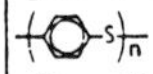 *(structure)*	12	AF	450
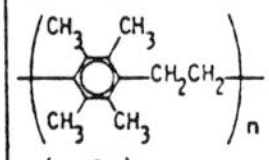 *(structure)*	13	AA	420
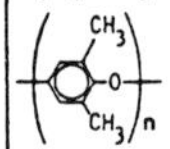 *(structure)*	14	AF	330
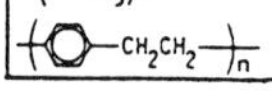 *(structure)*	15	AA	240

†Polymer decomposition temperature (TGA inflection temperature).
‡Starts to decompose at 270°C.
§Starts to decompose at 200°C.

A comparison of the effect of structure on the stability of some polybenzimidazoles is shown below,

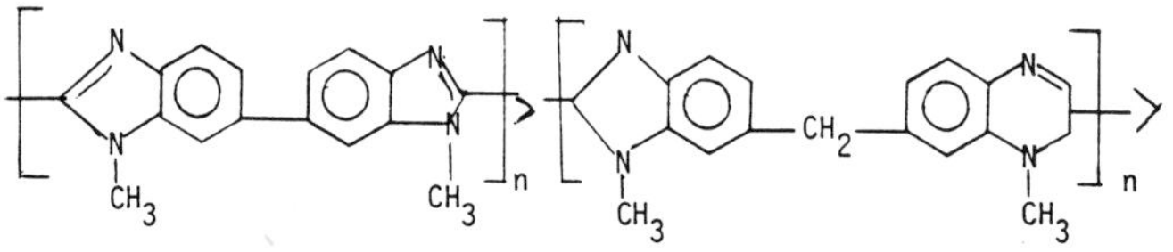

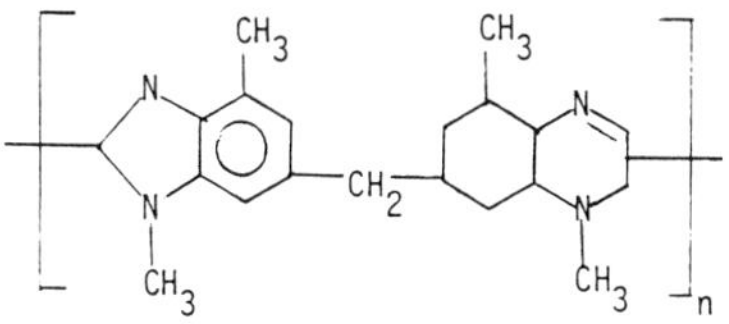

Most of the stabilities were measured by TGA.

(16) SURFACE DEGRADATION

Surface degradation is an important area of degradation. Changes of the surface properties of polymers can be caused unintentionally or intentionally.[38] The slightest oxidation of the uppermost surface layer can have profound effects on the adhesive properties of the polymer transforming the surface from a low to a high energy one. This is due to the generation of oxygenated groups on the surface. This was very difficult to investigate. I.R. reflectance methods, for instance, penetrate the surface to some extent. But new instrumental, non-destructive methods such as ESCA (electron spectroscopy for chemical analysis) have given a new impetus to surface science. The uppermost surface layer i.e. the first monolayer of a polymer surface can now be studied in detail.

Any reactions of polymers due to weathering (i.e. due to the sun's radiation in conjunction with oxygen and other chemicals) must start at the polymer surface and subsequently penetrate into the bulk polymer. Thus any deterioration will be at a maximum at the surface. The degree of degradation for random chain scission will be a function of the distance from the surface. α will decrease exponentially from the surface, e.g. for photodegradation,

$$\alpha_x = \alpha_0 e^{-at}$$

$$\alpha_0 = k_{ir} I_0 \frac{a}{m_1}$$

The molecular size distribution will be at a different stage in each succeeding polymer layer. α_0 is the degree of degradation at the surface $x = 0$, 'a' is an optical constant, I_0 is the incident light intensity and m_1 is the amount of monomer in the polymer at $t = 0$.

Oxygen uptake and emission of volatiles becomes frequently diffusion controlled in certain temperature ranges. Diffusion often plays an important role in degradation reactions. It can be recognized by relatively small energies of activation and by changing the thickness of polymer films.

The new instrumental methods leading to a very detailed characterization

J

of polymer surfaces often reveal that they have different characteristics from those of the bulk polymer. In the case of block-copolymers one type of block frequently lies preferentially in the polymer surface, which then shows the properties of that particular polymer.

A number of surface regression studies have been carried out but usually the apparatus has not been versatile enough for obtaining a clear picture of the process. Surface degradation is also referred to as surface regression or surface pyrolysis.

(17) MISCELLANEOUS

It would lead too far to discuss the multitude of experimental methods here. However, one recent method should be mentioned i.e. the factor-jump thermo-gravimetry.[39] Also NMR, ESR, x-ray, mass-spectrometry and various surface methods and weight loss methods etc. are used for studying degradation processes.[40]

A substantially abbreviated and non-exhaustive survey has been given here due to space limitation. However, it is hoped that an impression of the importance of research in degradation and stability of polymers has been conveyed.

References

1. K. H. Meyer, H. Hopff and H. F. Mark, *Ber.*, **62**, 1103 (1922); see also *J. Chem. Ed.*, **58**(7), 527 (1981).
2. W. Kuhn, *Ber.*, **63**, 1503 (1930); K. Freudenberg, W. Kuhn *et al.*, *Ber.*, **63**, 1510 (1930).
3. H. Staudinger and A. Steinhofer, *Ann.*, **517**, 35 (1935); see also H. Staudinger, Arbeitserinnerungen, Dr. Alfred Hüttig Verlag, G.M.H. Heidelberg, 1961.
4. A. V. Tobolsky, *Properties and Structure of Polymers*, John Wiley and Sons, Inc., New York (1960). (This is only one example).
5. L. A. Wall, *Pyrolysis*, Chap. V, Analytical Chemistry of Polymers, G. M. Kline Ed., Interscience, New York (1962); with M. Tryon, *Oxidation of Polymers*, Chap. 19, **2**, Autoxidation and Antioxidants, W. O. Lundberg, Ed., Interscience, New York (1962); L. A. Wall, Ed., *The Mechanism of Pyrolysis, Oxidation and Burning of Organic Materials*, N.B.S. Special Publication 357 issued June 1972; L. A. Wall Ed., *Fluoro-polymers*, Wiley-Interscience, New York (1972). (These are only a few examples).
6. H. H. G. Jellinek, *Degradation of Vinyl Polymers*, Academic Press Inc., New York (1955); H. H. G. Jellinek, Chapter 1 in Aspects of Degradation and Stabilization of Polymers, H. H. G. Jellinek, Ed., Elsevier, New York (1978).
7. Ref. 6 (1978), W. K. Bustfield, Chapter 2.
8. H. H. G. Jellinek and J. J. Lichorat, *Polymer J.*, **12**, 347 (1980); A. M. North, *The Collision Theory of Chemical Reactions in Liquids*, Methuen and Co. Ltd., London (1964).
9. W. Schnabel and J. Kiwi, Chapter 5, Aspects of Degradation and Stabilization of Polymers, H. H. G. Jellinek, Ed., Elsevier (1978); B. Ranby and J. F. Rabek, *Photodegradation, Photo-oxidation and Photo-stabilization of Polymers*, John Wiley & Sons, Inc. New York (1975), C. H. Bamford and C. F. H. Tipper, Eds., Chapter 3 (G. Geuskens) and 4 (J. F. Rabek), *Comprehensive Chemical Kinetics*, **14**, Elsevier (1975).

10. H. H. G. Jellinek Ref. 6, Chapter 1 (1978); A. Charlesby, *Atomic Radiation and Polymers*, Pergamon Press (1960).
11. W. Schnabel, Ref. 6, Chapter 4 (1978); *Comprehensive Chemical Kinetics*, **14**, Chapter 2 (C. David); A. Chapiro, Radiation Chemistry of Polymer Systems, Interscience Publ. (1962).
12. Yoshio Kamiya and Etsuo Niki, Ref. 6, Chapter 3 (1978); *Comprehensive Chemical Kinetics*, **14**, Chapter 4 (J. F. Rabek); C. Reich and S. S. Stivala, *Elements of Polymer Degradation*, McGraw Hill Book Co. (1971); W. Lincoln Hawkins, Ed., *Polymer Stabilization*, Wiley Interscience (1972).
13. E. H. Farmer *et. al.*, *Trans. Faraday Soc.*, **38**, 350 (1942); **42**, 228 (1946); *J. Chem. Soc.* (1942); 121, 139 (1943); 119, 122, 125 (1946), 10.
14. J. C. Bolland, *Trans. Far. Soc.*, **44**, 669 (1948); J. C. Bolland and G. Gee, **42**, 236, 2441 (1948); J. L. Bolland, *Quart. Rev.*, **3**, 1 (1949).
15. See Ref. 6 (1978), Chapter 3.
16. D. L. Allara *et al.*, *J. Coll. Interface Sci.*, **47**, 697 (1974); *Polym. Eng. Sci.*, **4**, 12 (1974); *J. Polym. Sci. (Chem. Ed)*, **14**, 93, 1977; *ibid.*, **14**, 1857 (1976).
17. H. H. G. Jellinek *et al.*, *J. Polym. Sci (Chem. Ed.)*, **17**, 1493 (1979); A. Czanderna et al., J. Vac. Sci. Techn., 14, 227, (1979).
18. J. Guillet, *Polymers and Ecological Problems*, Plenum Press, 1973.
19. See Ref. 9.
20. See Ref. 6 (1978), H. H. G. Jellinek, Chapter 1.
21. G. Geuskens and C. David, p. 113 in G. Geuskens Ed., *Degradation and Stabilization of Polymers*, [Halsted] J. Wiley and Sons (1975).
22. H. H. G. Jellinek, *Trans. Far. Soc. XL*, **40**, 266 (1944); *ibid.*, **41**, (1948).
23. Ref. 6 (1978), W. Schnabel and J. Kiwi, Chapter 5.
24. C. Reich and S. S. Stivala, Ref. 12.
25. H. Kachi and H. H. G. Jellinek, *J. Polym. Sci (Chem. Ed.)*, **17**, 2031 (1979).
26. H. H. G. Jellinek, *J. Polym. Sci (Chem. Ed.)*, **14**, 1249 (1976).
27. See Ref. 6, H. H. G. Jellinek, Chapter 9.
28. H. H. G. Jellinek and K. Takada, *J. Polym. Sci (Chem. Ed.)* **13**, 2709, (1975); *ibid.*, **15**, 2269 (1977); H. H. G. Jellinek and S. R. Dunkle, *J. Polym. Sci (Chem. Ed.)*, **18**, 1471 (1980); *ibid.*, **20**, 85 (1982); H. H. G. Jellinek and A. Das, *ibid.*, **16**, 2715 (1978).
29. H. H. G. Jellinek, A. Chauduri and K. Takada, *Polym. Journal*, **10**, 253 (1978).
30. Ref. 6, (1978), R. Akita, Chapter 10.
31. M. Lewis, S. M. Atlas, E. M. Pearce, Eds., *Flame-Retardant Polymeric Materials*, **1** and **2**, Plenum Press (1975, 1978).
32. C. F. H. Tipper, Ed., *Oxidation and Combustion Reviews*, **2**, Elsevier (1967).
33. H. Kachi, H. H. G. Jellinek and M. Hall, *J. Polym. Sci. (Phys. Ed)*, **19**, 1131 (1981).
34. See Ref. 6 (1978), K. Murakami, Chapter 7, see also ref. 4.
35. See Ref. 6 (1978), E. L. Straus, Chapter 11; R. F. Landel and A. Rembaum, Eds., *Chemistry in Space Research*, Elsevier, 1972.
36. A. H. Frazer, *High Temperature Resistant Polymers*, *Polymer Reviews*, **17**, Interscience Publ. (1968); P. E. Cassidy, *Thermally Stable Polymers*, Marcel Dekker Inc. (1980); C. Arnold Jr., *Macromol. Rev.*, **14**, *J. Polym. Sci.*, 265 (1979); W. W. Wright, *Develop. Polym. Degrad.*, **3**, 1 (1981); See also Ref. (Geuskens Ed.), W. W. Wright, Chapter 3.
37. C. S. Marvel, *J. Chem. Ed.*, **58**(7), 535 (1981).
38. H. H. G. Jellinek, *Surface Degradation*, Internatl. Symp. Polymer Surfaces, ACS Meeting, New York, Amherst 23–28 (1981), in press.
39. B. Dickens, *Thermochemica Acta*, **29**, 87 (1979), gives earlier references.
40. I. M. Kolthoff, P. J. Elving and F. H. Stross, Treatise on Analytical Chemistry, Pt. III, *Anal. Chem. in Industry*, **3**, H. L. Friedmann, Part IV, Section D-1, 395, J. Wiley & Sons (1976).

NMR Review

H. R. KRICHELDORF

Universität Hamburg, 2000 Hamburg 13, Edmund-Siemers-Allee, West Germany

TRENDS

While in the two decades before 1972, ^{1}H-NMR spectroscopy was nearly the only NMR method used for the characterization of polymers, ^{13}C-NMR spectroscopy has become the predominant method in the last ten years. Thus, ca. 80% of all papers dealing with NMR spectroscopy of synthetic oligomers and polymers report and discuss ^{13}C-NMR data.

Relatively inexpensive Fourier-Transform NMR spectrometers with iron magnet equipped for both ^{1}H- and ^{13}C-NMR measurements are nowadays widely available for routine NMR work. At low frequencies (20–25 MHz for ^{13}C, 80–100 MHz for ^{1}H) ^{13}C-NMR spectra in comparison with ^{1}H-NMR, allow better resolution and characterization of the primary structure of most polymers. Consequently, ^{13}C-NMR spectra were most widely used to characterize the primary structure of synthetic polymers. Primary structure means here all aspects of the chemical structure including endgroups, while the term sequence is limited to the composition of copolymers. This field of application comprises the structural characterization of polymers prepared by new synthetic methods, identification of products obtained by modification of polymers with functional groups and detection of branching points in branched or cross-linked polymers. An increasing number of papers also deals with ^{13}C-NMR sequence analyses of various copolymers, because ^{1}H-NMR spectra are rather useless for this purpose. In contrast, ^{1}H-NMR spectra have yielded in former years much useful information on the tacticity of stereocopolymers; yet, even in this area ^{13}C-NMR spectroscopy is now more widely used. A complete spectroscopic analysis of the primary structure requires identification of the end-group. Due to the much higher sensitivity (better signal-to-noise ratio) ^{1}H-NMR spectroscopy is more advantageous for this purpose. It is presumably the low sensitivity which has prevented a more intensive application of ^{13}C-NMR spectroscopy in this field, although identification and quantification of active or dead end-groups is useful for two purposes. First, it allows a good insight into reaction mechanisms; and second, it allows the determination of number molecular weights.

Less common nuclei such as ^{19}F, ^{31}P, ^{29}Si and ^{15}N can also be used for the characterization of polymer structures, yet their application is limited. The first three nuclei only occur in a few classes of polymers, while the ^{15}N nucleus has a very low natural abundance. However, increasing magnetic field strengths and the application of cross-polarization[1] or INEPT[2] pulse sequence can strongly enhance the signal-to-noise ratio and will certainly stimulate the application of ^{15}N-NMR spectroscopy.

A further focal point for the application of both ^{1}H- and ^{13}C-NMR spectroscopy are dynamic measurements. The determination of spin-lattice relaxation times (T_1) and its interpretation in terms of segmental mobility is mostly done by means of the ^{13}C- nucleus. This way of study the dynamic behavior of polymers in solution or in the gel-phase has become a classical method which will also find continuing interest in the following years, a newer method which attracts increasing interest of physico-chemists in the 'Pulse NMR/Field-Gradient' technique.[3] This method is a domain of ^{1}H-NMR spectroscopy. It is based on spin–spin relaxation time (T_2) measurements by means of spin–echo pulse sequences. Of particular interest are these T_2 measurements when they are combined with a periodically pulse gradient of the magnetic field. Then, very fast motions of small polymer segments are detectable, above all self-diffusion processes.

The fastest growing novel NMR technique in the course of the last five years is certainly the ^{1}H/^{13}C-CP/MAS method. The cross-polarization (CP) technique has the purpose of enhancing the signal intensity of the less sensitive ^{13}C nucleus (rare isotope) by magnetization-transfer from the more strongly polarized protons (abundant isotope). The magic angle spinning (MAS) reduces the line widths caused by the anisotropy of the chemical shift tensors of the nuclei under investigation. Instead of ^{13}C any other nucleus with low gyromagnetic ratio, such as ^{15}N, or ^{29}Si can be measured by this technique and the polarization transfer may be made from ^{19}F. However, since most polymers contain protons and carbons ^{1}H/^{13}C-NMR CP/MAS spectroscopy will be the most widely used version of all solid-state NMR measurements. Such measurements may be useful to characterize the primary structure of polymers that are insoluble in NMR solvents. Furthermore, they allow one to differentiate between the crystalline and amorphous (e.g. glassy) regions of polymer and they make dynamic studies of solid polymers accessible. CP/MAS spectra also allow one to identify and quantify different crystalline modifications of a polymer, they allow identification of the preferred conformations of solid polymers and detecton of temperature-dependent conformational changes. While ^{13}C-NMR CP/MAS studies of synthetic polymers have become numerous in the course of the last five years, the first CP/MAS investigations of biopolymers (e.g. lignin,[4,5] polypeptides[6] and cellulose[7] have been published in 1981.

In the field of biopolymers and biochemistry the leading role of ^{1}H-NMR spectroscopy is still unrivalled by ^{13}C-NMR spectroscopy. The reasons which account for this characteristic difference between the NMR spectroscopy of technical polymers and compounds of biological interest lay mainly in the high sensitivity of the ^{1}H nucleus. In particular, the use of high-field cryo-magnets allows measurements of small quantities (< 0.1 mg) of compounds which are difficult to isolate from natural sources. Two-dimensional NMR spectroscopy simplifies considerably the interpretation of complex ^{1}H-NMR spectra and provides additional information from coupling constants or NOE interactions. The published NMR literature demonstrates, furthermore, that ^{31}P-NMR spectroscopy has become a versatile tool in biochemistry. It is mainly applied for investigations of phospholipid-containing membranes and vesicles, for studies of association equilibria of mono- and oligo-nucleotides or for conformational analyses of polynucleotides. These areas of research and, thus, the application of NMR spectroscopy will further increase in interest.

^{13}C and to a lesser extent ^{15}N-NMR spectroscopy are mainly applied when synthetic models of biologically active compounds are to be studied, because the quantity of available substrate is then less limited. This is in particular true for the field of peptide chemistry. Thus, solvation and conformation of linear or cyclic oligopeptides are frequently investigated by means of ^{13}C- and also ^{1}H-NMR spectroscopy. Conformational changes and sequence analyses of copolypeptides were conducted by ^{13}C- and ^{15}N-NMR spectroscopy. ^{13}C-NMR studies of the primary and secondary structure of polysaccharides and polynucleotides were of minor interest. Concerning ^{1}H-, ^{13}C- and ^{31}P-NMR measurements of biopolymers in solution, the papers published presumably represent a rather stable trend which will continue in the near future. However, it is likely that the importance of ^{15}N-NMR measurements and of ^{13}C-NMR CP/MAS spectra of solid biopolymers will considerably increase in the following years.

A new dimension of NMR spectroscopy has developed in the last five years mainly for medical purposes, namely the so-called NMR Zeugmatography or Tomography. This novel NMR method is an imaging technique and does not yield spectra in the classical sense and is not directly comparable with the NMR method discussed above. Historically, the technique has been a form of low frequency (6–15 MHz) ^{1}H-NMR spectroscopy using large iron magnets so that samples up to the size of the human body can be studied. Strong field gradients and computer-controlled pulsing, data acquisition, and data processing are used to reconstruct the image (spin density map) of a large sample cross-section from several thousand accurately measured tiny sample volumes. A new step in the development of this technique for small samples is the adaption of conventional high-resolution wide-bore cryomagnets to imaging purposes. This limits the sample size to diameters of a few centimeters; yet higher field strengths and field gradients improve substantially both sensitivity and resolution. This means that nuclei other than protons may be used, and this new field of NMR mini imaging[8] will certainly find the interest of polymer chemists and biochemists.

References

1. B. S. Holmes, G. C. Chingas, W. B. Moniz and R. C. Ferguson, *Macromolecules*, **14**, 1785–1789 (1981).
2. G. A. Morris, *J. Am. Chem. Soc.*, **102**, 428 (1980).
3. *Bruker Report*, **1**, 8–9 (1981).
4. J. Schaefer, M. D. Sefcik, E. D. Steijskal, R. A. McKay and P. L. Hall, *Macromolecules*, **14**, 557–559 (1981).
5. G. E. Macid, J. D. O'Donnel, J. J. Ackermann, B. H. Hawkings and V. J. Bartuska, *Makromol. Chem.*, **182**, 2297–2304 (1981).
6a. D. Müller and H. R. Kricheldorf, *Polym. Bull*, **6**, 101–108 (1981).
6b. T. Taki, S. Yamashito, U. Sator, A. Shibata, T. Yamashito, R. Tabeta and H. Saito, *Chem. Letters* 1803–1806.
7. W. L. Earl and D. L. vander Hart, *Macromolecules*, **14**, 570–574.
8. *Bruker Report*, **1**, 2–5 (1981).

Chemical Structure and Sequence of Synthetic Polymers

^1H-*NMR spectroscopy* of polyesters: refs. 9–11, polyethers: ref. 12 and ureacondensates: ref. 13.

9. H. R. Kricheldorf, M. Dröscher and W. E. Hull, *Polym. Bull*, **4**, 547–554.
10. P. M. Henrichs and J. M. Hewitt, *Macromolecules*, **14**, 461–463.
11 G. A. Russell, P. M. Henrichs, J. M. Hewitt, H. R. Grashof and M. A. Sandu, *Macromolecules*, **14**, 1764–1770.
12. M. Gibas and Z. Jedlinski, *Macromolecules*, **14**, 1012–1015.
13. G. E. Myers, *J. Appl. Polym. Sci.*, **26**, 747–764.

^{13}C-*NMR spectroscopy* of vinylpolymers in refs 14–30, of polycondensates in refs. 31–36.

14. A. E. Tonelli and F. C. Schilling, *Macromolecules*, **14**, 74–76.
15. K. J. Ivin, L.-M. Lam and J. J. Rooney, *Makromol. Chem.*, **182**, 1847–1854.
16. M. Tsuhino and T. Kunitake, *Polym. J.*, **13**, 671–678.
17. K. J. Ivin, S. Pitchurnani, C. R. Reddy and S. Rajadurai, *Europ. Polym. J.*, **17**, 341–346.
18. D. Galini, I. P. Pascault and Q.-T. Pham, *Makromol. Chem.*, **182**, 2321–2342.
19. P. Fruhe, M.-F. Grenier-Loustalot and F. Metras, *Makromol. Chem.*, **182**, 2305–2320.
20. C. Pichot, M.-F. Llauro and Q.-T. Pham, *J. Polym. Sci. Chem. Ed.*, **19**, 2619–2633.
21. Y. Doi and T. Asakura, *Macromolecules*, **14**, 69–71.
22. F. Cavagna, *Macromolecules*, **14**, 215–216.
23. A. E. Tonelli, F. C. Shilling and F. A. Bovey, *Macromolecules*, **14**, 560–564.
24. H. Koinuma, T. Tanake and H. Hirai, *Macromolecules*, **14**, 883–885.
25. D. Braun, G. Holzer and J. Klein, *Polym. Bull.*, **5**, 367–372.
26. G. J. Earl, J. Spanswick, J. R. Knox and C. Seres, *Macromolecules*, **14**, 1323–1337.
27. K. Arita, T. Oktoma and Y. Tsurumi, *J. Polym. Sci. Letters Ed.*, **19**, 211–216.

28. G. S. Georico, *Polym. Bull.*, **6**, 29–34.
29. K. Matsuzaki, H. Morii, N. Inoue, T. Kawai, Y. Fujiwara and T. Higaskimero, *Macromolecules*, **14**, 1008–1012.
30. K. Matsuzaki, H. Morii, T. Kawai and Y. Fujiwara, *Macromolecules*, **14**, 1004–1008.
31. D. M. White, T. Tabekoshi, T. J. Williams, H. M. Relles, P. E. Donahue, H. J. Klopfer, G. R. Louks, J. S. Manello, R. O. Matthews and R. W. Schluenz, *J. Polym. Sci. Chem. Ed.*, **19**, 1635–1658.
32. R. A. Newmark, M. L. Runge and J. A. Chermark, *J. Polym. Sci. Chem. Ed.*, **19**, 1329–1336.
33. F. C. Schilling, W. U. Ringo, Jr, N. J. Scloane and F. A. Bovey, *Macromolecules*, **14**, 532–537.
34. E. A. Williams, P. E. Donahue and J. D. Cargioli, *Macromolecules*, **14**, 1016–1018.
35. C. Delides, R. A. Pethrick, A. V. Cunliffe and P. G. Klein, *Polymer*, **22**, 1205–1210.

^{15}N-*NMR spectroscopy* of polyurethanes: ref. 36 and ^{29}Si-*NMR spectroscopy* of polysiloxanes: ref. 37.

36. H. R. Kricheldorf and W. E. Hull, *Makromol. Chem.*, **182**, 1177–1196.
37. G. Engelhardt and H. Janke, *Polym. Bull.*, **5**, 577–584.

Tacticity of Vinyl Polymers

^{13}C-*NMR spectroscopic* tacticity analyses of vinylpolymers: refs. 38–47.

38. J. Ando and T. Asakura, *Makromol. Chem.*, **182**, 1243–1251.
39. K. Hatada and T. Kitayama, *Makromol. Chem.*, **182**, 1449–1458.
40. K. Matsuzaki, T. Kawai and Y. Aoki, *Makromol. Chem.*, **182**, 1027–1032.
41. J. Kaspercyk, A. Dworak and Z. Jedlinsky, *Makromol. Chem. Rapid. Comm.*, **2**, 663–666.
42. V.-D. Staudt and J. Klein, *Makromol. Chem. Rapid. Comm.*, **2**, 41–45.
43. A. C. Segre, F. Andruzzi, D. Lupinacci and P. L. Maggnini, *Macromolecules*, **14**, 1845–1847.
44. S. S. Huang, C. Mathis and T. E. Hagen-Esch, *Macromolecules*, **14**, 1802–1807.
45. R. E. Cais and J. M. Kometani, *Macromolecules*, **14**, 1346–1350.
46. H. N. Cheng, T. E. Smith and D. M. Vitus, *J. Polym. Sci. Letters Ed.*, **19**, 29–31.
47. H. N. Sang and J. H. Noggle, *J. Polym. Sci. Phys. Ed.*, **19**, 1593–1602.

Chemical Modification of Polymers

^{13}C-*NMR spectroscopy:* refs. 48–50, ^{1}H-*NMR spectroscopy:* ref. 51 and ^{15}N-*NMR spectroscopy:* ref. 52.

48. J. G. de la Campa and Q. T. Pham, *Makromol. Chem.*, **182**, 1415–1428.
39. J. H. Hodgkin, R. I. Willing and R. Eibl, *J. Polym. Sci. Chem. Ed.*, **19**, 1239–1249.
50. K. Yokota and T. Hirabayashi, *Macromolecules*, **14**, 1613–1615.
51. W. Kern, M. M. Bhavagat and V. Böhmer, *Makromol. Chem. Rapid. Comm.*, **2**, 557–562.
52. H. R. Kricheldorf, *J. Polym. Sci. Chem. Ed.*, **19**, 2195–2214.

End group Analyses (by ^{1}H and ^{13}C-*NMR spectroscopy*)

53. Q. T. Pham, *Macromol. Chem.*, **182**, 1167–1176.
54. L. Vancea and S. Bywater, *Macromolecules*, **14**, 1776–1778.
55. L. Vancea and S. Bywater, *Macromolecules*, **14**, 1321–1323.

Dynamic Measurements

^{1}H-*NMR measurements* of T_1 and T_2: refs. 56–61.

56. J. Jajicek, H. Pivcova and B. Schneider, *Makromol. Chem.*, **182**, 1369–1376.
57. G. Feio and J. P. Cohen-Addad, *Polym. Bull.*, **5**, 277–284.

58. S. Giowinkowski, K. Jurga and Z. Pajak, *Polym. Bull.*, **5**, 271–275.
59. J. S. Blicharski, H. Haraucyk and K. Strzalka, *Polym. Bull.*, **5**, 282–289.
60. L. Westmann and T. Lindström, *J. Appl. Polym. Sci.*, **26**, 2545–2559.
61. E. von Norwall and R. D. Ferguson, *J. Polym. Sci. Phys. Ed.*, **19**, 77–92.

^{13}C-*NMR spectroscopic* T_1 *measurements*: refs. 62–71.

62. P. M. Henrichs, J. M. Hewitt, G. A. Russell, M. A. Sandhu and H. R. Grashof, *Macromolecules*, **14**, 1770–1775.
63. L. W. Jelinski, *Macromolecules*, **14**, 1341–1345.
64. L. W. Jelinski, F. C. Schilling and F. A. Bovey, *Macromolecules*, **14**, 581–586.
65. W. T. Ford and T. Balakrishnan, *Macromolecules*, **14**, 284–288.
66. T. Asakura and T. Doi, *Macromolecules*, **14**, 72–74.
67. T. Asakura, K. Sazuki and K. Horie, *Makromol. Chem.*, **182**, 2289–2295.
68. K. Hatada, T. Kitayama, Y. Okamoto, K. Ohta, Y. Umemura and H. Yuki, *Makromol. Chem.*, **182**, 617–621.
69. M. Paci, S. Y. Fu and F. Andruzzi, *Polym. Bull.*, **6**, 81–85.
70. W. Gronski and R. Peter, *Polym. Bull.*, **6**, 113–119.

^{19}F-*NMR measurements* and ^{15}N-*NMR measurements:* refs. 62 and 71.

71. P. L. Rinaldi, Chin Yu, G. C. Levy; *Macromolecules*, 14, 551–554.

Line-Shape-Analyses

^1H- *and* ^2H *measurements:* refs. 72–77; ^{13}C-*NMR spectroscopy:* ref. 78.

72. Y. Hori, T. Tanigawa, S. Shimada and H. Kashiwabara, *Polym. J.*, **13**, 293–294.
73. J. P. Cohen-Addad and C. Roby, *J. Polym. Sci. Phys. Ed.*, **19**, 1395–1403.
74. P. T. Inglefield, A. A. Jones, R. P. Lubianez and J. F. O'Gara, *Macromolecules*, **14**, 228–292.
75. I. Loboda-Caikovič and H. Caikovič, *Polym. Bull.*, **5**, 291–298.
76. R. Hentschel, H. Sillescu, H. W. Spiess, *Polymer*, **22**, 1516–1521.
77. R. Hentschel, H. Sillescu, H. W. Spiess, *Macromolecules*, **14**, 1605–1607.
78. R. F. Boyer, J. P. Heeschen, J. K. Gillham, *J. Polym. Sci. Phys. Ed.*, **19**, 13–21.

Pulsed NMR

^1H *NMR* T_2 *measurements* in solution: refs. 79–83 and in the solid state: rafs. 83–86.

79. P. T. Callaghan, D. N. Pinder, *Polym. Bull.*, **5**, 305–309.
80. A. Charlesby, B. J. Bridges, *Europ. Polym. J.*, **17**, 645–656.
81. B. Nystrom, M. W. Mosely, W. Brown, J. Routs, *J. Appl. Polym. Sci.*, **26**, 3385–3394.
82. Y. K. Sung, D. E. Gregonis, M.-S. John, J. D. Andrade, *J. Appl. Polym. Sci.*, **26**, 3719–3728.
83. K. G. Barnett, T. Cosgrove, B. Vincent, D. S. Sissons, M. Cohen-Stuart, *Macromolecules*, **14**, 1018–1020.
84. R. A. Assink, G. C. Wilkes, *J. Appl. Polym. Sci.*, **26**, 3689–3698.
85. L. Kamel, A. Charlesby, *J. Polym. Sci. Phys., Ed.*, **19**, 803–814.
86. T. Nishi, T. Chikaraish, *J. Macromol. Sci. Phys.*, **1319**, 445–457.

Solid Polymers

^{13}C *NMR CP/MAS* studies of frozen conformations: refs. 87–89, of polymer blends: refs. 90, 91, of poly-imides): refs. 92, 93 and various polymers: refs. 94–102; ^{29}Si-NMR spectra: ref. 102.

87. M. Möller, H.-J. Cantow, *Polym. Bull.*, **5**, 119–124.
88. M. Möller, H.-J. Cantow, J. K. Krüger, H. Höcker, *Polym. Bull.*, **5**, 125–129.
89. W. Gronski, H. Hasenhindl, H. H. Limbach, M. Möller, H.-J. Cantow, *Polym. Bull.*, **6**, 93–100.

90. J. R. Havens, H. Ishida and J. L. Koenig, *Macromolecules*, **14**, 1327–1333.
91. A. C. Wong, A. V. Garsoway and W. M. Ritchey, *Macromolecules*, **14**, 832–836.
92. E. O. Stejskal, J. Schaefer, M. D. Sefcik and R. A. McKay, *Macromolecules*, **14**, 275–279.
93. J. Schaefer, M. O. Sefcik, E. O. Stejskal and R. A. McKay, *Macromolecules*, **14**, 188–192.
94. J. Schaefer, M. D. Steyskal and R. A. McKay, *Macromolecules*, **14**, 280–283.
95. B. Schröter and A. Posern, *Makromol. Chem.*, **182**, 675–680.
96. F. Sergot, E. Pauprêtre, C. Louis and J. Kirlet, *Polym.* **22**, 1150–1153.
97. B. Schröter, H. H. Hörhold and D. Raabe, *Makromol. Chem.*, **182**, 3185–3193.
98. A. Dilks, S. Kaplan and A. van Laeken, *J. Polym. Sci. Chem. Ed.*, **19**, 2987–2996.
99. H. T. Edzes and W. S. Veeman, *Polym. Bull.*, **5**, 255–261.
100. G. Hempel and H. Schneider, *Polym. Bull.*, **6**, 7–12.
101. T. J. Sanford, R. D. Allendoerfer, E. T. Kang, P. Ehrlich and J. Schaefer, *J. Polym. Sci. Phys. Ed.*, **19**, 1151–1152.
102. C. E. Maciel, M. J. Sullivan and D. W. Sindorf, *Macromolecules*, **14**, 1607–1608.

Polysaccharides

^1H- and ^{13}C-*NMR analyses* of substituted cellulose: refs. 103–105 of conformational changes: ref. 106.

103. D. T. Clark, P. J. Stephanson, and F. Heatly, *Polymer*, **22**, 1112–1117.
104. K. Kamide and K. Okajima, *Polym. J.*, **13**, 163–166.
105. K. Kamide and K. Okajima, *Polym. J.*, **13**, 127–133.
106. A. J. Stipanovic and E. S. Stevens, *Makromol. Chem. Rapid. Commun.*, **2**, 339–341.

^{23}Na-, ^{127}J-, ^{133}Cs- and ^1H-*NMR studies* of ion binding to polysaccharides: refs. 107–110.

107. J. Kunze, A. Ebert, B. Schröter, K. Frigge and B. Philipp, *Polym. Bull.*, **5**, 399–406.
108. H. Grasdalen and O. Smidsröd, *Macromolecules*, **14**, 1842–1845.
109. H. Grasdalen and O. Smidsröd, *Macromolecules*, **14**, 229–231.
110. R. Basosi, F. Laschi and C. Rossi, *Polym. Bull.*, **5**, 263–269.

Synthetic Oligo- and Polypeptides

^1H- and ^{13}C-*NMR studies* of conformation and solvation of oligopeptides: refs. 111–120.

111. T. Asakura, *Makromol. Chem.*, **182**, 1135–1145.
112. T. Asakura, *Makromol. Chem.*, **182**, 1097–1109.
113. Y. V. Venkatachalapathi and P. Balaram, *Biopolymers*, **20**, 625–628.
114. M. Goodman and R. P. Saltman, *Biopolymers*, **20**, 1929–1948.
115. L. M. Gierasch, C. M. Deber, V. Madison, C. H. Hice and E. R. Blout, *Biochemistry*, **20**, 4730–4738.
116. Y. V. Venkatachalapathi, G. M. Nair, M. Vijayan and P. Balaram, *Biopolymers*, **20**, 1123–1136.
117. Y. V. Venkatachalapathi and P. Balaram, *Biopolymers*, **20**, 1137–1145.
118. N. Higuchi, Y. Kyogoku and H. Yajima, *Biopolymers*, **20**, 2203–2212.
119. R. Deslauriers, J. J. Evans, S. L. Leach, Y. C. Meinwald, E. Minasian, G. Nemethy, J. P. Pau, H. A. Sheraga, R. L. Sonosjai, E. R. Stimson, J. W. van Nispen and R. W. Woody, *Macromolecules*, **14**, 985–996.
120. F. Migazawa and T. Higashijima, *Biopolymers*, **20**, 1949–1958.

^{13}C- and ^{15}N-*NMR investigations* of conformations of polypeptides: refs. 121–123.

121. S. Sasaki, Y. Yasumoto and I. Uematsu, *Macromolecules*, **14**, 1797–1801.
122. B. Perly, Y. Chevalier and C. Shachaty, *Macromolecules*, **14**, 969–975.
123. W. E. Hull, E. Büllesbach, H. J. Wieneke, H. Zahn and H. R. Kricheldorf, *Org. Magn. Resonance*, **17**, 92–96.

256 POLYMER YEARBOOK

^{13}C- and ^{15}N-NMR Sequence and tacticity analyses of polypeptides: ref. 124–127.

124. H. Pivcova, V. Saudek, J. Drobnik and J. Vlasak, *Biopolymers*, **20**, 1605–1614.
125. V. Saudek, H. Pivcova and J. Drobnik, *Biopolymers*, **20**, 1645–1623.
126. H. R. Kricheldorf and T. Mang, *Makromol. Chem.*, **182**, 3077–3098.
127. H. R. Kricheldorf, *Org. Magn. Resonance*, **15**, 162–177.

Naturally occurring Peptides and Proteins

^{1}H-NMR measurements of conformations and association equilibria: refs. 128–145.

128. Y. Kobayashi, Y. Kyogoku, J. Emura and S. Sakakiba, *Biopolymers*, **20**, 2021–2031.
129. M. T. Hinke, B. D. Sykes and C. M. Kay, *Biochemistry*, **20**, 3286–3294.
130. N. R. Krishna, D.-H. Huang, J. B. Vaughn, Jr., G. A. Heavner and G. Goldstein, *Biochemistry*, **26**, 3933–3940.
131. G. M. Smith and A. S. Mildran, *Biochemistry*, **20**, 4340–4346.
132. M. Igbal and P. Palaram, *Biochemistry*, **20**, 7278–7284.
133. H. Hakimi, J. Casver and P. H. Atkins, *Biochemistry*, **20**, 7314–7319.
134. N. C. Alma, B. J. Hermsen, W. E. Hull, G. van der Marvel, J. H. van Boom and C. W. Hilbers, *Biochemistry*, **20**, 441–442.
135. C. N. La Mar, R. P. Anderson, D. C. Budd, K. M. Smith, K. C. Langry, K. Gersonde and H. Sick, *Biochemistry*, **20**, 4426–4429.
136. P. Rösch, H. R. Kalbitzer, V. Schmidt-Aderjan and W. Hengstenberg, *Biochemistry*, **20**, 1599–1605.
137. J. P. Carver and A. A. Grey, *Biochemistry*, **20**, 6607–6616.
138. J. P. Carver, A. A. Grey, F. M. Winnik, J. Hakimi, C. Ceccarini and P. H. Atkimon, *Biochemistry*, **20**, 6600–6606.
139. F. Jordan and L. Polgar, *Biochemistry*, **20**, 6366–6370.
140. C. H. Pletchner, E. G. Boukoutsos-Brown, R. G. Bryant and G. L. Nelsestuen, *Biochemistry*, **20**, 6149–6155.
141. P. R. Rosevaer, P. Desmeules, G. L. Keynon and A. S. Mildvan, *Biochemistry*, **20**, 6155–6164.
142. H. R. Kalbitzer, J. Deutscher, W. Hengstenberg and P. Rösch, *Biochemistry*, **20**, 6178–6185.
143. A. A. Ribeiro, D. Wemmer, R. P. Bray and O. Jardetzky, *Biochemistry*, **20**, 3346–3356.
144. A. Guyon-Gruaz, J.-P. Demonte, M.-C. Fournie-Zaluski, A. Englert and B. P. Roques, *Biochemistry*, **20**, 6677–6683.
145. J. Feeney, B. Birdsall, J. P. Albrand, G. C. Roberts, A. S. Burger, P. A. Charlton and D. W. Young, *Biochemistry*, **20**, 1837–1842.

^{13}C-NMR studies of teichuronic acid: ref. 146 and various enzyme proteins: refs. 147–150.

146. S. D. Johnson, K. P. Lacher and J. S. Anderson, *J. Am. Chem. Soc.*, **20**, 4784–4785.
147. J. B. Wooten, J. S. Coben, J. Vig and A. Scheijter, *Biochemistry*, **20**, 5394–5402.
148. L. Cocco, J. P. Groff, C. Temple, Jr., J. A. Montgomery, R. E. London, N. A. Matwizoff and R. L. Blakley, *Biochemistry*, **20**, 3972–3978.
148. J. L. Zweier, J. B. Wooten and J. S. Cohen, *Biochemistry*, **20**, 3505–3510.
149. R. A. Halpin, D. G. Hegeman and G. L. Kenyon, *Biochemistry*, **20**, 1525–1533.
150. C. K. Kishwanat and K. R. Easwaran; *Biochemistry*, **20**, 2018–2023.

^{19}F-, ^{35}Cl- and ^{113}Cd-NMR spectroscopy of association equilibrium with enzymes: refs. 151–153.

151. C. A. Lewis, P. D. Ellis and R. B. Dunlap, *Biochemistry*, **20**, 2275–2285.
152. S. Subramanian, H. Skindo and B. T. Kaufman, *Biochemistry*, **20**, 3226–3230.
153. J. C. Everhoch, D. F. Boican and J. L. Sudmeier, *Biochemistry*, **20**, 4951–4952.

Oligo- and Poly(nucleotides)

^{1}H-NMR studies of oligo- and poly(nucleotides): refs. 154–159.

154. B. Gaugain, J. Markovits, J.-B. Peyu and B. P. Roques, *Biochemistry*, **20**, 303–3042.
155. M. P. Stone, D. L. Johnson and P. N. Borer, *Biochemistry*, **20**, 3604–3610.
156. T. A. Early, D. R. Kearns, W. Hillen and R. D. Wills, *Biochemistry*, **20**, 3756–3764.
157. J. Tropp and A. G. Redfield, *Biochemistry*, **20**, 2133–2140.
158. R. A. Bell, J. R. Everett, D. N. Hughes, D. Alkena, P. Hader, T. Neilson and P. J. Romanjilk, *Biopolymers*, **20**, 1383–1398.
159. L. A. Merkey, D. Patel and K. J. Breslauer, *Biochemistry*, **20**, 1427–1431.

^{31}*P-NMR studies of DNS and RNS conformations*

160. D. B. Lerner and D. R. Kearns, *Biopolymers*, **20**, 803–816.
161. J. M. Neumann and S.-Fran-Dinh, *Biopolymers*, **20**, 89–109.
162. B. T. Nell, P. Rothwell, J. S. Waugh and A. Rupprecht, *Biochemistry*, **20**, 1881–1887.
163. D. G. Gorenstein, E. M. Goldfield, R. Chen, Ken-Kovar and B. A. Luxon, *Biochemistry*, **20**, 2141–2150.

^{31}*P-NMR studies of* mononucleotide-protein complexes.

164. M. Brauer and D. B. Sykes, *Biochemistry*, **20**, 2060–2064.
165. J. W. Shriver and B. D. Sykes, *Biochemistry*, **20**, 6357–6362.
166. J. W. Shriver and B. D. Sykes, *Biochemistry*, **20**, 2004–2012.
167. S. G. Withers, N. B. Madsen and B. D. Sykes, *Biochemistry*, **20**, 1748–1756.

Lipid membranes and vesicles

^{31}*P-NMR measurements:* refs. 168–170; ^{2}*H- and* ^{13}*C-NMR measurements*, 171–172.

168. Y. Boulanger, S. Schreier and J. C. P. Smith, *Biochemistry*, **20**, 6824–2830.
169. A. M. Thayer and S. J. Kohler, *Biochemistry*, **20**, 6831–6834.
170. J. Seelig, L. Tamm, L. Hymel and S. Fleischer, *Biochemistry*, **20**, 3922–3932.
171. H. M. Gally, G. Pluschke, P. Overath and J. Seelig, *Biochemistry*, **20**, 1826–1831.
172. R. J. Witterbort, C. F. Schmidt and R. G. Griffin, *Biochemistry*, **20**, 4223–4228.

Polymer Drugs

L. G. DONARUMA, O. VOGL

Polytechnic Institute of New York, 333 Jay Street, Brooklyn, NY 11201, USA.

R. M. OTTENBRITE

Virginia Commonwealth University, Richmond, VA 23284

It is not surprising that polymers, and basic polymer science itself, has had a tremendous effect on medical treatment in recent years. Polymers have been used for a considerable amount of time for a variety of biomedical applications such as prosthetic devices, cosmetic implants, and, more recently, as drugs and for drug administration. The requirements for any materials for biological utilization are exceptionally demanding; some of the characteristics that are looked for in these materials are, bio-compatability, durability, nontoxicity, and biodegradability.

This review will be to introduce the reader to the scope of research and development in the area of polymeric drugs by a discussion of some important generalities and the citation of literature where more specific data can be found.

Drugs generally are distributed throughout an organism in the aqueous phase of the blood plasma or lipid phases of the body. Unless topically active, a drug must first enter the blood system. It will reach the tissues of an organ at a rate determined by blood flow through that organ and by the rapidity of passage of the drug molecule across the capillary bed and into the tissue cells of that particular organ. Within blood plasma, drug molecules can become bound to substances such as proteins and cell surfaces. Thus the amount of any drug present in tissues at a site of activity is a small part of the total. Most of the drug remains in various fluid compartments or is localized in subcellular particles, at macromolecular surfaces and in fat deposits by adsorptive or partition processes. Even with target tissue, cellular fractionation and radioautographic studies reveal that most drug molecules are associated with structures having nothing to do with the specific drug effect.

Consequently, one of the most difficult problems in drug administration is getting the agent in sufficient quantity to the desired site for the required period of time. With conventional delivery systems, such as oral injection methods, it is necessary to redose, as the whole body becomes initially infused with a high concentration of the drug which slowly dissipates. If dosages are spaced too widely, there will be periods when an insufficient amount of the drug is present and disease can recur. At the opposite side, reapplication can lead to build-up of drug to the point of exceeding toxic levels. Thus, the therapeutic utility of many drugs is often limited to short *in vivo* half-lives, lack of specificity, acute toxicity, and undesirable side effects.

Recent pharmacological attention has been directed to the development of therapeutic systems for controlled administration of drugs using synthetic polymeric materials for the regulated release and transport of drugs. A number of techniques have been devised to alleviate many of the problems inherent with repeated dosages associated with oral and injection methods.

The incorporation of polymers with drugs is of particular interest as it embodies several desirable features such as: (a) retarded absorption and excretion resulting in increased duration of drug activity, (b) variable solubility by altering the polymer lipophilic and hydrophilic character and thus influencing specific body distribution with regard to cell interaction, protein binding, and resorption, and (c) pharmacokinetic variability which affects the metabolic pathways, drug activity, and toxicity.

Several polymer systems that are presently being explored for pharmaceutical applications include (a) *polymeric drugs*: these are polymers or copolymers that are physiologically active as polymers themselves, (b) *drug-carrying polymers*: these are polymers that have active drugs bound to a parent polymer backbone, (c) *time-release drug polymers*: these are polymers used for contolled release of the drug. In this system the drug is encapsulated with water soluble polymer coatings which dissolve at different rates, releasing the drug or the drug is imbedded into a polymer matrix from which it diffuses at specific rates, (d) *site-specific drugs*: these are polymers that have special chemical groups attached to the polymer carrying the drug; these groups can combine with specific receptor sites on protein, cell surfaces or in the lipid areas.

(A) POLYMER DRUGS

Several synthetic polymers have been evaluated for biological activity and were found to elicit a number of physiological activities. The most important activities observed have been antibacterial, antifungal, antiparasital, antiviral, antineoplastic, and interferon inducing capabilities.[1] Other activities include appetite suppressing, anticlotting antihistaminic, enzyme inhibition and activation, and change in blood viscosity.[2] To be effective, most polymer drugs are water soluble and consequently consist of polyanions such as carboxylates, sulfonates, and phosphates; polycations, such as quaternary ammonium, and sulfonium salts as well as neutral polymers such as polyols.

Synthetic polyanions exhibit a large number of interesting biological responses. Those of major interest are antiviral, antitumor and immunological effects which were recently reviewed in detail.[1] The most thoroughly studied polyanion is a copolymer of divinyl ether and maleic anhydride which is known in the literature as PYRAN or DIVEMA and, more recently, as MVE. It is parenterally active against a variety of virus and carcinogen induced tumors. These activities include Lewis lung carcinoma, F_{16} melanoma, Rauscher leukemia, LSTRA sarcoma, and Friend leukemia, as well as many other viral and neoplastic diseases.[3]

The apparent mechanism of pyran's activity is through the activation of macrophages and enhanced phagocytosis.[4] The most important feature of the polymer is that cytotoxicity of activated macrophage is confined to tumor cells[5] with no effect on normal cells. In a recent study by Kaplan,[6] it was reported that animals pretreated with pyran and channeled with Ehrlich ascites cells had no detectable tumor cells in G_2M-phase of the cell life cycle and only a few in the S-phase after 2–6 days compared to pyran untreated mice which had 22% in the G_2M and 58% in the S-phases respectively. This study also showed that pyran activitated macrophage resulted in a high level of tumoricidal activity *in vivo* and *in vitro*, a shift of tumor cells from G_2M-phase to the resting G_1-Phase and that the tumor cell population had 50% less DNA content.[6]

Although clinical studies of pyran[7] initially were not encouraging, due to some severe toxicity problems, recent developments have shown that many of these toxicities can be eliminated by using discrete low molecular weight polymer fractions.[8] It was also recently demonstrated that the calcium salts were much less toxic than the sodium salts of pyran previously used and that the toxicity with both salts was molecular weight dependent.[9] Subsequently, Breslow[10] has developed a method of synthesizing pyran in low and narrow molecular weight fractions. The fractions in the molecular weight range of 10,000–15,000 have presently completed phase I clinic trials as the calcium salt.[11]

Many other polyanions have been prepared and show biological activity similar to pyran.[12] It was reported that copolymers of maleic anhydride such as styrene, allyl phenol, and dimethyl-1,3-dioxepin exhibit antitumor effects against Ehrlich Ascites as well as pyran.[12] In this study, several copolymers of 2,3-dicarboxynorborn-5-ene were also shown to be effective against the Ehrlich Ascites tumor. More recently it was found that maleic anhydride copolymers of methacrylic acid and 4-methyl-4-ene-2-pentanone and vinyl acetate, the copolymer of itaconitic acid-styrene and α-ethylacrylic acid homopolymer elicited macrophage activation that resulted in enhanced tumor cytotoxicity whereas maleic anhydride copolymers of styrene, ethylene and allyl urea were not effective.[13]

A new polyelectrolyte prepared by Monsanto has been shown to inhibit the growth of a number of solid tumors such as Lewis lung, Madison 109, and colon carcinomas as well as B_{16} melonoma and P815 mastocytoma.[14] The drug is first prepared as a low molecular weight (1000) copolymer of maleic anhydride and ethylene; the anhydride is converted to the half amide and then on heating 15–25% of these groups form an imide. This polymer drug, in contrast to the nuleic anhydride copolymers such as pyran, shows little toxicity, no anticoagulant activity, and no induced T-Cell, macrophage or interferon effects.[15,16] Clinical data for this polymer indicated increased life spans for patients with various neoplastic diseases as well as being effective in preventing post-surgical tumor recurrences.[16]

Another group of new synthetic polymers with antitumor activity containing 2-methylene-1,3-propanediol as a homopolymer and copolymer has been reported.[17] Investigations indicate that the nonionic polymers were better tolerated biologically than the ionic substances and neither toxicity nor activity was molecular weight dependent. Antitumor activity was observed in mice with carcinoma EO771, fibrosarcoma F1026, sarcoma 180, and leukemia P388. This activity was observed both therapeutically and prophylactically. Cytotoxicity and cytostaticity were not observed in cultures of normal cells or tumor cells. This latter result, along with the prophylatic effects, indicates a host-mediated mechanism for tumor inhibition by these polymers.

(B) DRUG CARRYING POLYMERS

Since polymeric biological systems diffuse slowly and are often absorbed at interfaces, the attachment of pharmaceutical moieties to the structure of an inactive macromolecular chain has been found to produce polymers with distinct pharmacological activity. It has become recognized that the binding of drugs to a polymer backbone can effect some desirable properties such as sustained therapy, slow drug release, prolonged activity, and drug latentiation as well as decreased drug metabolism and excretion.

A model for pharmacologically active polymer-drug carriers has been developed by Ringsdorf and others.[18,19] In this schematic representation four different groups are attached to a biostable or biodegradable polymer backbone. One group is a pharmacon or drug, a second is a spacing group, a third is a transport system, and a fourth is a group to solubilize the entire biopolymer system. The pharmacon is the agent which elicits the physiological response in the living system; it can be attached permanently by a stable bond between the drug and the polymer or it can be temporarily attached and removed by hydrolysis, ion exchange or by an enzymatic process. The transport systems for these soluble polymer-drug carriers can be made specific for certain tissue cells by the use of homing devices such as receptor-active components of pH-sensitive groups or they may be made nonspecific. Solubilizing groups such as carboxylates, quaternary amines and sulfonates, are added to increase the hydrophilicity of the complete polymer system in aqueous media while large alkyl groups adjust the solubility in lipid regions. Another important feature necessary in a polymer-drug carrier is to space the pharmacon away from the polymer backbone of other groups so that there is a minimal structural interference with the pharmacological action of the drug.[20]

The method of attachment of a drug to a polymer is dependent upon the ultimate use of the adduct. Further, the chemical reaction conditions for the attachment of the drug to the polymer should not adversely affect the biological activity of the drug. Temporary attachment of a pharmacon is necessary if the drug is active only in the free form, which is usually the case with agents that function intracellularly. Permanent attachment of the drug moiety is generally used when the drug exhibits activity in the attached form. The pharmacon is usually attached away from the polymer chain and other pendent groups by means of a spacer moiety to allow for drug-receptor interaction. For example, catecholamines were ineffective when bound directly to polyacrylic acid but did affect heart rates and muscle contractions when attached away from the backbone of the macromolecular carrier.[24] Similarly, isoproterenol was found to elicit a pharmacological response only when coupled to a polymer by pendant azo groups but not when directly attached to the polymer chain.[22]

A large number of synthetic polymers have been evaluated against several bacterial and parasitic species.[23,24] The polymer systems examined include sulfonamide-formaldehyde, tropolone-formaldehyde and N-methacrylyl-aminoadamantane-methacrylic acid and piperazine-dibasic acid copolymers. Many structure-activity relationships have been observed as well as effects of molecular weight, copolymer composition, stereochemical configuration, and other polymeric properties.[24] More recently, Donaruma et al.[25] have prepared a series of polythiosemicarbazides involving monomers not considered to be drugs. It was found that the polymers themselves were inactive as antibacterial agents, however, they selectively formed complexes with copper II and these complexes were active against B. subtilis and E. coli.[25]

The types of polymer carriers used are natural polypeptides and polysaccharides as well as synthetic polymers with drugs such as daunomycin,[26] methotrexate[27] and actinomycin.[28] It has been shown, for example, that methotrexate bound to albumin is as effective as the drug alone for treatment of murine L1210 and even more effective against Lewis lung.[29] Research has shown that the nature of the carrier's structure is important to the interaction with the cell surface.[30] For example, two of the more effective carriers for methotrexate are L- and D- forms of polylysine, which are more effective than the free drug against rat liver tumors.[31]

Polymeric substances are also being utilized to serve as tolerogenic carriers for a variety of clinically relevant heptans. According to theory, a polymeric antigen can render an immunocompetent cell refactory to an immunologic challenge by cross-linking its receptors.[32] This has been demonstrated experimentally; polyethylene glycol has been shown to be an effective tolerogenic carrier for heptans as well as for proteins[33] and, more recently, carbomethyl cellulose was reported acting as a powerful tolerogen *in vivo* to several heptans.[34]

(C) TIME-RELEASE DRUG POLYMERS

Conventional dosage forms (oral, injection, drop or ointment) all deliver medication at rates that are very high initially and then steadily decline; this is known as a first-order delivery rate. Ideally, once the drug concentration is at an effective level the drug should be delivered at a zero-order rate;[35] that is, it should be delivered at the same rate it is being consumed by the biological system. The present oral administratives are not consistent because human gastrointestinal tracts vary in acidity, contents and motality. Thus, intense research is presently being carried out to develop simple regimen oral dosage forms to more closely control drug concentrations in the blood.

Consequently, implanted drug delivery devices have been developed over the past few years with some very good successes.[36] Among these is OCUSERT®, an ocular therapeutic device in the form of a thin oval film that is worn under the eyelid for one week to deliver the antiglaucoma drug pilocarpine.[37] Another success is PROGESTASERT , which is a T-shaped device that provides intrauterine delivery of progesterone for birth control. Thus, it reduces a number of side effects usually associated with the 'pill'.[38] Both of these systems work on the principle of controlled diffusion of the drug through polymeric membranes. Both systems are available and have the advantage over previous delivery methods by decreasing the amount of drug administered for effective treatment and thus reducing the side-effects and the frequency of dose administration.

Minipumps have also been developed and used in small animals for drug delivery. These devices are capsular in size and shape with a volume of 280 ml. Most pumps consist of a flexible drug reservoir with an exterior orifice surrounded by an impermiable sleeve containing an osmotic agent inside a semipermiable membrane. The minipump is filled with the drug and implanted either subsutaneously or intraperitioneally.[39] Water from the surrounding tissues passes through the semipermiable membrane to the osmotic agent thus exerting pressure on the internal reservoir forcing the drug through the orifice at rate commensurate to the rate of osmosis. ALZET® was the first of the minipumps developed and is used primarily for small animals.[40] For human pharmacological application, OSMETTM was developed as an oral administration, giving continuous drug delivery for specific times up to 24 hours.[41]

Recently developed osmotic pump systems, EOP and OROS®,[42] do give predictable drug release rates for 12 or 24 hour periods. These devices use a solid drug, either alone or with an osmotic agent, surrounded by an aqueous permeable membrane with an external opening. Water from the area diffuses through the membrane, dissolving the drug, which is forced out of the orifice.

Another systemic method of treatment is the transdermal therapeutic system for release of drug through intact skin into the bloodstream. This system eliminates the GI tract which can alter a drug through acid-base conditions as well as enzymatic activity. One of the first marketed systems delivered scopolamine for motion sickness prevention.[43] Presently being developed is a system for the treatment of vertigo.[44]

Another important method for controlled drug administration is rate dependent diffusion of the drug from a polymer matrix. The central problem confronting this type of controlled release technology is that most vehicles display release rates that decay with time. The Higuchi model which has been used to develop approaches to achieve zero-order release kinetics[45] predicts that many designs, such as a slab, would have release kinetics that decreases with time. The reason is that the release rate is inversely proportional to the distance the drug must travel from within the matrix to the matrix surface. Since the diffusion distance increases with time, the release rate decreases. However, a new design by Hsieh and Langer[46] involving hemispheres has been reported to produce the desired zero-order release kinetics.

Biodegradable polymers are another polymer system currently receiving considerable interest for controlled drug release applications.[47,48] Poly(lactic acid) was the first biodegradable polymer used for the controlled release of therapeutic agents from an implanted site and current approaches are based on homopolymers and copolymers of poly(lactic acid), poly(glycolic acid) and aliphatic polyesters.[49] Zero-order kinetics can be achieved when these polymers are used as bioerodible membranes. Recent work with polyacetals and poly(orthoesters) demonstrated that nearly zero-order kinetics can be achieved by monolithic systems.[49] Homopolymers and copolymers of ε-caprolactone and σ-valerolactone have produced interesting properties as biodegradable material for zero-order drug release.[50] The degradation mechanism is based on hydrolysis of the ester linkage or by enzymatic attack at the polymer surface.

(D) SITE-SPECIFIC DRUGS

In the general design of a drug, the main emphasis has been the development of a substance that can be taken orally. Therefore, the drug must be able to penetrate from the gastrointestinal tract into the bloodstream. This penetrability, however, is common to other barriers as well, such as the blood-brain and cell membranes. Consequently, all tissues are penetrated and the whole organism is perfused with the drug, causing side effects which can complicate the drug's clinical utility. To overcome this problem, drugs must be designed to limit distribution and to localize themselves to desired biological sites.

It has been found that polymers, as drug carriers, can be targeted or deposited in specific areas of the animal body such as tissues or organs by means of pendent ligands that bind to the cell surface or protein at specific receptor sites. This type of treatment is designed to ensure that appropriate levels of drug are retained continuously at the required site. This localization is particularly important with cytotoxic drugs such as those used for intratumor chemotherapy so as to minimize the killing of normal cells. Goldberg[51] recently prepared several polymeric drugs with adriamycin, a broad spectrum antitumor agent, designed to maintain high local antitumor activity[52] in the tumor without dissipating to other areas and giving rise to toxic reactions. Pitha *et al.* have developed polysaccharide-based antagonists with affinity for beta-adrenergic receptors.[53] Pitha has also developed other polymers with alprenolomenthane residues which interact specifically with beta-adrenergic receptors.[54]

SUMMARY

Although polymers already play a large role in drugs and drug administration, it is evident that the use of these substances for these purposes is only in its infancy. Many new polymer drugs are needed and much more has to be understood about toxicities, specificity, structure, molecular weight and other physical properties. More and better drug carriers are required to suit the delicate balance of the biological system. Further research in pharmaco-kinetics of polymers for controlled drug release and longer duration of release are just beginning to bear fruit. Probably the most fascinating area of all is the site-specific polymer drugs that are targeted to discrete biological locations, thus limiting the area of drug perfusion. In conclusion, we see before us an exciting and unique challenge to be met, requiring the resources of several disciplines for outstanding rewards.

References

1. R. M. Ottenbrite, *Biologically Active Polymers*, C. E. Caraher and C. G. Gebelein, Eds., American Chemical Society, Washington, D.C., 1982.
2. W. Regelson, *J. Polym. Sci.*, **66**, 483 (1979).
3. A. M. Kaplan, P. S. Morahan and W. Regelson, *J. Natl. Cancer Inst.*, **52**, 1919 (1974).
4. J. D. Stinnett and J. A. Majeski, *J. Surg. Oncol.*, **14**, 327 (1980).
5. A. G. Currie and C. Basham, *Br. J. Cancer*, **38**, 653 (1978).
6. A. M. Kaplan, K. M. Connolly, W. Regelson, *The Host Invader Interplay*, H. Yan Vendogosshe, ed., Elsevier, Holland (1980).
7. W. Regelson, *Anionic Polymer Drugs*, L. G. Donaruma, R. M. Ottenbrite, and O. Vogl, Eds., John Wiley and Sons, New York (1980).
8. R. M. Ottenbrite, *Anionic Polymeric Drugs*, L. G. Donaruma and O. Vogl, Eds., John Wiley and Sons, New York (1980).
9. A. E. Munson, D. White and P. Klykken, *Cancer Res.*, **16**, 329 (1981).
10. D. S. Breslow, *Polymer Preprints*, **22**, 24 (1981).
11. M. L. Powell, E. M. Hursh, J. U. Gatterman, A. R. Zander, L. Granati, L. Alexander, G. Hortobagyl and S. G. Murphy, *Abs. Am. Ass. Cancer Res.*, **22**, 189 (1981).
12. R. M. Ottenbrite, *Polymer Preprints*, **21**, (1981).
13. K. Kuss, R. M. Ottenbrite and A. M. Kaplan, Federation of American Experimental Biology, New Orleans Meeting (April 1982).
14. J. E. Fields, S. S. Asculai and J. H. Johnson, U.S. Pat. 4,255,537 (March 10, 1981).
15. R. E. Falk, L. Makowka, N. A. Nossal, J. Falk, J. Rotstein, J. E. Fields and S. S. Asculai, *J. Surg. Res.*, **88**, 120 (1980).
16. R. E. Falk, L. Makowka, N. Nossal, J. A. Falk, J. E. Fields and S. S. Asculai, *Brit. J. Surg.*, **66**, 861 (1979)
17. R. Burling, G. D. Wolf and B. Bömer, *Naturwissenschaften*, **67**, 367 (1980).
18. H. Ringsdorf, *J. Polym. Sci.*, **51**, 135 (1975)
19. E. G. Goldberg, Polymeric Affinity Drugs, *Polymeric Drugs*, L. G. Donaruma and O. Vogl, Eds., Academic Press, New York, 239 (1978).
20. A. Zaffaroni and P. Bonsen, *Polymeric Drugs*, L. G. Donaruma and O. Vogl, Eds., Academic Press, New York (1978).
21. C. M. Samour, *Chem. Tech.*, 494 (1978).
22. J. C. Venter, *et al.*, *Proc. Nat. Acad. Sci.*, **69**, 1141 (1972).
23. L. G. Donaruma, *Polymeric Drugs*, L. G. Donaruma and O. Vogl, Eds., Academic Press, New York (1978).
24. L. G. Donaruma, *Progress in Polymer Science*, **4**, Pergamon Press (1974).
25. J. A. Brierley, L. G. Donaruma, S. Lockwood, R. Mercogliano, S. Kitoh, R. Warner, J. Depinto and J. Edzward, *Biomedical Polymers*, E. P. Goldberg and A. Nakajima. Eds., Academic Press, New York, 1980, p. 425.
26. E. R. Hurwitz, R. Maron, A. Bernotein, M. Wilchek, M. Sela and R. Arnon, *Int. J. Cancer*, **21**, 747 (1978).
27. W. Fung, M. Przybylski, H. Ringsdorf and D. S. Zaharko, *J. Nat. Cancer Inst.*, **62**, 1261 (1979).

28. M. Szekerke and J. S. Driscoll, Europ. *J. Cancer.*, **13**, 529 (1977).
29. J. M. Whiteley, Z. Nimec and J. Galvin, *Organic Coatings and Plastic Chemistry*, **44**, 127 (1981).
30. B. C. Chu and J. M. Whiteley, *J. Natl. Cancer Inst.*, **62**, 79 (1979).
31. B. C. Chu and J. M. Whiteley, *Mol. Pharmacol.*, **17**, 382 (1980).
32. E. Diener and M. Feldman, *Transplant. Rev.*, **8**, 76 (1972).
33. U. Y. Lee and A. H. Sehon, *Nature*, **267**, 618 (1977).
34. U. Dimer and E. Diener, *J. Immunol.* **111**, 1886 (1979).
35. J. Urquhart, *Natl. Acad. Sci.*, 329 (1979).
36. A. Zaffaroni, *Poly. Sci. and Techn.*, **14**, 293 (1980).
37. J. Urquhart, *Ophthalemic Drug Delivery Systems*, J. Robinson, Ed., *Acad. of Pharm. Soc.*, 180 (1980).
38. B. B. Pharriss, *J. Reprod. Med.*, **20**, 155 (1978).
39. F. Theeuwes and B. Eckenhoff, *Sixth International Symposium on Controlled Release of Bioactive Materials* (1981).
40. B. Eckenhoff, *Controlled and Topical Release Session of the ALCHE Meeting Philadelphia, PA* (1980).
41. F. Theeuwes and W. Bayne, *J. Pharm. Sci.*, **66**, 1388 (1977).
42. A. Zaffaroni, *Polym. Sci and Tech.*, **14**, 293 (1981).
43. J. Shaw and J. Urquhart, *Trends Pharmacol. Sci.*, **1**, 208 (1980).
44. N. Price, L. G. Schmitt ad J. E. Shaw, *Clin. Ther.*, **2**, 258 (1979).
45. N. M. Weinshenker, *Polymeric Drugs*, L. G. Donaruma and O. Vogl, Eds., Academic Press, New York (1978).
46. D. S. T. Hsieh and R. Langer, *Controlled Release of Bioactive Materials*, D. H. Lewis Ed., Plenum Press, New York (1981).
47. J. Heller, *Biomaterials*, **1**, 15, (1980).
48. R. Langer, *Bioavailability and the Pharmokinetic Control of Drug Response*, V. Smolen Ed., John Wiley and Sons, New York (1981).
49. J. Heller, *Polymer Reprints*, **24**, 540 (1982).
50. A. Schindler and C. C. Pitt, *Polymer Reprints*, **24**, 544 (1982).
51. J. Pitha, J. Zjawiony, R. J. Lefkowitz and M. G. Caron, *Proc. Natl. Acad. Sci., USA*, **77**, 2219 (1980).
52. J. Pitha, J. Zjawiony, N. Narin, R. J. Lefkowitz and M. G. Caron, *Life Sci.*, **27**, 1791 (1980).
53. J. Pitha, Organic Coatings and Plastic Chemistry, **44**, 74 (1981).
54. E. P. Goldberg, Organic Coatings and Plastics Chemistry, **44**, 132 (1981).

Interrelation of Rheological and Thermodynamic Solution Properties

B. A. WOLF

Institut für Physikalische Chemie, Johannes-Gutenberg-Universität Mainz, D-65 Mainz, Jakob-Welder-Weg 15, West Germany.

New knowledge and new aspects concerning the interrelation named in the title have accumulated during the past years, so that it seemed worthwhile to attempt a critical survey. The accent in the present report is on the essentials, no attempt being made to cover all phenomena or to present a comprehensive review of the literature. The discussion is mainly based on our own experiments carried out with solutions of polystyrenes (PS) with narrow molecular weight distributions in various single solvents. The first part of the present report deals with the question how the thermodynamic conditions influence the flow behavior. In the second part, the question is reversed, i.e. whether the thermodynamic behavior of a polymer solution changes when it is subjected to laminar flow.

THERMODYNAMIC INFLUENCES ON THE FLOW BEHAVIOR

Assume that the viscosity of the fluid constitutes a function of state so that the differential of the viscosity coefficient η can be written as

$$d \ln \eta = \left(\frac{\partial \ln \eta}{\partial c}\right)_{\omega,P,D} dc + \left(\frac{\partial \ln \eta}{\partial(1/T)}\right)_{c,P,D} d\left(\frac{1}{T}\right) + \left(\frac{\partial \ln \eta}{\partial P}\right)_{c,\omega,D} dP$$

$$+ \left(\frac{\partial \ln \eta}{\partial \ln D}\right)^{c,\omega,P} d \ln D \tag{1}$$

where c (g.cm^{-3}) is the polymer concentration and D (s^{-1}) the rate of shear. The individual partial differentials are related to the following characteristic quantities:

The concentration dependence of η at vanishing (small values of) D yields the generalized Staudingerindex[1]

$$\left(\frac{\partial \ln \eta}{\partial c}\right)_{D \to 0} \equiv \{\eta\}, \tag{2}$$

which approaches the normal $[\eta]$ at infinite dilution.

From the T and p dependence of η one obtains the well known apparent energies and volumes of activation according to

$$\frac{\partial \ln \eta}{\partial(1/T)} \equiv \frac{E^{\neq}}{R} \quad \text{and} \quad \frac{\partial \ln \eta}{\partial p} \equiv \frac{V^{\neq}}{RT}. \tag{3}$$

Finally, the dependence of η on the rate of shear gives the slope m of the flow curves ($\log \eta$ versus $\log D$)

$$\frac{\partial \ln \eta}{\partial \ln D} \equiv m, \tag{4}$$

which changes from zero (Newtonian region) to a constant of the order of minus unity (power-law region) when D is raised. It will now be discussed, how a variation of the thermodynamic quality of the solvent (expressed in terms of the Flory–Huggins parameter χ or the second osmotic virial coefficient A_2) influences the above partial differentials.

Variable c[2,3] If we only consider Newtonian behavior, the question becomes one of how $\{\eta\}$, the specific hydrodynamic volume of the polymer at a given concentration, depends on χ. For most polymer solutions, the Martin equation can well reproduce the dependence of the zero-shear viscosity η_0 on the reduced polymer concentration $\tilde{c} = c[\eta]$; hence

$$\eta_0 = \eta_s(1 + \tilde{c}e^{k_H\tilde{c}}) \tag{5}$$

η_s being the viscosity of the solvent and k_H the Huggins constant (typically < 1). From Equations (5) and (2) one obtains

$$\{\eta\} = [\eta]\frac{(1 + k_H\tilde{c})e^{k_H\tilde{c}}}{1 + \tilde{c}e^{k_H\tilde{c}}} \tag{6}$$

which yields the following limiting cases

$$\{\eta\} \infty [\eta] \text{ for } k_H\tilde{c} \ll 1 \text{ and } \{\eta\} \infty k_H[\eta] \text{ for } k_H\tilde{c} \gg 1.$$

Looking at the thermodynamic influences in terms of an increase of χ, the growth of the number of intersegmental contacts associated with such a deterioration of the solvent will influence $\{\eta\}$ in a very different way at low and at high concentrations. In the case of sufficiently dilute systems, practically only the intramolecular effects (manifesting themselves in $[\eta]$) will be operative, i.e. the coils shrink and $\{\eta\}$ decreases. If, however, there already exists a considerable overlap of polymer coils, the intermolecular effects (manifesting themselves in k_H) will result in a coupling of neighbouring molecules and in a rise of $\{\eta\}$. Going from isolated coils to high values of $\tilde{c}$, the flow mechanism changes from non-draining to free draining behavior.

Variable T and P[5-8]   The activation energies and volumes of the pure low molecular weight solvents ($E_s^{\neq}$ and $V_s^{\neq}$) typically amount to $\frac{1}{4}$ of the corresponding molar heats of vaporization and molar volumes. In the case of liquid polymers, the above fraction is much less and independent of molecular weight if a certain number of segments is exceeded; these observations have led to the concept of independently moving flow units, the size of which expresses itself in the measured parameters of activation. The question therefore becomes, how the contributions stemming from the polymer, i.e. $E^{\neq} - E_s^{\neq}$ and $V^{\neq} - V_s^{\neq}$, vary with the quality of the solvent.

In order to guarantee reasonable effects, we have confined ourselves to moderate polymer concentrations ($1 < \tilde{c} < 10$); in addition, tert.-butyl-acetate (TBA) was chosen as the solvent for *PS*, since it allows the entire range from endo- to exothermal demixing to be covered. Figure 1(a) shows how A_2 varies with T[4] and Figure 1(b), how $V_s^{\neq}$ and $V^{\neq}$ depend on T[5]. From a comparison of these results it becomes obvious that $V^{\neq} - V_s^{\neq}$ (and within experimental error also $V^{\neq}$) reaches a minimum when A_2 runs through its maximum. The pressure dependence of the viscosity of the solution is therefore smallest for optimum thermodynamic conditions. The worse the solvent becomes (the smaller A_2), the more the polymer raises $V^{\neq}$

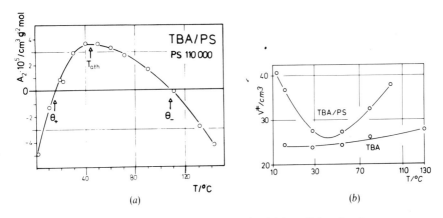

(a)   (b)

Figure 1   (a) T dependence of the second osmotic virial coefficient for the system tert.-butylacetate/polystyrene (TBA/PS).
(b) T dependence of the volume of activation (Equation (3)) for TBA/PS ($M = 110\,000$) and pure TBA.

over $V_s^{\neq}$: This behavior results from the increasing preference of interseg-mental contacts and the corresponding thermodynamic 'pull-along effect' which increases the size of the flow unit. $V^{\neq} - V_s^{\neq}$ is, however, not an unequivocal function of the Gibbs energy of mixing. It also depends on the enthalpy of mixing, which in turn governs the T dependence of $V^{\neq} - V_s^{\neq}$[6]. From a comparison of the P dependence of the viscosity in different solvents at theta-conditions it is known[6] that $(V^{\neq} - V_s^{\neq})_{c,\omega=\theta}$ (like $V_s^{\neq}$ itself) is largely determined by the shape of the solvent molecules; high values resulting from a bulky and low values from a smooth appearance.

All the above considerations hold true for $E^{\neq} - E_s^{\neq}$ analogously. As the critical temperature T_c is approached, the effects become particularly pronounced. For the thermodynamic critical concentration one obtains[7]

$$E^{\neq} = RT^2\left(\alpha + \frac{\beta}{T - T_c}\right); \ \alpha, \beta \ \text{constants} > 0 \tag{7}$$

where the second term, accounting for the critical extra effects, leads to a rapid increase of $E^{\neq}$ over $E_s^{\neq}$ as the solvents become extremely poor. In the case of a lower critical solution temperature, this term causes the increase in viscosity ($E^{\neq} < 0$) observed when T is raised towards T_c[8]. Normally, the critical contributions to $E^{\neq}$ are treated in terms of the vanishing concentration dependence of the chemical potential. In the light of the present reasoning they reflect the extraordinary increase in the size of the flow unit associated with the approach to T_c—the demixing itself simply marking the point at which thermal and flow agitations become unable to keep the molecules from clumping together and forming a second phase (cf. the discontinuity in $\eta(D)$ of Figure 2).

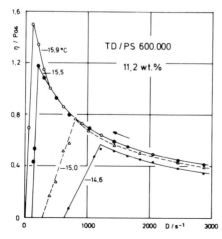

Figure 2 D dependence of the viscosity of a solution of PS ($M = 600\,000$, 11.2 wt%) in trans-decalin at the indicated constant temperatures.

Variable D[2,9]   The question regarding the thermodynamic influences on the non-Newtonian behavior is equivalent to the question how m depends on χ at a given shear rate. For sufficiently high D values, a direct answer can be given, since m is a constant within the power-law regime. For good solvents one obtains m ca. -0.8 in accord with entanglement theories[2], and ca. -0.5 for theta solvents. The reason for the reduced shear-thinning effect in poor solvents lies in the occurrence of a frictional contribution which does not depend on D[9]. Maintaining the reasoning of the preceding sections, this extra friction in poor solvents can be attributed to the high number of non-entangling intersegmental contacts.

To answer the question concerning the thermodynamic influences on η for m values close to zero (Newtonian behavior), τ_0, the characteristic relaxation time associated with the onset of shear rate dependence turns out helpful. In this region the following equation holds true, practically regardless of the solvent quality[9]:

$$\eta = \eta_0 - k_1(\tau_0\eta D)^3 \text{ if } \tau_0 D < \tfrac{2}{3}. \tag{8}$$

Figure 3 shows the experimental results[9] for solutions of *PS* in toluene (good solvent) and in trans-decalin (theta solvent). The dramatic increase in τ_0 associated with the approach to the demixing temperature T_D reflects the increasing preference of intersegmental contacts and the corresponding reduction of the molecular mobility.

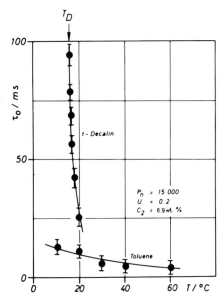

Figure 3 T dependence of the characteristic relaxation time associated with the onset of shear rate dependence of η for solutions of PS ($DP = 15\,000$, $6.9\,\text{wt}\%$) in toluene and in trans-decalin (T_D = demixing temperature).

INFLUENCE OF LAMINAR FLOW ON THE THERMODYNAMIC BEHAVIOR

Changes in the Gibbs energy of mixing brought about in solutions of chain molecules on streaming primarily result from an orientation and/or a deformation of the solute. At high dilution the effects of shearing are small for binary systems in both, good[10] and poor[11] solvents. They can, however, become quite pronounced in the case of two (incompatible) macromolecular solutes and a common solvent, where the homogeneous region may be extended by the order of 10°C by shearing[12]. For moderate polymer concentrations ($1 < \tilde{c} < 10$) and the binary systems of present interest, experimental information only exists in regard of poor solvents, to the knowledge of the author. It stems from two sources that will be discussed separately in the following, namely turbidimetric and viscometric investigations.

Light transmittance[13,14]   Normally the turbidity of polymer solutions close to demixing decreases as the liquids are sheared[13]. Such a shear-clearing is discernible from Figure 4, where the T dependence of the transmittance is plotted for different constant D values. Examples are, however, also known for systems that show shear-darkening[13,14]. As could be proved by viscometric experiments, these phenomenon appears within the homogeneous region and does not mean shear-demixing, i.e. no production of phase separation by the shear stress.

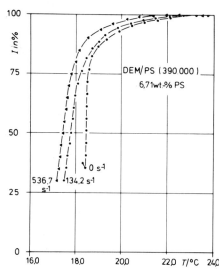

Figure 4 T dependence of the transmittance of a solution of PS ($M = 390\,000$, 6.7 wt%) in diethylmalonate at the indicated shear rates.

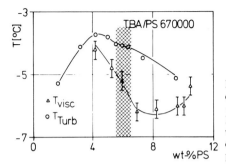

Figure 5   Comparison of the demixing curve of the system tert.-butylacetate/PS ($M = 670\,000$) at rest (O) and under shear ($\triangle$). For experimental reasons the D values at which the viscometric demixing points were determined fall from ca. 500 to $50\,\mathrm{s}^{-1}$ as the polymer concentration is raised.

Viscosity break-down[15-17]   For most polymer solutions the entrance into the two-phase region via a variation of T[15], P[16] or D[17] can be determined very accurately from the discontinuity in η associated with the formation of a suspension of the more concentrated coexisting phase in the matrix of the more dilute. Figure 2 demonstrates the situation for a low temperature demixing (16.0°C at rest) and variable D[17]: Solutions that would already be demixed at $D = 0$ can be held homogeneous by flow; the lower T, the higher the D value that must be exceeded to prevent the solution from segregating into a second phase. The extent of shear-dissolving depends on c[5]; it is largest in the region of the thermodynamic critical concentration and fades out on both sides of it, as is shown in Figure 5.

Theoretical considerations[18]   The interpretation of the turbidimetric information turns out to be difficult. Although it should in principle be possible to explain the phenomena of shear-clearing and shear-darkening in terms of the influence of D on the concentration dependence of the chemical potential, no such calculation has been performed so far. At present, it is not even possible to ascribe beyond doubt the demixing condition to a certain characteristic point on a turbidimetric curve like that shown in Figure 4.

The viscometric information, on the other hand, yields unambiguous demixing points for flowing systems. Furthermore, the observed influences of D can be interpreted in a comparatively simple manner. It suffices to calculate the conditions for which the system, already demixed at rest, can be completely homogenized by shearing. Equating the shear stress (which tends to disrupt the droplets of the suspended second phase) to the sum of the stress of curvature and the elastic stress (which both counteract such a subdivision) one obtains[18] the following relation for the stationary state:

$$k_2(\eta D) = \frac{2\sigma}{R} + k_3(\eta' D)^2 \qquad (9)$$

where η and η' are respectively the viscosity coefficients of the matrix and of the fluid contained in the droplets, σ = interfacial tension, R = radius of the droplet, k_2, k_3: constants.

Knowing that σ evolves from zero at T_c according to $\sigma = k_4 |T - T_c|^{3/2}$ and assuming that R must be reduced down to the mean end-to-end distance of the polymer coils for complete homogenization, it is possible to calculate the amount of shear-dissolving quantitatively. In view of the simple approach, the agreement with the experimental observations[5] showed surprisingly well[18].

OUTLOOK

In this short review it was attempted to summarize the present information concerning the interrelation of rheological and thermodynamic solution properties, and to raise some new perspectives. It seems promising to the author to extend the present approach to the question of kinetic consequences of thermodynamic conditions. In the case of the thermodynamically induced shear degradation, this concept has already proved successful[19].

References

1. H. L. Frisch, R. Simha, *Rheology, Theory and Applications*, **1**, F. Eirich, Ed., Academic Press, New York, 1956, Chapter 12.
2. W. W. Graessley, *Adv. Pol. Sci.*, **16**, 1 (1974) and references therein.
3. U. Brand, *Examensarbeit Mainz* (1980) and J. R. Schmidt, *Thesis Mainz* (1981), to be published.
4. B. A. Wolf and H. J. Adam, *J. Chem. Phys.*, **75**, 4121 (1981).
5. J. R. Schmidt and B. A. Wolf, *Coll. Polymer Sci.*, **257**, 1188 (1979).
6. H. Geerissen, J. R. Schmidt and B. A. Wolf, *J. Appl. Pol. Sci.*, in press.
7. P. Debye, B. Chu and D. Woermann, *J. Pol. Sci.*, **A1**, 249 (1963).
8. J. K. Rigler, B. A. Wolf and J. W. Breitenbach, *Angew. Makromol. Chem.*, **57**, 15 (1977).
9. M. Ballauff, *Thesis Mainz* (1981); M. Ballauff and B. A. Wolf, *J. Pol. Sci.*, **A2** (Physics Ed.) submitted for publication.
10. J. V. Champion and I. D. Davis, *J. Chem. Phys.*, **52**, 381 (1970), and references therein.
11. H. Tjakraatmadja, R. Hosemann and J. Springer, *Coll. Pol. Sci.*, **258**, 1145 (1980).
12. A. Silberberg and W. Kuhn, *Nature*, **170**, 450 (1952); *J. Pol. Sci.*, **13**, 21 (1954).
13. H. Krämer, *Diplomarbeit Mainz* (1980); B. A. Wolf and H. Krämer, *J. Pol. Sci., Pol. Letters Ed.*, **18**, 789 (1980).
14. G. Ver Strate and W. Philippoff, *J. Pol. Sci., Pol. Letters Ed.*, **12**, 267 (1974).
15. B. A. Wolf and M. C. Sezen, *Macromolecules*, **10**, 1010 (1977).
16. B. A. Wolf and R. Jend, *Macromolecules*, **12**, 732 (1979).
17. M. Klimiuk, *Diplomarbeit Mainz* (1981).
18. B. A. Wolf, *Makromol. Chem., Rapid Comm.*, **1**, 231 (1980).
19. J. W. Breitenbach, J. K. Rigler and B. A. Wolf, *Makromol. Chem.*, **164**, 353 (1973); M. Ballauff and B. A. Wolf, *Abstracts of the 27th Int. Symp. Macromol.*, **2**, 659 Strasbourg (1981).

Polyelectrolytes—Study on Solute Distribution in Solutions

N. ISE

Department of Polymer Chemistry, Kyoto University, Kyoto, Japan.

The solute distribution in solutions reflects the solute–solute, solvent–solute and solvent–solvent interactions present in the systems. Therefore, it is essential to determine the solute distribution, theoretically or experimentally[1] if a complete understanding of the solution properties is to be obtained. The distribution of simple ionic species in very dilute solutions can adequately be described by the Debye-Hückel theory.[2] This has been possible solely because the theory is 'complete'. Another example, in which the solute distribution has been established, is dilute solutions of electrically charged polymer latex particles. Taking advantage of the relatively large dimensions, direct microscopic observation was possible for these solutions.[3] These two cases are really exceptional; at higher concentrations of the simple ions, in the entire concentration range of biological or synthetic macroions or in the solution of ionic micelles, theoretical analyses are far from being satisfactory. Thus, experimental elucidation of the distribution has to be undertaken. It seems that recent technical advancement begins to allow us to tackle this formidable task.

Earlier light scattering measurements on salt-free polyelectrolyte solutions show a very low intensity of scattered light, suggesting that concentration fluctuations in polyelectrolyte solutions can not be independent, because of the necessity of maintaining net electrical neutrality (according to Doty and Steiner).[4] This fact led Fuoss,[5] for example, to the statement that 'long range of Coulomb forces tends to establish an ordered distribution'. The ordering of macroions was inferred on the basis of the dependence of the mean activity coefficients of polyelectrolytes on the cube-root of the concentration.[6] The unusually sharp decrease of the single-ion activity of hydrophilic macroions with increasing concentration was also in line with such an ordered distribution.[7] In 1970's, the neutron scattering technique was introduced in this field; Cotton and Moan reported an intensity peak at higher degrees of neutralization of a polymethacrylic acid, and attributed it to an ordering of macroions.[8] Moan found that the peak vanished progressively with increasing salt concentration and that the peak intensity decreased with increasing charge density.[9] Rinaudo and Domard carried out the neutron scattering of poly-L-glutamic acids of rather small molecular weights and the observed peak was attributed to an interchain ordering.[10] Maret et al. studied the influence of high magnetic field on flexible polypeptides in solution and found a local parallel ordering of the macroions with intermolecular distances up to several hundred angstroms.[11]

X-ray small angle scattering is also an interesting tool. Dusek et al. and Ise et al. have applied this technique for the study of dilute solutions of synthetic and bio-polyelectrolytes[12–14] such as polyacrylate, polymethacrylate, polystyrene sulfonate, poly-L-lysine and bovine serum albumin (BSA). Early X-ray analyses by Bernal et al. on concentrated solutions and gel states of plant virus,[15] by Oster et al. on concentrated solutions of proteins[16] and by Harkins et al. on concentrated ionic micelle solutions,[17] reported a single scattering peak. The peak disappeared when a large amount of neutral salt was added or when the polymer concentration was lowered.[12–14] Furthermore no peak was observed when the macromolecules were not electrically charged, namely, at the isoelectric point for the protein,[18] at low degree of neutralization of weak polyions,[12,13] or for neutral polymers.[14] Ise et al. noticed[13,14] that the scattering behaviors depended on the molecular weight of the linear macroions and that mixing of two fractions of different molecular weights caused disappearance of the original peaks and gave a new single peak between the mother peaks. From this and other results, the present writer strongly believes that the ordering, which gave rise to the scattering peak, is of an intermolecular nature. Thus, the intermacroion distance ($2D_{exp}$) was estimated by assuming the validity of Bragg's relation for these solution systems. $2D_{exp}$ was found to be fairly large (for example, 70 Å–160 Å for polyacrylates, polymethacrylate, and poly-L-lysine) and it decreased with increasing polymer concentration, with decreasing salt concentration, and with increasing degree of neutralization, and with decreasing degree of polymerization. For BSA, $2D_{exp}$ was found to be 105, 111, and 116 Å at 10°, 25°, and 40°C at 0.072 g/ml and pH = 3.9 ~ 3.8.[18]

Comparison of $2D_{exp}$ with the average interparticle distance ($2D_0$) revealed an interesting feature. For the linear polyelectrolytes, $2D_{exp}$ was smaller than $2D_0$ by a factor of 2 under the experimental conditions employed. This suggests that there exist simultaneously an ordered region and a disordered region in the solution although the volume ratio of these two regions is not yet determined; in other words, a two-state structure is maintained. The underlying idea which emerges is that macroions attract each other. Because of the repulsion between macroions themselves, this attraction should arise from the presence of counterions existing in the space between the macroions. Qualitatively, the situation is close to the stabilization of triple ions, quadruples and higher aggregates, which is a familiar phenomenon in the physical chemistry of simple electrolyte solutions.[19] The potential energy (W_3) of a triple ion, for example ($\oplus \!-\! \ominus \!-\! \oplus$), is definitely lower than that (W_2) of an ionpair and a free ion ($\oplus \!-\!\cdots\! \oplus$) with the intercharge distance r. (The same argument holds for higher aggregates.) For monovalent ions, $W_3 = -1.5\,e^2/r$ and $W_2 = -e^2/r$, where e is the charge on the ions. W_3 is further lowered with increasing valency of the constituent ions. In other words, the two cations 'attract' each other (through the intermediary of the anion) and even more strongly with increasing charge valency. If this explanation applies to the macroion systems, the intermacroion distance ($2D_{\exp}$) should decrease with increasing degree of neutralization or charge density and with decreasing salt concentration, as a result of shielding effect. This is what is experimentally observed.

The X-ray method is useful but cannot be claimed to be finally conclusive. Thus, a similar charged system was sought. Polymer latex particles have been known to form an ordered structure in dilute solutions, when they are monodisperse and free from ionic impurities. Hachisu et al.[3] obtained the direct evidence of the ordering by visual observation using metallurgical microscope. By this method, the interparticle distance ($2D_{\exp}$) can be determined without assumptions. Though overlooked by the authors, the data show that $2D_{\exp}$ was approximately equal to the average interparticle distance $2D_0$ for Hachisu's sample which had a relatively low charge density (diameter $\cong 3000\,\text{Å}$, analytical valency $\cong 10^3$). According to the explanation mentioned above, $2D_{\exp}$ is expected to become smaller than $2D_0$ when the charge density of the particles is raised. Thus, monodisperse latex particles of copolymers of styrene and styrene sulfonate (diameter $3020\,\text{Å}$, valency $\simeq 10^5$) were studied by the microscopic observation.[20] It turned out, as expected, that $2D_{\exp} \cong 2D_0$ at higher concentrations (3–10 vol%) whereas $2D_{\exp} < 2D_0$ below 1% (e.g. $2D_{\exp} = 1.3 \times 10^4\,\text{Å}$ and $2D_0 = 2.3 \times 10^4\,\text{Å}$ at 0.15%); a two-state structure is maintained in the dilute solutions of latex as well as polyelectrolytes.

Independently, Chu et al. claimed an ordering in salt-free and low-salt solutions of tRNA by means of light and X-ray scatterings.[21] The latter method showed a clear peak, as was the case for macroions considered above. The Bragg distance was in rough agreement with the average distance: $2D_{\exp} \cong 2D_0$. This equality is quite understandable in light of the relatively low charge density of the polynucleotide. BSA solutions were found to show thixotropic properties, which suggest the presence of some sort of ordering.[22] Elastic and inelastic light scatterings reinforced this conclusion. Most interestingly, these authors arrived at the same conclusion as the present writer that an ordered region coexists with a disordered region. Lysozyme was also pointed out to form a structure in a concentration range between $10^{-4}\%$ and 30%.[23] The time constant characterizing the building up of the structure was determined for lysozyme to be $\sim 10^{-5} \sec^{-1}$. It should be remembered that the ordered structure in question is fluctuating with time as was directly confirmed for the first time for polymer latex solutions by Hachisu et al.[3] The time constant must depend sensitively on the properties of charged entities and the experimental conditions, although direct measurements have not yet been tried on macroions.

Because of the close relevancy to macroions, we finally consider simple electrolyte ions. These charged solutes have generally much higher charge densities than the macroions, proteins and latex particles, as a result of the small dimension. Thus, the consideration on the ion association mentioned above predicts that the interionic distance in simple electrolyte solutions must be much smaller than the average distance; namely $2D_{exp} \ll 2D_0$. It was most recently noticed on the basis of EXAFS data reported by Sadoc et al.[24,25] that this prediction was correct. The data of $ZnBr_2$ show that the distances between Zn^{2+} and Br^- in aqueous dilute solution (down to $0.09\,M$) and in ethyl acetate solution (down to $0.05\,M$) are $2.37\,\text{Å}$, respectively, which are surprisingly close to the value in the solid crystal $(2.40\,\text{Å})$. Since the $2D_0$ is roughly $40\,\text{Å}$ at these low concentrations, it can be claimed that there must exist a two-state structure, a much 'tighter' one than was experienced for the macroions and latex systems.

Finally two points have to be mentioned. The first is the problem of whether the single, broad X-ray or neutron peak observed can be an evidence for the ordering. In this respect, we note that light scattering experiments have furnished also a broad peak for latex particles[26] probably because of the lattice being distorted by thermal motion, although the ordering is almost unquestionably proven as mentioned above. It may be proposed that the ordering in solutions is not as 'perfect' as in solid crystals and is expected to be more 'perfect' at lower temperatures and dielectric constants. The second is the presence of the attraction, as discussed here, which has been almost completely overlooked in previous theoretical treatments. An exception is that due to Kirkwood and Mazur.[27-28] Even when an ordering of macroions was noted, the intermacroion repulsion was claimed to be responsible.[4] The ordering of latex particles has been attributed to repulsion[3] and this conclusion was erroneously justified on the basis of the fact that $2D_{exp}$ was close to $2D_0$. Obviously, the interpretation in terms of the repulsion is not widely valid because $2D_{exp}$ can be smaller than $2D_0$ for solutes of high charge density such as linear macroions, highly charged latex, and simple ions.

Thus, the ordering and the two-state structure appear to be a most significant aspect of electrically charged solutes in general in dilute solutions, which has not been discussed in previous theoretical models. Further detailed and quantitative experimental elucidation of the solute distribution and theoretical treatments of such a basic feature, if possible, are necessary for better and unfragmented understanding not only of poly-electrolyte and colloidal solutions but also ionic solutions in general.

References

1. For a review of this topic, see N. Ise and T. Okubo, *Acc. Chem. Res.*, **13**, 303 (1980).
2. P. Debye and E. Hückel, *Physik, Z.*, **24**, 185 (1923).
3. A. Kose, M. Ozaki, K. Takano, Y. Kobayashi and S. Hachisu, *J. Coll. Interface Sci.*, **44**, 330 (1973).
4. P. Doty and R. F. Steiner, *J. Chem. Phys.*, **17**, 743 (1949).
5. R. M. Fuoss and D. Edelson, *J. Polymer Sci.*, **6**, 767 (1951).
6. N. Ise and T. Okubo, *J. Phys. Chem.*, **70**, 1930 (1966).
7. N. Ise and T. Okubo, *J. Phys. Chem.*, **70**, 2407 (1966).
8. J. P. Cotton and M. Moan, *J. Phys. (Paris)*, **37**, L-75 (1976).
9. M. Moan, *J. Appl. Cryst.*, **11**, 519 (1978).
10. M. Rinaudo and A. Domard, *Polymer Letters Edition*, **15**, 411 (1977).
11. G. Maret, J. Tobert, E. Senechal, A. Domard, M. Rinaudo and H. Milas, *Symposium Europ. Chem. Societies* (September 1978).
12. J. Plestil, J. Mikes and K. Dusek, *Acta Polym.*, **30**, 29 (1979).
13. N. Ise, T. Okubo, *et al.*, *J. Am. Chem. Soc.*, **101**, 5836 (1979); **102**, 7901 (1980).
14. N. Ise, T. Okubo, K. Yamamoto, H. Matsuoka, H. Kawai, T. Hashimoto and M. Fujimura, *J. Chem. Phys.*, **78**, 541 (1983).
15. J. D. Bernal and I. Fankuchen, *J. Gen. Physiol.*, **25**, 111 (1941).
16. D. P. Riley and G. Oster, *Discuss. Faraday Soc.*, **11**, 107 (1951).
17. R. W. Mattoon, R. S. Stearns and W. D. Harkins, *J. Chem. Phys.*, **16**, 644 (1948).
18. N. Ise, T. Okubo, S. Kunugi, K. Yamamoto and H. Matsuoka, publication in preparation.
19. For example, see H. S. Harned and B. B. Owen, *The Physical Chemistry of Electrolytic Solutions*, Reinhold, New York (1963).
20. N. Ise, T. Okubo, M. Sugimura, K. Ito and H. J. Nolte, *J. Chem. Phys.*, **78**, 536 (1983).
21. A. Patkowski, E. Gulari and B. Chu, *J. Chem. Phys.*, **73**, 4178 (1980).
22. R. Giordano, G. Maisano, F. Mallamace, N. Micali and F. Wanderlingh, *J. Chem. Phys.*, **75**, 4770 (1981).
23. R. Giordano, M. P. Fontana and F. Wanderlingh, *J. Chem. Phys.*, **74**, 2011 (1981).
24. P. Lagarde, A. Fontaine, D. Raoux, A. Sadoc and P. Migliardo, *J. Chem. Phys.*, **72**, 3061 (1980).
25. A. Sadoc, A. Fontaine, P. Lagardo and D. Raoux, *J. Am. Chem. Soc.*, **103**, 6287 (1981).
26. P. A. Hiltner and I. M. Krieger, *J. Phys. Chem.*, **73**, 2386 (1969).
27. J. G. Kirkwood and J. Mazur, *J. Polymer Sci.*, **9**, 519 (1952).
28. I. Sogami and N. Ise successfully formulated a theory, which predicts the presence of attraction at large interparticle distances and repulsion at small separations. For a preliminary communication, see I. Sogami, *Physics Letter* in press.

Review Article Source Selection

Each year this section will present a resumé of review articles in certain areas of polymer science. This years selection is rather larger than usual and contains references to a number of important early reviews published before (1980). Next year will cover reviews appearing in the period (1979–1982).

I POLYMERIZATION PROCESSES

a) Co-ordination complexes in polymerization

T. Keii, *Kinetics of Zeigler Natta Polymerization*, Chapman and Hall, London (1972).
I. Pasquon and L. Porri, Macromolecular Science, *MTP International Review of Science*, **8**, 159 (1972).
Coordination Polymerization; A Memorial to K. Ziegler, Ed. J. C. W. Chien, Academic Press, London, New York (1975).
W. Cooper, *Comprehensive Chemical Kinetics*, Ed. C. H. Bamford and C. F. H. Tipper, Elsevier, Amsterdam, 133 (1976).
E. J. Vandenberg and B. C. Repka, *High Polymers*, Ed. C. E. Schildknecht and I. Skeist, Wiley, New York, **29**, 337 (1977).
A. D. Cant, *Catalysis*, Ed. C. Kemball (Specialist Periodical Report), The Chemical Society, London, **1**, 234 (1977).
P. J. T. Tait, *Macromolecular Chemistry* (Specialist Periodical Reports), The Chemical Society, London, **1**, 3 (1980).
P. J. T. Tait, *Developments in Polymerization*, Ed. R. N. Haward, Applied Science Publishers, **2**, 81 (1979).

b) Cationic polymerization

K. E. Russell and G. J. Wilson, *High Polymers,* **29**, 306 (1977).
D. J. Dunn and J. M. Rooney, *Macromolecular Chemistry* (Specialist Periodical Reports), The Chemical Society, London, **1**, 22 (1980).

c) Anionic polymerization

R. N. Young, *Macromolecular Chemistry* (Specialist Periodical Reports), The Chemical Society, London, **1**, 34 (1980).

d) Radical polymerization

J. C. Bevington, *Macromolecular Chemistry* (Specialist Periodical Reports), The Chemical Society, London, **1**, 45 (1980).

e) Template polymerization

Polymerization of Organized Systems, Ed. H.-G. Elias, Midland Macromolecular Monographs, Gordon and Breach, **3**, (1977).

C. H. Bamford, *Developments in Polymerizations*—2, Ed. R. N. Haward, Applied Science Publishers (1979).

C. M. Bamford, *Macromolecular Chemistry* (Specialist Periodical Reports), The Chemical Society, London, **1**, 52 (1980).

f) Emulsion polymerization

D. C. Blackley, *Emulsion Polymerization: Theory and Practice,* Applied Science Publishers, London (1975).

J. Ugelstad and F. K. Hansen, *Rubber Chem Technology,* **49**, 536 (1976); **50**, 639 (1977).

J. L. Gardon, *Emulsion Polymerization: Theory and Practice, Ency. of Polymer Sci. & Tech.,* Supplement, **1**, 238, Interscience, New York (1976).

Emulsion Polymerization, Ed. I. Piirma and J. L. Gardon (ACS Symposium Series No 24) American Chemical Society, Washington DC -1976),

V. I. Eliseeva, S. S. Ivanchev, S. I. Kuchancy and A. V. Lededev, *Emul'sionnaya Polimerizatsiya i ee Primenenie v Promyshlennosti,* Khimiya, Moscow, USSR (1976).

J. Ugelstad, F. K. Hansen and K. H. Kaggarud, *Faserforsch u Textitech,* **28** 309 (1977).

g) Electrochemical polymerization

J. W. Breitenbach, O. F. Olaj and F. Sommer, *Adv. Polymer Science,* **9**, 47 (1972).

B. L. Funt and J. Turner, *Techniques of Electroorganic Synthesis,* Part 2, Ed. N. Weinberg in *Techniques of Chemistry,* Ed. G. A. Weissberger, Wiley, Chichester, **5**, 559 (1975).

G. S. Shapoval and T. E. Lipatova, Electrokhimicheskoe Initsiirovanie Polimerizatsii (Electrochemical Initiation of Polymerization) Kier, Naukova Dumka (1977).

B. M. Tidswell, *Macromolecular Chemistry* (Specialist Periodical Reports), The Chemical Society, London, **1**, 74(1980).

h) Step growth polymerization

R. H. Still, *Macromolecular Chemistry* (Specialist Periodical Reports), The Chemical Society, London, **1**, 81 (1980).

i) Copolymerization and multi-component polymerization reactions

T. Alfrey, J. J. Bohrer and H. Mark, *Copolymerization,* Interscience, New York (1952).

C. H. Bamford, W. G. Barb, A. D. Jenkins and P. F. Onyon, *The Kinetics of Vinyl Polymerization by Radical Mechanisms,* Butterworths, London (1958).

C. S. Marvel, *An Introduction to the Organic Chemistry of High Polymers,* Wiley, New York (1959).

W. J. Burland and A. S. Hoffman, *Block and Graft Polymers,* Reinhold, New York (1960).

R. J. Ceresa, *Block and Graft Copolymers,* Butterworths, London (1962).

Copolymerization, Ed. G. E. Ham, Interscience, Wiley, New York (1964).

G. E. Ham, *Copolymerization, Ency. of Polymer Sci. & Tech.,* **4**, Wiley, New York (1966).

H. Mark, B. Immergut, E. H. Immergut, L. J. Young and K. I. Benyon, *Polymer Handbook,*

Ed. J. Brandrup and E. H. Immergut, 11–142, Wiley, New York (1966).

M. L. Miller, *The Structure of Polymers*, Reinhold, New York (1966).

A. M. North, *The International Encyclopedia of Physical Chemistry and Chemical Physics; Topics* 17 *Macromolecules*, Ed. C. E. H. Bawn (1966).

H. A. J. Battaerd and G. W. Tregear, *Graft Copolymers*, Wiley, New York (1967).

R. W. Lenz, *Organic Chemistry of Synthetic High Polymers*, Wiley, New York (1967).

M. Szwarc, *Carbanions, Living Polymers and Electron Transfer Processes*, Interscience, New York (1968).

Block Copolymers, Ed. J. Maocanin, G. Holder and N. W. Tschoegel, *J. Polymer Sci.*, Polymer Symp. Ser. No. 20 (1969).

Block Polymers, Ed. S. L. Aggarwal, Plenum, New York (1970).

G. Odian, *Principles of Polymerization*, McGraw-Hill, New York (1970).

F. W. Billmeyer, *Textbook of Polymer Science*, Wiley-Toppan, New York (1971).

D. J. Williams, *Polymer Science and Engineering*, Prentice-Hall, New York (1971).

Kinetics and Mechanism of Polymerusation Series, Ed. G. E. Ham, 1–3, Marcel Dekker, New York (1972).

Block Copolymers, Ed. D. C. Allport and W. H. Janes, Applied Science Publishers, London (1973).

Reactivity, Mechanism and Structure in Polymer Chemistry, Ed. A. D. Jenkins and A. Ledwith, Wiley, London (1974).

D. B. V. Parker, *Polymer Chemistry*, Applied Science (1974).

P. E. M. Allan and C. R. Patrick, *Kinetics and Mechanisms of Polymerisation Reactions*, Ellias Horwood, London (1974).

M. P. Stevens, *Polymer Chemistry*, Addison-Wesley (1975).

Cationic Polymerization, Ed. J. P. Kennedy, *J. Polymer Sci.*, Polymer Symp. Ser. No. 56 (1976).

A. Noshay and J. E. McGrath, *Block Copolymers—Overview and Critical Survey*, Academic Press, New York (1977).

Polymerization Processes, Ed. C. E. Schildkneckt and I. Skeist, Wiley Interscience (1977).

H.-G. Elias, *Macromolecules—Structure and Properties*, Plenum, New York (1977).

V. V. Korshak, *Copolymers*, Nauka Moscow, USSR (1977).

Yu L. Spirin, *Polymerization Reactions*, Naukova Dumka, Kiev, USSR (1977).

H. S. Kaufman and J. J. Falcetta, *Introduction to Polymer Science and Technology*, Wiley, New York (1977).

A. F. Johnson and D. G. Catton, *Macromolecular Chemistry* (Specialist Periodical Reports), The Chemical Society, London, 1, 105 (1980).

II POLYSACCHARIDES AND GLYCOPROTEINS

R. J. Sturgeon, *Macromolecular Chemistry* (Specialist Periodical Reports), The Chemical Society, London (1980).

III NATURAL POLYMERS: PROTEINS AND ENZYMES

Amino-Acids Peptides and Proteins, Ed. G. T. Young (1–4) and R. C. Sheppard (5–8) (Specialist Periodical Reports), The Chemical Society, London (1969–1977).

Amino-Acids Peptides and Proteins, Ed. R. C. Sheppard (Specialist Periodical Reports), The Chemical Society, London, 9 (1978).

C. J. Gray, *Macromolecular Chemistry* (Specialist Periodical Reports), The Chemical Society, London, 1 (1980).

IV INORGANIC POLYMERS

D. L. Venezky, *Encyl. Polymer Sci Tech.*, 7, 664 (1967).

W. Noll, *The Chemistry and Technology of the Silicones* 2nd (Edn), Academic Press, New York (1968).

M. Zeldin, *Polymer News*, **3**, 65 (1976).

Homoatomic Rings, Chains and Macromolecules of Main Group Elements, Ed. A. L. Reingold, Elsevier Scientific Publishing Co., Amsterdam (1977).

Organometallic Chemistry Reviews—Annual Surveys; Silicon–Tin–Lead, Ed. D. Seyferth and R. B. King, Elsevier Scientific Publishing Co., Amsterdam (1977–78).

C. E. Carraher Jun., J. E. Sheats and C. U. Pittman Jun., *Conf. Proceedings Amer. Chem. Soc.* Published in *Organometallic Polymers*, Academic Press, New York (1978).

N. H. Ray, *Inorganic Polymers*, Academic Press, New York (1978).

K. M. Roch, *Macromolecular Chemistry* (Specialist Periodical Reports), The Chemical Society, London, **1**, 208 (1980).

V CONFIGURATIONS

P. J. Flory, *Principles of Polymer Chemistry*, Cornell U.P., Ithaca, New York (1953).

P. J. Flory, *Statistical Mechanics of Chain Molecules*, Wiley Interscience, New York, 15 (1969).

H. Yamakawa, *Modern Theory of Polymer Solutions*, Harper and Row, New York (1971).

S. B. Ross-Murphy, *Macromolecular Chemistry* (Specialist Periodical Reports), The Chemical Society, London, **1**, 222 (1980).

VI NUCLEAR MAGNETIC RESONANCE SPECTROSCOPY

F. A. Bovey, *Structural Studies of Macromolecules by Spectroscopic Methods—1974*, Ed. K. J. Ivin, Wiley, Chichester, England, 181 (1976).

A. R. Katrizky and D. E. Weiss, *Chem. In Britain*, **13**, 45 (1976).

J. R. Ebdon, *Structural Studies of Macromolecules by Spectroscopic Methods—1974*, Ed. K. J. Ivin, Wiley, Chichester, England, 241 (1976).

J. Schaefer, *Structural Studies of Macromolecules by Spectroscopic Methods—1974*, Ed. K. J. Ivin, Wiley, Chichester, England, 201 (1976).

A. Johnsen, E. Klesper and T. Wirhlin, *Macromol Chem*, **177**, 2397 (1976).

J. C. Randell, *Polymer Sequence Determination Carbon-13 nmr Method*, Academic Press, New York (1977).

K. Nagasawa, *Toyota Chuo Kenkyushu R and D Rebyu*, **13**, 87 (1977).

J. Urbanski, *Handbook of Analysis of Synthetic Polymers and Plastics*, Ellias Horwood, Chichester, 142 (1977).

N. A. Plate and L. B. Stroganov, *Polymer Sci USSR*, **18**, 1087 (1977).

R. A. Komoroski and L. Mandelkern, *Applications of Polymer Spectroscopy*, Ed. E. G. Brame, Academic Press, New York, 57 (1978).

D. Doskocilova and B. Schneider, *Adv. Colloid Interface Sci.*, **9**, 63 (1978).

F. Heatley, *Macromolecular Chemistry* (Specialist Periodical Report), The Chemical Society, London, **1**, 234 (1980).

VII POLYMER CHARACTERIZATION

P. E. Slade, *Polymer Molecular Weights*, Parts I and II, Dekker, New York (1975).

N. C. Billingham, *Molar Mass Measurements in Polymer Science*, Kogan Page, London (1977).

N. C. Billingham, *Macromolecular Chemistry* (Specialist Periodical Reports), The Chemical Society, London, **1**, 282 (1980).

VIII THERMODYNAMICS OF SOLUTIONS AND MIXTURES

G. Rialdi and R. L. Biltonen, *Int. Rev. Sci. Phys. Chem Ser.* **2**, **10**, 147 (1975).
P. L. Privalov, *Pure and Applied Chem.*, **47**, 293 (1976).
F. Kohler, *Ber Bunsengeschellschaft Phys Chem.*, **81**, 1037 (1977).
J. Grover, NATO Adv. Study Inst Ser C30 293 (1977).
R. F. Blamks, *Polymer Plast. Technol Eng.*, **8**, 13 (1977).
Structure-Solubility Relationships in Polymers, Ed. F. W. Harris and R. B. Seymour, Academic Press (1977).
S. Nagarajan, *Proc Royal Aust. Chem Inst.*, **44**, 151 (1977).
Chemical Thermodynamics, Ed. M. L. McGlashan, (Specialist Periodical Reports), The Chemical Society, London, **2** (1978).
L. E. Nielsen, *Predicting the Properties of Mixtures*, Marcel Dekker (1978).
J. W. Kennedy, *Macromolecular Chemistry* (Specialist Periodical Reports), The Chemical Society, London, **1**, 296 (1980).

IX POLYMER ENGINEERING

P. van Heerden, *Properties of Polymer*, Elsevier, Amsterdam (1972).
C. C. Chamis, *Composite Materials v Fracture and Fatigue*, Academic Press, 94 (1974).
G. P. Sendeckyi, *Mechanics of Composite Materials*, Ed. G. P. Sendeckyi, Academic Press, London and New York, 46 (1974).
Composite Materials, Ed. L. J. Broutman and R. H. Cook, Academic Press, London, **5** (1974).
T. L. Smith, *J. Polymer Eng Sci.*, **17**, 128 (1977).
C. B. Bucknall, *Toughened Plastic*, Applied Science, London (1977).
Developments in Polyurethane, Ed. J. M. Bruist, Applied Science, London (1977).
L. C. R. Blackman, *Advancing Technologies*, Mechanical Engineering Publications, London, 136 (1977).
Science and Technology of Polymer Processing, Ed. N. P. Suh and N. Sung, MIT Press, Cambridge, USA (1979).
S. F. Bush, *Macromolecular Chemistry* (Specialist Periodical Reports), The Chemical Society, London, **1**, 331 (1980).

X POLYMER DEGRADATION

Developments in Polymer Degradation, Ed. N. Grassie, Applied Science Publishers, London (1977).
F. H. Winslow, *Pure Appl Chem*, **49**, 495 (1977).
F. C. DeSchryver, N. Boens and J. Put, *Adv. Photochem*, **10**, 359 (1977).
K. Tsuju, *Polymer Plast Technology Eng*, **9**, 1 (1977).
Aspects of Degradation and Stabilization of Polymers, Ed. H. H. G. Jellinek, Elsevier, Amsterdam (1978).
N. S. Allan and J. F. McKeller, *Photochemistry of Man Made Polymers*, Applied Science Publishers, London (1978).
G. Scott, *Polymer Plast Technology Eng.*, **11**, 1 (1978).
D. Braun, *J. Macromol Sci. Chem A*12, 379 (1978).
J. R. McCallum and W. W. Wright, *Macromolecular Chemistry* (Specialist Periodical Reports), The Chemical Society, London, **1**, 370 (1980).

Publications in Polymer Science

In this first issue, the compilation contains many titles which have been published during the last ten years in polymer science and technology. In future issues this section will only contain books published in the preceding three years. The titles are sub-divided into five groupings:

I   —Textbooks and Research Monographs.
II —Conference Proceedings.
III —Data Compilations.
IV —Technological Tects.
V   —Review Texts.

There is no particular significance in the order of appearance of a particular title in the following lists. The addresses of the publishers and distribution agents are to be found at the end of this section.

I TEXTBOOKS AND MONOGRAPHS

Authors	Title	Details
H. Batzer and F. Lohse	Introduction to Macromolecular Chemistry, 2nd Edition	ISSN: 0471 99645 9 312pp (1979) John Wiley & Sons Ltd
J. F. Rabek	Experimental Methods in Polymer Chemistry	ISSN: 0471 27604 9 888pp (1980) John Wiley & Sons Ltd
Hans-Georg Elias	Macromolecules, Vol 1—Structure and Properties, Vol 2—Synthesis and Materials	ISSN: 0471 99486 3 1131pp (1977) John Wiley & Sons Ltd
R. T. Bailey, A. M. North, and R. A. Pethrick	Molecular Motion in High Polymers	ISBN 019 851333 393pp (1981) Oxford University Press
A. Ziabicki	Fundamentals of Fibre Formation The Science of Fibre Spinning and Drawing	ISSN: 0471 98220 2 504pp (1976) John Wiley & Sons Ltd
I. M. Ward	Mechanical Properties of Solid Polymers	ISBN: 0 471 91995 0 375pp (1974) John Wiley & Sons Ltd
D. B. V. Parker	Polymer Chemistry	ISBN: 0 85334 571 6 251pp (1974) Applied Science
P. G. de Gennes	Scaling Concepts in Polymer Physics	ISBN: 0 8014 1203 320pp (1979) Cornell University Press
K. E. J. Barret	Dispersion Polymerization in Organic Media	ISBN: 0471 05418 6 338pp (1975) John Wiley & Sons Ltd
R. J. Ceresa	Block and Graft Copolymerization, Vol 1	ISBN: 0471 14227 1 390pp (1973) John Wiley & Sons Ltd
G. M. Bartenev	Relaxation Phenomena in Polymers	ISBN: 0706 51485 8 (1975) John Wiley & Sons Ltd

Authors	Title	Details
R. S. Asquith	Chemistry of Natural Protein in Fibres	ISBN: 0 471 99518 5 438pp (1977) John Wiley & Sons Ltd
F. W. Billmeyer	Textbook of Polymer Science 3rd Ed.	ISBN: 0 471 03196 8 598pp (1979) John Wiley & Sons Ltd
T. M. Birshtein and O. B. Ptitsyn	High Polymers Conformations of Macromolecules	ISBN: 0 470 39325 4 350pp (1966) John Wiley & Sons Ltd
E. A. Collins, J. Bares and F. W. Billmeyer	Experiments in Polymer Science	ISBN: 0 471 16585 9 530pp (1973) John Wiley & Sons Ltd
E. M. Fettes	Macromolecular Synthesis	ISBN: 0 471 05891 2 (1979) John Wiley & Sons Ltd
P. J. Flory	Statistical Mechanics of Chain Molecules	ISBN: 0 470 26495 0 432pp (1969) John Wiley & Sons Ltd
A. D. Jenkins and A. Ledwith	Reactivity, Mechanism and Structure in Polymer Chemistry	ISBN: 0 471 44155 4 613pp (1974) John Wiley & Sons Ltd
J. A. Moore	Macromolecular Synthesis	ISBN: 0 471 61451 3 710pp (1978) John Wiley & Sons Ltd
H. Moraweitz	Macromolecules in Solution, 2nd ed	ISBN: 0 471 02131 5 549pp (1975) John Wiley & Sons Ltd
J. D. Ferry	Viscoelastic Properties of Polymers, 3rd ed	ISBN: 0 471 25774 5 671pp (1979) John Wiley & Sons Ltd
D. M. Brewis	Surface Analysis and Pretreatment of Plastic and Metals	ISBN: 0 85334 992 4 266pp (1981) Applied Science

Authors	Title	Details
L. T. Butt and D. C. Wright	Use of Polymers in Chemical Plant Construction	ISBN: 0 85334 914 2 148pp (1981) Applied Science
C. W. Evans	Practical Rubber Compounding and Processing	ISBN: 0 85334 901 0 204pp (1981) Applied Science
N. M. Bikales	Characterization of Polymers— Encyclopedia Reprints Series	ISBN: 0 471 07230 3 1239pp (1972) John Wiley & Sons Ltd
M. Morton	Anionic Polymerization	ISBN: 0 12 508080 8 (1982) Academic Press
E. A. Turi	Thermal Characterization of Polymeric Materials	ISBN: 0 12 703780 2 960pp (1981) Academic Press
K. Murakami and K. Ono	Chemorheology of Polymers	ISBN: 0 444 41831 8 211pp (1979) Elsevier
M. Bohdanecky and J. Kovar	Viscosity of Polymer Solutions	ISBN: 0 444 42066 5 279pp (1982) Elsevier
D. C. Bassett	Principles of Polymer Morphology	ISBN: 0 521 23270 8 247pp (1981) Cambridge Univ. Press
B. W. Cherry	Polymer Surfaces	ISBN: 0 521 23082 9 156pp (1982) Cambridge Univ. Press
D. Hull	An Introduction to Composite Materials	ISBN: 0 521 23991 5 244pp (1982) Cambridge Univ. Press
D. C. Blakley	Emulsion Polymerization	ISBN: 0 85334 627 5 555pp (1975) Applied Science

Authors	Title	Details
M. Kurata	Thermodynamics of Polymer Solutions	ISSN: 0275-7265 291pp (1982) Gordon and Breach
A. Rudin	The Elements of Polymer Science and Engineering	ISBN: 012 601680 1 512pp (1982) Academic Press
D. A. Seanor	Electrical Properties of Polymers	ISBN: 0 12633680 6 400pp (1982) Academic Press

II CONFERENCE PROCEEDINGS

Editors	Title	Publication	Details
D. T. Clark and W. J. Feast	Polymer Surfaces		ISBN: 0 471 99614 9 441pp (1978) John Wiley & Sons Ltd
J. P. Kennedy	Cationic Craft Copolymerization	J. Applied Polymer Science	ISBN: 0 471 04426 196pp (1979) John Wiley & Sons Ltd
J. P. Kennedy	Fourth International Symposium on Cationic Polymerization	J. Polymer Science C56	ISBN: 0 471 04639 6 507pp (1979) John Wiley & Sons Ltd
M. Lewin	Fibre Science	J. Applied Polymer Science—31	ISBN: 0 471 04563 2 418pp (1978) John Wiley & Sons Ltd
J. P. Mercier and R. Legras	Recent Advances in the Field of Crystallization and Fusion of Polymers	J. Polymer Science	ISBN: 0 471 04425 3 146pp (1979) John Wiley & Sons Ltd
M. S. El-Aasser and J. W. Vanderhoff	Emulsion Polymerization of Vinyl Acetate		ISBN: 0 85334 971 1 285pp (1981) Applied Science
E. D. Feit and C. W. Wilkins	Polymeric Materials for Electronic Applications	ACS No 184	ISBN: 0 8412 0715 1 250pp (1928) ACS

Editors	Title	Details	
C. E. Carraher and C. G. Grebelein	Biological Activities of Polymers	ACS No 186	ISBN: 0 8412 0719 4 285pp (1982) ACS
D. N. S. Hon	Graft Copolymerization of Lignocellulosic Fibres	ACS No 187	ISBN 0 8412 07216 370pp (1982) ACS
J. G. Montalvo	Cotton Dust: Controlling an Occupation Health Hazard	ACS No 189	ISBN: 0 8412 0716 X 330pp (1982) ACS
J. E. Mark and J. Lal	Elastomers and Rubber Elasticity	ACS No 193	ISBN: 0 8412 0729 1 550pp (1982) ACS
J. M. Honig and C. N. R. Rao	Preparation and Characterization of Materials		ISBN: 0 12 355040 8 624pp (1982) Academic Press
F. C. Frank	Organization of Macromolecules in the Condensed Phase	Faraday Discussions No 88	ISBN: 0 85186 988 2 517pp (1979) RSC
R. J. Kostelnik	Polymeric Delivery Systems	MMI Monographs	ISBN: 0 677-15940 4 311pp (1978) Gordon and Breach
R. F. Boyer and S. E. Keinath	Molecular Motion in Polymers by ESR	MMI Monographs	ISSN: 0195 3966 321pp (1980) Gordon and Breach
K. Solc	Polymer Compatibility and Incompatibility	MMI Monographs	ISSN: 0195 3966 449pp (1982) Gordon and Breach
R. L. Miller	Flow Induced Crystallization in Polymer Systems	MMI Monographs	ISSN: 0141 0342 361pp (1979) Gordon and Breach
R. W. Lenz and F. Ciardelli	Preparation and Properties of Stereoregular Polymers	NATO ASI C51	ISBN: 90 277 1055 4 459pp (1978) D. Reidel

Authors	Title	Details	
R. A. Pethrick and R. W. Richards	Static and Dynamic Properties of the Polymeric Solid State	NATO ASI C94	ISBN: 90 277 1481 9 469pp (1981) D. Reidel

III DATA COMPILATIONS

Author	Title	Details	
A. F. M. Barton	Solubility Parameters and other Cohesion Parameters	ISBN: 0 8493 3295 8 624pp (1983) CRC Press	
C. T. Lynch	Handbook of Materials Science Vol 1 General Properties Vol 2 Metals, Composites and Refractory Materials Vol 3 Nonmetallic Materials and Applications	ISBN: 0 87819 231 X 760pp (1974) 448pp (1975) 642pp (1975)	
R. Summitt	Vol 4 Wood	472pp (1980) CRC Press	
D. D. Pollock	Physical Properties of Materials for Engineers	ISBN: 0 8493 6203 2 774pp (1981) CRC Press	
T. J. Henman	World Index of Polyolefine Stabilizers	ISBN: 0 85038 462 1 347pp (1983) RSC	

IV TECHNOLOGICAL TEXTS

Author	Title	Details	
C. A. Brighton, G. Pritchard, and G. A. Skinner	Styrene Polymers: Technology and Environmental Aspects	ISBN: 0 85334 810 3 281pp (1979) Applied Science	
R. P. Brown	Physical Testing of Rubbers	ISBN: 0 85334 788 3 330pp (1979) Applied Science	
R. H. Burgess	Manufacture and Processing of PVC	ISBN: 0 85334 972 X 275pp (1981) Applied Science	

Authors	Title	Details
G. Butters	Plastics Pneumatic Conveying and Bulk Storage	ISBN 0 85334 983 5 295pp (1981) Applied Science
L. T. Butt and D. C. Wright	Use of Polymers in Chemical Plant Construction	ISBN: 0 853314 2 148pp (1981) Applied Science
K. O. Calvert	Polymer Latices and their Applications	ISBN: 0 85334 975 4 256pp (1981) Applied Science
C. W. Evans	Developments in Rubber and Rubber Composites—1	ISBN: 0 85334 892 8 184pp (1980) Applied Science
C. W. Evans	Practical Rubber Compounding and Processing	ISBN: 0 85334 901 0 204pp (1981) Applied Science
C. W. Evans	Hose Technology	ISBN: 0 85334 830 8 230pp (1979) Applied Science
C. Hepburn and R. J. W. Reynolds	Elastomers: Criteria for Engineering Design	ISBN: 0 85334 809 X 365pp (1979) Applied Science
N. C. Hilyard	Mechanics of Cellular Plastics	ISBN: 0 85334 982 7 395pp (1981) Applied Science
R. S. Lenk	Polymer Rheology	ISBN: 0 85334 765 4 375pp (1978) Applied Science
G. D. Parfitt	Dispersion of Powders in Liquids	ISBN: 0 85334 990 8 510pp (1981) Applied Science
R. G. Weatherhead	FRP Technology	ISBN: 0 85334 886 3 460pp (1980) Applied Science

Authors	Title	Details
A. Whelan and J. L. Craft	Developments in Injection Moulding—2	ISBN: 0 85334 968 1 345pp (1981) Applied Science
A. Whelan and J. L. Craft	Developments in Injection Moulding—1	ISBN: 0 85334 798 0 283pp (1978) Applied Science
A. Whelan and J. L. Craft	Developments in PVC Production and Processing—1	ISBN: 0 85334 741 7 228pp (1977) Applied Science
A. Whelan and K. S. Lee	Developments in Rubber Technology—2	ISBN: 0 85334 949 5 274pp (1981) Applied Science
A. Whelan and K. S. Lee	Developments in Rubber Technology—1	ISBN: 0 85334 862 6 280pp (1979) Applied Science
K. W. Allen	Adhesion 5	ISBN: 0 85334 929 0 161pp (1981) Applied Science

V REVIEW TEXTS

Authors	Title	Details
Ed N. S. Allen	Developments in Polymer Photochemistry Vol 2	ISBN: 0 85334 936 3 350pp (1981) Applied Science
Ed. N. S. Allen	Developments in Polymer Photochemistry Vol 1	ISBN: 0 85334 911 8 222pp (1980) Applied Science
Ed. N. S. Allen and J. F. McKellar	Photochemistry of Dyed and Pigmented Polymers	ISBN: 0 85334 898 7 296pp (1980) Applied Science
Ed. E. H. Andrews	Developments in Polymer Fracture—1	ISBN: 0 85334 819 7 348pp (1979) Applied Science

Authors	Title	Details
Ed. A. D. Jenkins and J. F. Kennady	Specialist Periodical Report Macromolecular Chemistry Vol 1	ISBN: 0 85186 840 1 450pp (1979) Royal Soc. Chemistry
Ed. J. M. Bruist	Developments in Polyurethanes—1	ISBN: 0 85334 756 5 275pp (1978) Applied Science
Ed. J. V. Dawkins	Developments in Polymer Characterization Vol 2	ISBN: 0 85334 909 6 240pp (1980) Applied Science
Ed. J. V. Dawkins	Development in Polymer Characterization Vol 1	ISBN: 0 85334 789 1 283pp (1978) Applied Science
Ed. C. W. Evans	Developments in Rubber and Rubber Composites—1	ISBN: 0 85334 892 8 184pp (1980) Applied Science
Ed. I. Goodman	Developments in Block Copolymers—1	ISBN: 0 85334 145 1 355pp (1982) Applied Science
Ed. N. Grassie	Developments in Polymer Degradation Vol 3	ISBN: 0 85334 942 8 321pp (1981) Applied Science
Ed. N. Grassie	Developments in Polymer Degradation Vol 2	ISBN: 0 85334 854 5 214pp (1979) Applied Science
Ed. N. Grassie	Developments in Polymer Degradation Vol 1	ISBN: 0 85334 739 5 284pp (1977) Applied Science
Ed. R. N. Haward	Developments in Polymerization—2	ISBN: 0 85334 821 9 283pp (1979) Applied Science
Ed. R. N. Haward	Developments in Polymerization—1	ISBN: 0 85334 822 7 206pp (1979) Applied Science

Author	Title	Details
J. K. McKellar and N. S. Allen	Photochemistry of Man-made Polymers	ISBN: 0 85334 799 9 308pp (1979) Applied Science
Ed. G. Pritchard	Developments in Reinforced Plastics—1	ISBN: 0 85334 919 3 283pp (1980) Applied Science
Ed. G. Scott	Developments in Polymer Stabilization—4	ISBN: 0 85334 920 7 283pp (1981) Applied Science
Ed. G. Scott	Developments in Polymer Stabilization—3	ISBN: 0 85334 890 1 195pp (1980) Applied Science
Ed. G. Scott	Developments in Polymer Stabilization—2	ISBN: 0 85334 885 5 245pp (1980) Applied Science
Ed. G. Scott	Developments in Polymer Stabilization—1	ISBN: 0 85334 838 3 334pp (1979) Applied Science
P. Dreyfuss	Poly(tetrahydrofuran)	ISSN: 0275 5777 305pp (1982) Gordon and Breach
R. A. Wessling	Polyvinylidene chloride	ISBN: 0 677 01700 6 183pp (1977) Gordon and Breach
W. D. Comper	Heparin (and related polysaccharides)	ISBN: 0275 5777 260pp (1981) Gordon and Breach
J. M. Pearson	Poly(N-vinylcarbazole)	ISSN: 0275 5777 163pp (1981) Gordon and Breach

Publishers and distributors

Academic Press	PO Box 733, New York, NY 10113, USA 24/26 Oval Road, London, NW1 7DX, UK
Applied Science Publishers	22 Rippleside Commercial Estate, Ripple Road, Barking, Essex, UK
Cambridge University Press	The Pitt Building, Trumpington Street, Cambridge, CB2 1RP, UK
Cornell University Press	124 Roberts Place, Ithaca, New York, 14850, USA
CRC Press Inc.	2000 Corporate Blvd., N.W. Boca Raton, Florida 33431, USA
Elsevier Scientific Publishing Company	335 jan van Galenstraat, PO Box 211, 1000 AE Amsterdam, The Netherlands
Gordon and Breach	One Park Avenue, New York, NY 10016, USA
Oxford University Press	Walton Street, Oxford, OX2 6DP, UK
D. Reidel Publishing	PO Box 17, 3300 AA Dordrecht, Holland
Royal Society of Chemistry	Distribution Centre, Blackhorse Road, Letchworth, Herts, SG6 1HN, UK
John Wiley & Sons Ltd	Baffins Lane, Chichester, Sussex, PO19 1UD, UK

Compilation of Journals in the Area of Macromolecular Science

Journal	Editors	Address	Publishers	Code
Carbohydrate Polymers	Dr J. M. V. Blanshard, J. Mitchell	Department of Applied Biochemistry and Nutrition, School of Agriculture, University of Nottingham, Sutton Bonington, Loughborough, UK	Applied Science Publishers, Ripple Road, Barking, Essex, IG11 0SA	ISSN: 0144-8617 F—4 S—6½ × 9¼ P—1981 A—1, 4, 8
Polymer Degradation and Stability	Prof N. Grassie	Department of Chemistry, University of Glasgow, Glasgow, C2, Scotland, UK	Applied Science Publishers, Ripple Road, Barking, Essex, IG11 0SA	ISSN: 0141-3910 F—6 S—6½ × 9¼ P—1981 A—1, 7, 9
Polymer Photochemistry	Dr N. S. Allen	Department of Chemistry, John Dalton Faculty of Technology, Manchester Polytechnic, Chester Street, Manchester, MA1 5GD, UK	Applied Science Publishers, Ripple Road, Barking, Essex, IG11 0SA	ISSN: 0144-2880 F—6 S—6½ × 9¼ P—1981 A—2, 4, 9
Polymer Testing	Dr. R. Brown	RAPRA, Shawbury, Salop, UK		ISSN: 0142-9418 F—4 S—6½ × 9¼ P—1979 A—2, 4, 7
Advances in Urethane Science and Technology	Dr K. C. Frisch, Dr S. L. Reegan		Technomic Publishing Co. Inc., 265 Post Raoda West, Westport, CT 06880, USA	ISBN: 0 87762 240-X A—3, 4, 7

Journal	Editors	Address	Publishers	Code
Urethane Plastics and Products	Dr M. Kuhudic		Technomic Publishing Co. Inc., 265 Post Raoda West, Westport, CT 06880, USA	A—3, 4, 7, 8
Polymer Physics	Prof R. S. Stein		Academic Press Inc, 111 Firth Ave., New York, NY 10003, USA	A—2
Journal of Polymer Science:				
Polymer Chemistry	C. G. Overberger	Department of Chemistry, University of Michigan, Ann Arbor, Michigan 48104, USA	John Wiley Sons Inc., 605 Third Avenue, New York, NY 10016, USA	ISSN: 0360-6376 A—3
Polymer Physics	E. F. Casassa	Department of Chemistry, Carnegie-Mellon University, 4400 Fifth Avenue, Pittsburgh, Pennsylvania 15213, USA	John Wiley Sons Inc., 605 Third Avenue, New York, NY 10016, USA	ISSN: 0098-1273 A—2
Polymer Letters	C. G. Overberger, E. F. Casassa	As above: Chemistry, Physical Organic Chemistry Physics, Physical Chemistry	John Wiley Sons Inc., 605 Third Avenue, New York, NY 10016, USA	ISSN: 0360-6384 A—2, 3
Polymer Symposia	Prof H. Mark	Polytechnic Institute of New York, 333 Jay Street, Brooklyn, New York 11201, USA	John Wiley Sons Inc., 605 Third Avenue, New York, NY 10016, USA	ISSN: 0360-8905 A—2, 3, 6
Polymer Reviews	Prof H. Mark, Prof E. H. Immergut	Polytechnic Institute of New York, 333 Jay Street, Brooklyn, New York 11201, USA	John Wiley Sons Inc., 605 Third Avenue, New York, NY 10016, USA	A—6

Journal	Editors	Address	Publishers	Code
Journal of Macromolecular Science: Part A: Chemistry	George E. Ham	G. E. Ham Associates, 284 Pine Road, Briarcliff Manor, New York 10510, USA	Marcel Dekker Journals, 270 Madison Avenue, New York, NY 10016, USA	Library of Congress card number: 74-9191 A—3
Part B: Polymer Physics	Philip H. Geil	Polymer Group, University of Illinois, 1304 W Green Street, Urbana, Illinois 618001, USA	Marcel Dekker Journals, 270 Madison Avenue, New York, NY 10016, USA	ISSN: 0022-2348 A—2
Part C: Reviews in Macromolecular Chemistry	G. B. Butler, K. F. O'Driscoll, G. L. Wiles	Department of Chemistry, University of Florida, Gainsville, Florida 32601, USA; Department of Chemical Engineering, University of Waterloo, Waterloo, Ontario, Canada; Polymer Materials and Interface Laboratory, Virginia Polytechnic Institute and State University, Blacksburg, Virginia 24061, USA	Marcel Dekker Journals, 270 Madison Avenue, New York, NY 10016, USA	ISSN: 0022-2356 A—2, 3, 6
Polymer Plastics Technology and Engineering	Dr Louis Naturman	Business Connection Co, 9 Viaduct Road, Box 2070 C, Stamford, Connecticut 06906, USA	Marcel Dekker Journals, 270 Madison Avenue, New York, NY 10016, USA	ISSN: 0360 2559 A—4, 7

Journal	Editors	Address	Publishers	Code
European Polymer Journal	Prof J. C. Bevington	Department of Chemistry, University of Lancaster, Lancaster, UK	Pergamon Press Inc., Journals Division, Maxwell House, Fairview Park, Elmsford, New York, USA	ISSN: 0014-3057 A—2, 3
Polymer Science USSR	Dr W. Cooper	217 Little Aston Road, Aldridge, Walsall, WS9 O9A, West Midlands	Pergamon Press Inc., Journals Division, Maxwell House, Fairview Park, Elmsford, New York, USA	ISSN: 032-3950 A—2, 3
Progress in Polymer Science	Prof A. D. Jenkins, Prof V. T. Stannett	School of Molecular Sciences, University of Sussex, Falmer, Brighton, BH1 9QJ, UK The Graduate School, North Carolina, State University at Raleigh, PO Box 5335, Raleigh, NC 27650, USA	Pergamon Press Inc., Journals Division, Maxwell House, Fairview Park, Elmsford, New York, USA	A—2, 3, 6
Angewandte Makromolekulare Chemie	Dr Dietrich Braun	Deutsches Kunstoff-Institute, Schlossgartenstrasse 6R, D-6100, Darmstadt, Germany	Huethig and Wepf Verlag, Eisengasse 5, CH-4001, Basel, Switzerland	ISSN: 0003-3146 A—2, 3, 6
Makromolekulare Chemie	Prof D. W. Kern	Die Makromolekulare Chemie, Hegelstrasse 45, D-6500, Meinz, Federal Republic of Germany	Heuthig and Wepf Verlag, Im Weiher 10, D-6900, Heidelberg, Federal Republic of Germany	ISSN: 0025-116X A—2, 3, 6
Makromolekulare Chemie-Rapid Communications	Prof Werner Kern	Die Makromolekulare Chemie, Hegelstrasse 45, D-6500, Meinz, Federal Republic of Germany	Heuthig and Wepf Verlag, Room 213, 611 Broadway, New York 10012, USA	ISSN: 0173-2803 A—2, 3, 6

Journal	Editors	Address	Publishers	Code
Macromolecules	Dr F. H. Winslow	Bell Laboratories, Murray Hill, New Jersey 07974, USA	American Chemical Society, 1155 16th Street, NW Washington, DC 20036, USA	ISSN: 0024-9297 A—2, 3, 6
Polymer Preprints			American Chemical Society, Division of Polymer Chemistry, c/o Dr F. Dammont, Box 20453, Newark, NJ 07101, USA	A—6
Chemical and Polymer Times	Prof R. K. Gupta		Small Business Publications, Box 2131, 4/45 Roop Naggar, Delhi 110007, India	A—4, 5, 7
Polymer India	K. S. S. Kaghavan		Sevak Publications, B-26 Royal Industrial Estate, Naigaum Cross Road, Wadala, Bombay 400031, India	A—4, 5, 7
Polymer Applications Kobunshi Kako	Hitoshi Okuda		High Polymer Publishing Association, Kubunshi Kankokai/Chiekon-Sagura, Marutamachi, Kamikyuku, Kyoto 602, Japan	A—4, 5, 7
Polymer Journal	Prof Shigehura Onogi		Society of Polymer Science, Kubunshi Gakkai, Honshu Bldg, 5-12-8 Ginza, Chuo-Ku, Tokyo 104, Japan	ISSN: 0032-3896 A—2, 3, 4

Journal	Editors	Address	Publishers	Code
Vinyls and Polymers/Enbi to Purima	Shugo Miyamoto		Institute of Polymer Industry Inc., Purima Kugyo Kenkyusho, CPO Box 1176, Tokyo 100-91, Japan	A—4, 5
Polymer Engineering and Science	Prof R. S. Porter	Polymer Science and Engineering, University of Massachusetts, Amherst, Massachusetts 01003, USA	Society of Plastics Engineers Inc., 14 Fairfield Drive, Brookfield Centre, CT 06808, USA	ISSN: 0032-3888 A—4, 5, 7
Polymer Engineering Reviews	Prof J. L. White		Elsevier Sequoia S.A. Box 851. Ch-1001, Lausanne 1, Switzerland	A—4, 5, 7
Additives for Polymers			Yarsley Research Laboratories Ltd, The Street, Ashtead, Surrey, KT21 2AB	A—5, 7
Polimery			Zamenhofa 2/4, 61-120 Poznan, Poland, Distributers: ARS Polona-Ruch, Krakowshie, Przedmieschie 7, Warsaw, Poland	A—2, 3
Colloid and Polymer Science Kolloid-Zeitschift und Zeitschift für Polymere	H. G. Kilian, A. Weiss		Dr Dietrich Steinkopff Velag, Saalbaustr 12, Postfach 11, 1008, 6100 Darnstadt 11, West Germany	A—2, 3

Journal	Editors	Address	Publishers	Code
Polymer News	G. S. Kirshenbaum	Celanese Plastics Co., PO Box 1000, Summit, New Jersey 07901, USA	Gordon and Breach, Scientific Publishers, 42 William IV Street, London, WC2N 4DE, UK	ISSN: 0032-3918 F—12 S—$8 \times 10\frac{3}{4}$ A—1, 2, 3, 4
International Journal of Polymeric Materials	H. S. Kaufman	Vice-President for Development, Ramapo College of New York, PO Box 542, Mahwah, NJ 07430, USA	Gordon and Breach, Scientific Publishers, 42 William IV Street, London, WC2N, 4DE, UK	ISSN: 0091-4037 F—4 S—5×9 A—2, 3, 4, 7
Advances in Polymer Technology	Dr P. Hold		Van Nostrand Reinhold Co., Professional Journals and Periodicals Division, 135 West 50th Street, New York, NY 10020, USA	A—4, 7
Polymer	Prof C. H. Bamford	Biomedical Engineering and Medical Physics Unit, Faculty of Medicine, Duncan Building, University of Liverpool, Liverpool, L69 3BX, UK	Butterworth Scientific Ltd, Journals Division, Box 63, Westbury House, Bury Street, Guildford, Surrey, GU2 5BH, UK	ISSN: 0032-361 A—2, 3
Rubber, Plastics and Fibre	Sadanori Itonori		Taiseisha Ltd, Publishing Division, 1–5 Kyobashi, Chou-Ku, Tokyo 104, Japan	A—4, 7, 8
Polymers, Monomers Enantiomers Abstracts	Ruth G. Shoemaker	SMS Publications, 30 Sprinborn Center, Enfield, CT 06082, USA	Taiseisha Ltd, Publishing Division, 1–5 Kyobashi, Chou-Ku, Tokyo 104, Japan	A—5, 7

Journal	Editors	Address	Publishers	Code
Mechanics of Composite Materials (Formerly Polymer Mechanics)			Consultants Bureau, Subsidiary of Plenum Publishing Corp., 233 Spring Street, New York, NY 10013, USA	A—4, 5, 7, 8
Polinerim ve-Homarim Plastiyim Polymers and Plastic Materials	S. Kenig		Israel Plastics Society, c/o Centre for Industrial Research, POB 311, Haifa, Israel	A—1, 2, 3
Polymer Bulletin	Prof H. J. Cantow	Makromolekulare Chemie, Universitaat Freiburg, Stefan Meier Strasse 31, d-7800, Freiburg, Germany	Springer Velag, 175 5th Avenue, New York, NY 10010, USA	A—2, 3
Polymer Science Library	Prof A. D. Jenkins	University of Sussex, The School of Molecular Science, Falmer, Brighton, BN1 9QJ, UK	Elsevair Scientific Publishing Co., Box 211, 1000 Ae Amsterdam, Netherlands	ISSN: 0444-41832-6 F—1 S—$6\frac{1}{4} \times 9\frac{3}{4}$ P—1979 A—6
Polymer Monographs	Prof H. Morawetz		Gordon and Breach, Science Publishers, One Park Avenue, New York, NY 10016, USA	A—6

L

Journal	Editors	Address	Publishers	Code
Cellular Polymers	Dr J. M. Buist	Kingslea, Cliff Side, Wilmslow, Cheshire, SK9 4AF, UK	Applied Science Publishers, 22 Rippleside Commercial Estate, Ripple Road, Barking, Essex, UK	ISSN: 0262-4893 F—4 S—$6\frac{1}{4} \times 9\frac{1}{4}$ P—1982 A—2, 4, 5
Fibre Science and Technology	Prof G. S. Holister	Faculty of Technology, The Open University, Walton Hall, Milton Keynes, MK7 6AA, UK	Applied Science Publishers, 22 Rippleside Commercial Estate, Ripple Road, Barking, Essex, UK	ISSN: 0015-0568 F—8 S—$6\frac{1}{4} \times 9\frac{1}{4}$ P—1974 A—1, 3, 8
Plastics and Rubber Processing and Applications	Dr P. L. Clegg	ICI Limited, Petrochemical and Plastics Div., Bessemer Road, Weywn Garden City, Herts, A17 1HD, UK	Applied Science Publishers, 22 Rippleside Commercial Estate, Ripple Road, Barking, Essex, IG11 0SA, UK	ISSN: 0144-6045 F—4, S—$8\frac{1}{4} \times 11\frac{1}{2}$ P—1981 A—4, 5, 7
British Polymer Journal	Dr M. Sherwood	Society of Chemical Industry, Belgrave Square, London, UK	Royal Society of Chemistry, Blackhorse Road, Letchworth, Herts, SG6 1HN, UK	A—1, 2, 3
Progress in Rubber Technology	S. H. Morell	RAPRA, Shawbury, Salop, UK	Applied Science Publishers, 22 Rippleside Commercial Estate, Ripple Road, Barking, Essex, IG11 0SA, UK	ISSN: 0306-3542 F—1 S—$6\frac{1}{4} \times 9\frac{1}{4}$ P—1937 A—5, 6

Journal	Editors	Address	Publishers	Code
Polymer Journal	Prof Takeshi Takahiko	The Society of Polymer Science Japan, Hon Bldg 5-12-8 Ginza, Chou-Ku, Tokyo 104, Japan	International Academic Printing Co., Tokyo 160, Japan Agents: U.S. Asiatic Co. Ltd, 1-13-12 Shinbashi Minato-Ku, Tokyo 105, Japan	ISSN: 0032-3896 A—2, 3
J Elastomers and Plastics	Prof C. J. Hilado	University of San Francisco, San Francisco, California, USA	Technomic Publishing Company, 265 Post Road West, West Port, Conn 06880, USA	ISSN: 0095-2443 A—2, 5, 7
Plastics and Rubber Weekly			PO Box Maclauren House, Scarbrook Road, Croydon, CR9 1QH, UK	A—5, 6, 7
Plastics	Dr N. W. Hastings	1704 Colorado Avenue, Santa Monica, California 90404, USA		A—5, 6, 7
International Polymer Science and Technology	Translation of Chemicky Prumysl etc.		RAPRA, Shawbury, Salop, SY4 4NR, UK	ISSN: 0307-173X A—5, 6, 7
J. Applied Polymer Science	H. Mark	Polytechnic Institute of New York 333 Jay Street, Brooklyn, New York 11201, USA	John Wiley Sons Inc., 605 Third Avenue, New York, NY 10016, USA	ISSN: 0021-8995 A—2, 3, 6
J. Applied Polymer Science: Symposia			John Wiley Sons Inc., 605 Third Avenue, New York, NY 10016, USA	ISSN: 0570-4898

Journal	Editors	Address	Publishers	Code
Macromolecular Reviews	A. Peterlin	National Bureau of Standards, Washington, DC, USA	John Wiley Sons Inc., 605 Third Avenue, New York, NY 10016, USA	ISBN: 0 471 08889 7 A—2, 3
Advances in Plastics Technology	S. Levy	Sidney Levy P.E. & Associates, 6621 North Blosser Street, Fresno, California 93711, USA	Van Nostrand Reinhold, 135 West 50th Street, New York, NY 10020, USA	ISSN: 0 272-9504 A—2, 3, 6
Acta Polymerica	W. Bobeth, H. Klare, B. Philipp, C. Ruscher	Akademie der Wissenschaften der DDR, Institute for Polymerenchemie, DDR-1530 Teltow, Kantstrasse 55, E. Germany	Akademie Verlag, DDR-1086, Leipziger Strasse 3-4, E. Berlin	ISSN: 0323-7648 A—2, 3, 6
Rubber Chemistry and Technology	A. Coran	Rubber Chemical Division, Monsanto Co., 260 Springside Drive, Akron, OH 44313, USA	R. H. Gerster, Rubber Division, American Chemical Society, University of Akron, Akron, OH 44325, USA	ISSN: 0035-9475 A—3, 5, 6

Code

F—frequency of publication—digit indicates number of issues per year.
S—size of journal.
P—first published.
ISSN: International Science Citation Number.
A—Topics published—1) Natural Biopolymers.
   2) Polymers Physics.
   3) Polymer Chemistry
   4) Polymer Technology.
   5) Rubber Technology.
   6) Reviews and Abstracts.
   7) Plastics Processing.
   8) Fibres and Composites.
   9) Polymer Degradation.

Compilation of Dissertation Abstracts in Polymer Science

The following titles are available from University Microfilms International and are listed in Dissertation Abstracts International. The reference code refers to the microfilm text and is indicated by the volume number/issue. The titles listed below were selected from January 1980 to August 1982 (issues 40/07B to 43/02B). The full texts are available from:

University Microfilms International,
30–32 Mortimer Street,
London, W1N 7RA,
UK

Title	Students Name	Thesis	University	Reference	Code
Toughening of Epoxy Resin by Acrylic Elastomer	Gazit, Shmuel	PhD (1980)	The University of Connecticut	271pp 41/09B, p3520	RRW81-06735
Phenomenological Study on Reinforcement of Elastomers	Kakarala, Srimannarayana	D.Engr (1980)	University of Detroit	170pp 40/12B, p5748	RRW80-13692
Structure-Property Relationships in Segmented Elastomers: Small Angle X-Ray Scattering, Thermal, and Mechanical Studies	Van Bogart, John William Chapman	PhD (1981)	The University of Wisconsin-Madison	566pp 42/05B, p1969	RRW81-17545
An Experimental Study and Theoretical Analysis of the Injection Molding of Thermoplastic Polymers	Abdalla, Samir Z.	PhD (1980)	State University of New York at Buffalo	289pp 42/12B, p4869	RRW82-04034
Mold Filling and Curing Studies for the Polyurethane RIM Process	Castro, Jose Mario	PhD (1980)	University of Minnesota	269pp 41/11B, p4194	RRW81-09399
The Injection Molding of Phenolic Resin	Chung, Tai-Shung	PhD (1981)	State University of New York at Buffalo	308pp 42/01B, p292	RRW81-14659
A Theoretical Analysis and Experimental Study of Extrusion Blow Molding	Dutta, Anit	PhD (1981)	State University of New York at Buffalo	422pp 42/01B, p293	RRW81-14668

Title	Students Name	Thesis	University	Reference	Code
Several Aspects of the Chromatography of Macromolecules on Porous Support	Ivory, Cornelius Francois	PhD (1980)	Princeton University	286pp 41/01B, p277	RRW80-15117
Conformation and Branching Studies of Polymers: Using Gel Permeation Chromatography Coupled with On-line Low Angle Laser Light Scattering Photometry	Jenkins, Robert Francis	PhD (1980)	University of Massachusetts	188pp 41/08B, p3113	RRW81-01338
Characterization of Reversed-Phase Chromatographic Systems and Displacement Chromatography	Nahum, Avi	PhD (1981)	Yale University	215pp 42/09B, p3763	RRW81-25691
An Experimental Study and Theoretical Analysis of the Injection Molding of Thermoplastic Polymers	Abdalla, Samir Z.	PhD (1980)	State University of New York at Buffalo	289pp 42/12B, p4869	RRW82-04034
Thermodynamic Characterization of Miscible Polymer Blends	Barnum, Robert Sheppard	PhD (1981)	The University of Texas at Austin	270pp 42/03B, p1098	RRW81-19255
Mechanical Behaviors of Particulate Solids and its Ramification to Polymer Processing	Cheng, Chia Yung	PhD (1980)	Stevens Institute of Technology	236pp 41/08B, p3112	RRW81-00851
Compatibilty of Polymer Blends: Blends of "Polyepsilon-Caprolactone (PCL)" and "Styrene acrylonitrile Copolymer (SAN)"	Chiu, Shao-Cheng	PhD (1981)	University of Maryland	178pp 42/08B, p3346	RRW82-01640

Title	Students Name	Thesis	University	Reference	Code
The Behaviour of Polymer Solutions in Non-uniform Flows	Cohen, Yoram	PhD (1981)	University of Delaware	466pp 42/11B, p4498	RRW82-10121
Interfacial Phenomena Occurring in Drying Surfactant Droplets and Polymer Latex Films	Durbin, Daniel Paul	PhD (1981)	Lehigh University	348pp 41–11B, p4195	RRW81-08193
Excimer Fluorescence as a Probe of Energy Migration and Segmental Diffusion in Polymers	Fitzgibbon, Patrick David	PhD (1981)	Stanford University	400pp 41/11B, p4196	RRW81-08924
Polymerization Kinetics and Ionization Equilibria in Aqueous Silica Solutions	Fleming, Bruce Allan	PhD (1981)	Princeton University	183pp 42/09B, p3759	RRW82-03239
I: Dynamics of Flowing Polymer Solutions. II: The Measurement of Velocity Gradients by Homodyne Light Scattering Spectroscopy	Fuller, Gerald Gendall	PhD (1980)	California Institute of Technology	377pp 41/04B, p1439	RRW80-16366
Factors Affecting Polymer–Polymer Miscibility: The Relative Importance of Entropy and Enthalpy.	Harris, James, Elmer	PhD (1981)	The University of Texas at Austin	283pp 42/12B, p4872	RRW82-08179
Strain-Induced Crystallization of Polymers	Hong, Kuo-Zong	PhD (1981)	The University of Michigan	282pp 42/05B, p1964	RRW81-22777

Title	Students Name	Thesis	University	Reference	Code
Flow of Polymer Solutions is Porous Media and Related Flow Fields	Hong, Seong Ahn	PhD (1981)	The Pennsylvania State University	180pp 43/01B, p192	RRW82-13314
On the Use of a Truncated-Cone and Plate Apparatus for the Measurements of Normal Stress Differences Generated by Polymer Solutions Under Steady Shear Flow	Hou, Tan-Hung	PhD (1980)	The University of Wisconsin-Madison	344pp 41/09B, p3521	RRW80-23414
Effect of Solvents on the Radiation-Induced Polymerization of Ethyl and Isopropyl Vinyl Ethers	Hsieh, Wen-Chyi	PhD (1981)	North Carolina State University at Raleigh	112pp 42/05B, p1964	RRW81-21731
A Study of the Dynamic and Transient Properties of Polymer Solutions	Vrentas, Christine Mary Jarzebski	PhD (1981)	Northwestern University	211pp 42/05B, p1970	RRW81-25028
The Study of the Rheological Properties of Polymeric Solutions in Steady Shear Flow using a Slit Rheometer	Ybarra, Robert Michael	PhD (1980)	Purdue University	387pp 41/08B, p3118	RRW81-02725
The Synthesis and Characterization of Chlorinated Polyethylene-g-styrene and its Application as a Blend Modifier	Youg, Teng-Shau	PhD (1980)	The University of Oklahoma	196pp 41/10B, p3841	RRW81-07970

Title	Students Name	Thesis	University	Reference	Code
Oscillatory and Transient Sorption Studies of Diffusion in Polyvinyl Acetate	Ju, Shiaw-Tzuu	PhD (1981)	The Pennsylvania State University	101pp 42/04B, p1531	RRW81-20441
Flow Properties of Surfactant Solutions in Porous Media and Polymer-Surfactant Interactions	Kalpakci, Bayram	PhD (1981)	The Pennsylvania State University	259pp 42/10B, p4131	RRW82-05927
Polymer Bonding in Never-Dried Fiber Assemblages	Krumpos, John David	PhD (1980)	University of Washington	127pp 41/05B, p1845	RRW80-26260
Application of Advanced Process Control Techniques to Continuous Emulsion Polymerization	Leffew, Kenneth Wayne	PhD (1981)	University of Louisville	333pp 42/07B, p2921	RRW81-29577
Thermodynamics and Diffusion in Polymer Solutions Containing Associating Species	Lin, Joe Su-Shien	PhD (1980)	Michigan State University	165pp 42/02B, p700	RRW81-17251
Diffusion Controlled Vinyl Polymerization	Soh, Sung Kuk	PhD (1981)	University of New Hampshire	254pp 42/07B, p2924	RRW81-29271
Absolute Rate Constants for the Free Radical Polymerization of Ethylene in the Supercritical Phase	Takahashi, Tsutomu	PhD (1980)	State University of New York at Buffalo	220pp 41/08B, p3117	RRW81-04244

Title	Students Name	Thesis	University	Reference	Code
Effect of Capillary Diameter on the Rate of Radiation Induced Polymerization of Acrylamide in Saline Solutions Containing Dissolved Phosphorus-32	Verma, Saty Ajit	PhD (1980)	The Louisiana State University and Agricultural and Mechanical College	199pp 41/11B, p4199	RRW81-10428
Random and Ordered Copolyesters from Sterically Hindered Diols by Modification of PET and Liquid-Crystalline Copolyesters from Sterically Hindered Diacids	Prasadarao, Meka	PhD (1981)	Polytechnic Institute of New York	124pp 42/03B, p1103	RRW81-18892
Sorption, Transport and Dilatometric Studies in Glassy Polymers	Ranade, Gautam Ramchandra	PhD (1980)	North Carolina State University at Raleigh	168pp 41/04B, p1441	RRW80-20543
The Dynamics of Continuous Emulsion Polymerization Reactors	Schork, Francis Joseph	PhD (1981)	The University of Wisconsin-Madison	328pp 42/08B, p3348	RRW81-26653
Fracture Locus Studies in a Polymethyl Methacrylate-Polystyrene System	Irani, Jamsheed P.	PhD (1981)	State University of New York at Buffalo	118pp 42/09B, p3761	RRW82-04071
Conformation and Branching Studies of Polymers: Using Gel Permation Chromatography Coupled with On-Line Low Angle Laser Light Scattering Photometry	Jenkins, Robert Francis	PhD (1980)	University of Massachusetts	188pp 41/08B, p3113	RRW81-01338

Title	Students Name	Thesis	University	Reference	Code
A Study of Thermodynamics and Molecular Diffusion in Polymer-Solvent Systems	Liu, Han-Tai	PhD (1980)	The Pennsylvania State University	134pp 41/07B, p2688	RRW80-24470
Interpretation of the Rheological Properties of Carbon Black Reinforced Polymer Melts	Lobe, Vincent Michael	PhD (1980)	The University of Tennessee	170pp 41/08B, p3114	RRW81-04604
High Stress Viscometry of Dilyte Polymer Solutions, and Internal Viscosity Model Predictions	McAdams, Joseph Edward	PhD (1981)	University of California, Berkeley	733pp 42/12B, p4875	RRW82-12038

Calendar of Meetings in Polymer Science and Related Topics

Date	Title of Meeting	Location	Contact
9–11 May	1st AFP/SME European Conference on Radiation Curing	Lausanne, Switzerland	Susan E. Burn, Administration Technical Activities, The Association for Finishing Processes of SME, 1 SME Drive, PO Box 930, Dearborn, Michigan 48128, USA
11 May	Multiphase Polymer Systems	SCI, London, UK	Dr G. C. Eastmond, Dept. of IPI Chemistry, University of Liverpool, PO Box 147, Liverpool, L69 3BX
11 May	Structure & Properties of Foam	PRI, London, UK	Mr J. N. Ratcliffe, Plastics & Rubber Institute, 11 Hobart Place, London, SW1W 0HL
18 May	Polymer Molecular Weight Characterisation Aqueous Gel Permeation Chromatography	Shawbury, UK	Dr S. R. Holding, PSCC, RAPRA, Shawbury, Shrewsbury, Shropshire, SY4 4NR
19–20 May	Scandinavian Rubber Conference 1983	Mondal, Norway	The Programme Committee SRC-83, PO Box 101, 4501 Mandal, Norway
24–26 May	Transition & Relaxation in Polymer Materials	Melbourne, Australia	Dr D. R. G. Williams, Chemical Engineering Dept., Adelaide University, GPO Box 498, Adelaide 50001, Australia
30 May–3 June	Phase Relationships & Properties of Multicomponent Systems	Naples, Italy	Prof R. Palumbo, ITPR-CNR Via Toiano, 6-80072 Arco Felice, (Napoli), Italy

Date	Title of Meeting	Location	Contact
1–3 June	Stabilisation & Controlled Degradation of Polymers	Lucerne, Switzerland	Dr N. C. Billingham, School of Molecular Sciences, University of Sussex, Falmer, Brighton, BN1 9QJ
2–3 June	Polymer Photochemistry	Berlin, W. Germany	Dr N. S. Allen, Dept. of Chemistry, John Dalton Faculty of Technology, Manchester College of Technology, Manchester
6–9 June	Milestones & Trends in Polymer Science, Tribute to T. Alfrey	Midland, Michigan, USA	Dr R. Boyer, MMI, 1910 W St Andrews Road, Midland, MI 48640, USA
4–12 June	IUPAC Congress of Pure & Applied Chemistry	Federal Republic of Germany	Dr W. Fritsche, c/o Gesellschaft Deutscher Chemiker, PO Box 9000440, Frankfurt/M90, Federal Republic of Germany
8–10 June	Polyethylenes 1933–83 Golden Jubilee Conference	London, UK	Mr J. N. Ratcliffe, PRI, 11 Hobart Place, London, SW1W 0HL
13–17 June	Aromatic Heterocycles	Ystad, Sweden	EUCHEM Conference, Organic Chemistry 1, Chemical Center, Box 740, S-220 07 Lund 7, Sweden
14 June	RIM & Coupling Agents	SCI, London, UK	Dr E. F. T. White, Dept. of Polymer Science & Technology, UMIST, PO Box 88, Manchester, M60 1QD

Date	Title of Meeting	Location	Contact
15–18 June	Oil & Colour Chemicals Association Conference	York, UK	Mr R. H. Hamblin, Secretary OCCA, Priory House, 967 Harrow Road, Wembley, Middlesex
15–25 June	Modulated Structure Materials	Crete, Greece	Dr T. Tsakalakos, Dept. Materials Science, Rutgers Univ., Piscataway, NJ 08854, USA
15–30 June	Energy Transfer Processes in Condense Matter	Erice, Italy	Prof B. Di. Bartolo, Dept. Physics, Boston College, Chestnut Hill, MA 02167, USA
20–22 June	Symposium on Chromatography & Mass Spectrometry in Nutrition and Food Safety	Montreux, Switzerland	Dr A. Frigerio, Italian Group for Mass Spectrometry in Biochemistry and Medicine, Via Eritrea 62 1-20157, Milano, Italy
26 June–1 July	Colloquium Spectroscopicum	Amsterdam, The Netherlands	23rd CSI, c/o Organisatie Bureau Amsterdam BV, Europaplein, 1078 GZ, Amsterdam, The Netherlands
July	EUCHEM Conference on High Resolution Electron Microscopy in Solid State Chemistry	Sweden	Prof L. Kihlborg, Dept. of Inorganic Chemistry, Arrhenius Laboratory, University of Stockholm, S-106 91 Stockholm, Sweden
4–16 July	Engineering of Surface Modification of Materials NATO ASI	Les Arcs, France	Dr R. Kossowsky, Westinghouse R & D Center, Pittsburgh, PA 15235, USA

Date	Title of Meeting	Location	Contact
11–13 July	Mossbauer Discussion Group	Manchester, UK	Dr F. J. Berry, Dept. of Chemistry, University of Birmingham, PO Box 363, Birmingham, B15 2TT, England
11–14 July	Copolymers, Structure and Solution Properties	Prague, Czechoslovakia	Dr P. Kratochvil, Institute of Macromolecular Chemistry, Czechoslovak Academy of Sciences, Heyrovskeho nam. 2, 162 06, Prague 616, Czechoslovakia
11–15 July	International Meeting on NMR Spectroscopy	Edinburgh, UK	Dr D. Shaw, Oxford Research Systems, Ferry Hinksey Road, Oxford, OX2 0DL, England
11–15 July	Gums and Stabilizers for the food Industry. II International Conference Hydrocolloids	Connan's Quay, Deeside	Conference Secretariat, Research Division, The NE Wales Institute, Kelsterton College, Connan's Quay, Deeside, Clwyd, CH5 4BR
18–21 July	Polymer Stability & Processing Prague Microsymposium	Prague, Czechoslovakia	Dr J. Pospil, Institute of Macromolecular Chemistry, Czechoslovak Academy of Sciences, Heyrovskeho, n 2 162 06, Prague 616, Czechoslovakia
18–21 July	SPI-RP International Conference on Acoustic Emission of Reinforced Plastics (CARP)	Holiday Inn Golden Gateway, San Francisco, CA, USA.	Wilda Roman, The Society of the Plastics Industry, 355 Lexington Avenue, New York, New York 10017, USA

Date	Title of Meeting	Location	Contact
26 July– 7 August	NATO ASI on NMR of Liquid Crystals	San Miniato, Italy	Dr J. W. Emsley, Department of Chemistry, The University, Southampton, SO9 5NH, England
Aug	8th European Crystallograhic Meeting, Liege, Belgium	Liege, Belgium	Prof J. Toussaint, Institute de Physique B5, Universite de Liege au Sart-Tilman, B-4000, Liege, Belgium
1–11 Aug	NATO ASI Non-Equilibrium Cooperative Phenomina in Physics and Related Fields	El Escorial (Madrid), Spain	Dr M. G. Verlarde, UNED- Ciencias Apdo, Correos 50 487, Madrid, Spain
7–12 Aug	8th Canadian Symposium on Theoretical Chemistry	Halifax, Nova Scotia, Canada	Dr R. J. Boyd, Department of Chemistry, Dalhousie University, Halifax, Nova Scotia, B3H 4J3, Canada
21–26 Aug	9th International Congress of Heterocyclic Chemistry	Tokyo, Japan	Yuichi Kanaoka, Faculty of Pharmaceutical Sciences, Hokkaido University Sapporo, 060 Japan
22–26 Aug	5th International Symposium on Olefin Metathesis	Graz, Austria	Professor K. Hummel, Institut fur Chemische Technologie, Der Technischen Universitat Graz, Graz, Austria
26 Aug–2 Sept	Solvent Extraction	Denver, CO, USA	ISEC'83, American Institute of Chemical Engineers, 345 East 47th Street, New York, New York 10017, USA

Date	Title of Meeting	Location	Contact
28 Aug–2 Sept	ACS Meeting, Washington. Rubber Modified Thermoset Resins	Washington, DC, USA	C. K. Riew, B. F. Goodrich Co., R & D Centre, 9921 Brecksville Ohio 44141 J. Gillham, Department of Chemical Engineering, Princeton University, Princeton, NJ 08544
	Polyurethanes: Chemistry and Applications		K. Frisch and D. Klempner, Polymer Institute, University of Detroit, 4001 W McNichols Road, Detroit, Mich. 48221, USA
	Low Temperature Curing of Coatings by Photon and Electron Beams		G. Gruber, PPG Industries, PO Box 9, Allison Park, PA 15101, USA
	Case Histories of Industry-Academe Cooperation in Polymer Orientated Fields: Plastics, Coatings, Rubber, Textiles and Paper		E. M. Pearce, Department of Chemistry, Polytechnic Institute of New York, 333 Jay Street, Brooklyn, New York 11201, USA
	Advances in Organometallic Polymers		C. E. Carraher Jr., Department of Chemistry, Wright State University, Dayton, OH 45431, USA J. Sheats, Chemistry Department, Rider College, Lawrenceville, NJ 08648, USA C. U. Pittman, Department of Chemistry, University of Alabama, University, AL 35486, USA

Date	Title of Meeting	Location	Contact
	Reaction Injection Moulding		J. E. Kresta, Polymer Institute, Univ. of Detroit, 4001 W McNichols Road, Detroit, MI 48221, USA
	Polymers and Fibres		Michael Jaffe, Celanese Corp, Summitt, NJ, USA
	Forensic Chemistry of Coatings		R. Holsworth, Glidden Coatings & Resins, Division of SCM Corporation, 16651 Sprague Road, Strongsville, Ohio 44136, USA
28 Aug–1 Sept	2nd IUPAC Symposium on Organometallic Chemistry directed toward Organic Synthesis	Dijon, France	Prof J. Tirouflet, Universite de Dijon, Boite, Postale 138, 21004 Dijon Cedex, France
29 Aug–2 Sept	Danube Symposium on Chromatography and 7th International Symposium "Advances and Application of Chromatrography in Industry"	Bratislave, Czechoslovak	Dr Jan Remen, The Analytical Section of the Czechoslovak Scientific and Technical Society, Slovnaft, 823 00 Bratislave, CSSR
30 Aug–2 Sept	Cationic Polymerization & Related Processes	Ghent, Belgium	Prof E. Goethals, Institute of Organic Chemistry, University of Ghent, Krijgslaan 281, B-9000 Ghent, Belgium
30 Aug–2 Sept	6th International Symposium on Cationic Polymerization and Related Processes	Ghent, Belgium	Prof E. Goethals, Institute of Organic Chemistry, Rijksuniversiteit-Ghent, Krijglaan 281 (S-4) B-9000, Ghent, Belgium

Date	Title of Meeting	Location	Contact
September	Polymerization of Heterocycles International Symposium on Polymerization of Small Heterocycles and Aldehydes	Zabrze, Poland	Polymer Institute, POB 49, Curie-Sklodowskiej 34, PL-41 800 Zabrze, Poland
5–9 Sept	International Conference on Fourier Transform Spectroscopy	Durham, UK	J. R. Birch, Division of Electrical Science, National Physical Laboratory, Teddington, Middlesex, TW11 0LW, UK
5–9 Sept	29th International Symposium on Macromolecules	Bucharest Romania	Acad. Cristofor Simionescu, Institutul de Chimie, Macromoleculara Petru Poni Aleea Grigore Ghica Voda Nr. 41A Iasi, Romania
5–9 Sept	3rd European Symposium on Organic Chemistry	Canterbury, UK	Prof R. F. Hudson, University of Kent, Canterbury, Kent, CT2 7NH, UK
5–9 Sept	International Conference on Phosphorous Chemistry	Nice, France	Prof J. G. Reiss, Laboratoire de Chimie Minerale Moleculaire, Universite de Nice, Parc Valrose 06034, Nice, France
12–14 Sept	Biotechnology in the Pulp & Paper Industry	London, UK	D. Attwood, Director, Paper and Board Division, PIRA, Randells Road, Leatherhead, Surrey, KT22 7RU

Date	Title of Meeting	Location	Contact
12–16 Sept	European Congress on Molecular Spectroscopy	Sofia, Bulgaria	Dr B. Jordanov, Institute of Organic Chemistry, Bulgarian Academy of Sciences, BG-1113 Sofia, Bulgaria
14–16 Sept	Polymer Physics Group Biennial Conference	Reading, UK	Dr J. V. Champion, Department of Physics, City of London Polytechnic, 31 Jewry Street, London, FC3N 2EY
14–16 Sept	Concentrated Colloidal Dispersions Faraday Discussion No 76	Loughborough, UK	Prof R. H. Ottewill, School of Chemistry, University of Bristol, Cantock's Close, Bristol, BS8 1TS
21–23 Sept	2nd International Conference on Drug Adsorption: Rate Control in Drug Therapy	Edinburgh, UK	Secretariat, 2nd International Conference on Drug Adsorption, Centre for Industrial Consultancy and Liaison, University of Edinburgh, 16 George Square, Edinburgh, EH8 9LD, UK
3–5 Oct	5th International Conference PRP-Automation	Antwerp, Belgium	BIRA Jan van Rijswijcklaan 58, B-2000 Antwerp, Belgium
5–12 Oct	K '83 International Trade Fair Plastics & Rubber	Dusseldorf, West Germany	Dusseldorf Messegesellschaft mbH, NOWEA, Postfach 320203 D-40000, Dusseldorf 30, West Germany
4–6 Oct	Polypropylene Fibres & Textiles	York	Mr J. N. Tadcliffe, Plastics and Rubber Institute, 11 Hobart Place, London, SW1W 0HL, UK

Date	Title of Meeting	Location	Contact
5–7 Oct	VIth Australian Rubber Technology Convention	Canberra, Australia	L. R. Smith, Chairman, Technical Papers Committee, Contential Carbon Australia Pty Ltd. Private Bag, Cronulla, NSW 2230, Australia
9–14 Oct	Joint Symposium on Fundamental Aspects of Corrosion Protection by Surface Modification	Washington, DC, USA	R. P. Frankenthal, Chairman, Corrosion Division of Electrochemical Society, Bell Laboratories, 600 Mountain Avenue, Murray Hill, NJ 07974, USA
16–21 Oct	EUCHEM Conference on Organic Free Radicals	Schloss Elmau, Germany	Prof R. Sustmann, Universitaat Essen- Gesamthochschule, Universitatsstrasse 5, 4300 Essen 1, Federal Republic of Germany
30 Oct–4 Nov	SPI-Polyurethane Division, 6th International Technical Conference	San Diego Convention Centre, CA, USA	P. Toner, The Society of the Plastics Industry, 355 Lexington Avenue, New York, New York 10017, USA
10–11 Nov	Flame Retardents	London, UK	Diane Varley. Plastics and Rubber Institute, 11 Hobart Place, London, SW1W 0HL, UK
7th Dec	The Crosslinking of Polyolefins: Recent Advances	London, UK	Dr J. V. Champion, Dept. of Physics, City of London Polytechnic, 31 Jewry Street, London, EC3N 2EY, UK

Date	Title of Meeting	Location	Contact
16–20 July	Cellucon 84	Wrexham, UK	Conference Secretariat, Cellucon 84, Research Division, NE Wales Institute, Kelsterton College, Deeside, Clwyd, CH5 4BR, UK
17–20 July	Chemistry of Carbanions	Durham, UK	Secretary, Perkin Division, Royal Society of Chemistry, Burlington House, Picadilly, London, W1V 0BN, UK
7th-Aug	IUPAC Conference on Physical Organic Chemistry	Auckland, New Zealand	Prof B. R. Davis, Chemistry Department, University of Auckland, Private Bag, Auckland, New Zealand
20–24th Aug	New Developments in Polymer Science & Engineering	Kyoto, Japan	Secretariat, The Society of Polymer Science, Japan, Hon Building, 5-12-8 Ginza, Chuo-ku, Tokyo 104, Japan
26–30 Aug	5th International Conference on Organic Synthesis	Freiburg, Germany	Gesellschaft Deutscher Chemiker, Postfach 90 04 40, D-6000 Frankfurt/Main 90, Federal Republic of Germany
26–31 Aug	188th Meeting of the American Chemical Society	Philadelphia, USA	A. T. Winstead, American Chemical Society, 1155 16th Street NW, Washington, DC 20036, USA

Date	Title of Meeting	Location	Contact
	Multicomponent Polymer Materials: Polymer Blends, Grafts, Blocks and Interpenetrating Polymer Networks		L. H. Sperling, Materials Research Center, Coxe Laboratory, 32, Lehigh University, Bethleham, PA 18015, USA D. R. Paul, Department of Chemical Engineering, University of Texas at Austin, Austin, Texas, USA
	Coulombic Interactions in Macromolecular Systems		Adi Eisenberg, Department of Chemistry, McGill University, Montreal, PQ. Canada, H-3A-2K6
	Polymers in Medication		F. E. Bailey, Union Carbide Corp. So. Charleston, W. Va. 25303, USA C. G. Gebelein, Department of Chemistry, Youngstown, OH 44555, USA
	Anticorrosion Barrier, Chemistry and Applications		Henry Leidheiser Jr, Center for Surface and Coatings Research, Sinclair Memorial Laboratory 7, Lehigh University, Bethleham, PA 18015, USA
12–14 Sept	Adhesion & Adhesives: Science, Technology & Applications	Nottingham, UK	Diana Varley, Plastics and Rubber Institute, 11 Hobart Place, London, SW1W 0HL, UK
Sept	Conference on Polymers of Small Heterocycles and Aldehydes	Poland	Prof Z. Jodinski, Polymer Institute, POB 49, Curie Sklodowskiej 34, PL-41 800 Jabrze, Poland
Nov/Dec	7th International Biotechnology Symposium	New Delhi, India	Prof T. K. Ghose, Chairman, National Organisation Committee & Head Biochemical Enginering Research Centre, Indian Institute of Technology, Hauz Khas, New Delhi 110029, India

Date	Title of Meeting	Location	Contact
8–9th Dec	Molecular & Microstructural Basis of Viscoelasticity and Related Phenomena	Cambridge, UK	Dr M. Lal, Unilever Research, Port Sunlight Laboratory, Bebington, Wirral, L63 3JW, UK
1984 Jan	2nd Chemical Congress of the Federation of Asian Chemical Societies	India	Dr M. Singh, Rubber Research Institute of Malaysia, PO Box 150, Kuala Lumpur, Malaysia
Jan	3rd International Conference on Solid Films and Surfaces	Sydney, Australia	Prof D. Haneman, Department of Physics, University of NSW, Kensington, NSW 2033, Australia
12–16 Feb	14th Australian Polymer Symposium	Ballarat, Australia	RACI, Polymer Division, PO Box 224, Belmont, Victoria 3216, Australia
28–30 March	Radiation Processing for Plastics & Rubber II	Canterbury, UK	Mr J. N. Radcliffe, The Plastics & Rubber Institute, 11 Hobart Place, London, SW1W 0HL, UK
1–6 April	Corrosion '84 International Conference	New Orleans, LA, USA	Conference Cordinator, National Association of Corrosion Engineers, POB 218340 Houston, TX, USA
8–13 April	American Chemical Society -St Louis Meeting Recent Advances in Characterization of Polymers	St Louis, USA	A. T. Winstead, ACS 1155 16th Street, Washington, DC 20036, USA C. Craver, Chemir Laboratories, 761 Kirkham, Glendale, MO 63122, USA

Date	Title of Meeting	Location	Contact
	Wear of Polymers and its Control		L.-H. Lee, Xerox Webster Research Center, Building 114, 800 Phillips Road, Webster, NY 14580, USA
	Reactive Oligomers		F. W. Harris, Chem. Dept. Wright State University, Dayton, OH 45435, USA
	Mechanisms of Ring Opening Polymerization		Prof J. McGrath, Dept. of Chemistry, VPI and Su Blacksburg, VA 24061, USA
	Polymeric Precursors to Inorganic Materials		K. Wynne, Office of Naval Research, Arlington, VA 22217, USA
16–17 April	Health & Safety in the Plastics & Rubber Industry II	York, UK	Mr J. N. Radcliffe, Plastics and Rubber Institute, 11 Hobart Place, London, SW1W 0HL, UK
26 April–1 May	2nd World Congress on Biomaterials	Washington, DC, USA	Dr S. R. Pollack, Department of Bioengineering, 285 Town Building D3, University of Pennsylvannia, Philadelphia, PA 19104, USA
2–6th July	8th International Congress on Catalysis	Berlin, Germany	DECHEMA, PO Box 970146, D-6000 Frankfurt am Main 97, Federal Republic of Germany
10–14 July	4th International Symposium on Organic Free Radicals	St Andrews, Scotland	Secretary, Perkin Division, Royal Society of Chemistry, Burlington, Piccadilly, London, W1V 0BN, UK

Date	Title of Meeting	Location	Contact
16–21 Dec	International Congress of Pacific Basin Societies		A. T. Winstead, 1155-16th St. N.W. Washington, DC 20036, USA
1985 11–14th Feb	Polymer 85-Characterization and Analysis of Polymers	Melbourne, Australia	Dr J. H. O'Donnell, Polymer & Radiation Group, Department of Chemistry, University of Queensland, Brisbane, 4067, Australia
25–29 March	Corrosion '85 International Conference	Boston, MA, USA	Conference Coordinator, National Association of Corrosion Engineers, POB 218340, Houston, TX 77218, USA
April	Polymer Liquid Crystals Faraday Discussion		Dr J. V. Champion, Dept. of Physics, City of London Polytechnic, 31 Jewry Street, London, EC3N 2EY, UK
7–13 July	14th International Conference on Medical and Biological Engineering and 7th International Conference on Medical Physics	Helsinkii, Finland	Dr N. Saranummi, Finnish Society for Medical Physics and Medical Engineering, PO Box 27, 33231 Tampere 23, Finland
9–13 Sept	IUPAC Congress	Manchester, UK	Secretary, Royal Society of Chemistry, Burlington House, Picadilly, London, W1V 0BN, UK

Index